Advances in Experimental Medicine and Biology

This series of volumes focuses on concepts, techniques and recent advances in the field of proteomics, interactomics, metabolomics and systems biology. Recent advances in various 'omics' technologies enable quantitative monitoring of myriad various biological molecules in a high-throughput manner, and allow determination of their variation between different biological states on a genomic scale. Now that the sequencing of various genomes, from prokaryotes to humans, has provided the list and linear sequence of proteins and RNA that build living organisms, defining the complete set of interactions that sustain life constitutes one of the key challenges of the postgenomic era. This series is intended to cover experimental approaches for defining protein-protein, protein-RNA, protein-DNA and protein-lipid interactions; as well as theoretical approaches dealing with data analysis, integration and modeling and ethical issues.

More information about this series at http://www.springer.com/series/5584

Flavia Vischi Winck
Editor

Advances in Plant Omics and Systems Biology Approaches

Springer

Editor
Flavia Vischi Winck
Center for Nuclear Energy in Agriculture (CENA)
University of São Paulo
Piracicaba, São Paulo, Brazil

ISSN 0065-2598 ISSN 2214-8019 (electronic)
Advances in Experimental Medicine and Biology
ISSN 2730-6216 ISSN 2730-6224 (electronic)
Proteomics, Metabolomics, Interactomics and Systems Biology
ISBN 978-3-030-80351-3 ISBN 978-3-030-80352-0 (eBook)
https://doi.org/10.1007/978-3-030-80352-0

This Springer imprint is published by the registered company Springer Nature Switzerland AG
The registered company address is: Gewerbestrasse 11, 6330 Cham, Switzerland

I dedicate this book to my daughter and husband. In gratitude for changing my life forever and for their endless love and support.

Preface

This book was written with the goal of reaching ungraduated and graduate students from different areas related to plant sciences who look for introductory current knowledge about plant omics and systems biology and those scientists and professionals who wants to follow the evolution of the field.

The authors that contributed to this book are expert researchers that are currently developing their main research related to the topic they contributed to the book chapters.

This book had gathered authors from different countries, who contributed to reviewing the latest advances on the areas of functional genomics, including transcriptomics, proteomics, metabolomics, and data analysis and discussed the challenges ahead and the achievements of plant omics that contributed to the development of plant biotechnology applications.

The book is organized in ten chapters which describe several recent approaches and technical aspects of the omics analysis in plants, as follow:

1. Introduction: Advances in Plant Omics and Systems Biology
2. Modern Approaches for Transcriptome Analyses in Plants
3. Plant Proteomics and Systems Biology
4. Subcellular Proteomics as a Unified Approach of Experimental Localizations and Computed Prediction Data for Arabidopsis and Crop Plants
5. The Contribution of Metabolomics to Systems Biology: Current Applications Bridging Genotype and Phenotype in Plant Science
6. Interactomes: Experimental and In Silico Approaches
7. Probabilistic Graphical Models Applied to Biological Networks
8. Cataloging Posttranslational Modifications in Plant Histones
9. Current Challenges in Plant Systems Biology
10. Contribution of Omics and Systems Biology to Plant Biotechnology

The information contained in this book contributes to describe and discuss current methodologies applied in plant omics and systems biology and how these approaches led to novel biological data and knowledge. It brings a great opportunity for a straightforward reading of important aspects of the current field in a single book. It also includes discussions on technical and methodological issues and successful or advantageous strategies for achieving a system view of a biological problem.

Piracicaba, São Paulo, Brazil Flavia Vischi Winck

Acknowledgments

This book could not be done without the valuable contributions of the many authors that dedicated efforts, even in adverse times, to write their book chapters and share their knowledge within the pages of this book. I would like to thank all contributing authors Diego Mauricio Riaño-Pachón, Hector Fabio Espitia-Navarro, John Jaime Riascos, Gabriel Rodrigues Alves Margarido, André Luis Wendt dos Santos, Maria Juliana Calderan-Rodrigues, Cornelia M Hooper, Ian R Castleden, Sandra K Tanz, Sally V Grasso, A Harvey Millar, Marina CM Martins, Valeria Mafra, Carolina C. Monte-Bello, Camila Caldana, Luíza Lane de Barros Dantas, Marcelo Mendes Brandão, Natalia Faraj Murad, Ericka Zacarias, J. Armando Casas-Mollano, Danilo de Menezes Daloso, Thomas C.R. Williams, Ronaldo J. D. Dalio, Celso Gaspar Litholdo Junior, Gabriela Arena, Diogo Magalhães, Marcos A. Machado, Lucca de F. R. Monteiro, and Glaucia Mendes Souza. In many ways we have also to thank the Institute of Chemistry and Center for Nuclear Anergy in Agriculture of the University of São Paulo and the Brazilian research foundation agencies the São Paulo research foundation (FAPESP), Coordination for the Improvement of Higher Education Personnel (CAPES), and National Council for Scientific and Technological Development (CNPq) for supporting several scientific projects discussed here and the fellowships for the students and researchers that contributed to this book.

The editor also thanks all anonymous colleagues, academics, and professionals that contributed to improve the book chapters by giving valuable suggestions.

The editor thanks Prof. Dr. Alexander Henning Ulrich, from the Institute of Chemistry at the University of São Paulo, Brazil, for his valuable comments and suggestions on the creation and organization of this book project. The editor would like to thank Prof. Dr. Michael Hippler from the University of Münster for valuable comments and suggestions.

The authors thank the São Paulo Research Foundation (FAPESP) for financial support (grant #2016/06601-4, São Paulo Research Foundation (FAPESP); grant #2014/50921-8, São Paulo Research Foundation (FAPESP); grant #2015/21075-4, São Paulo Research Foundation (FAPESP); and grant #2017/01284-3, São Paulo Research Foundation (FAPESP)). ALWS thanks Dr. Eny IS Floh and Alexandre Junio Borges Araújo (Department of Botany, University of São Paulo) for the valuable collaboration.

A special thanks from the editor to the Springer Nature publisher and its associated editors for their support along the way on the development of this book, especially Noreen Henson, Anu Pradhaa Subramonian, Deepak Devakumar, Patrick Carr, and Larissa Albright for their kindness and understanding during all the steps of the development of this book.

The development of this book could not be possible without the understanding, endless support, and encouragement of the editor's husband and daughter, even under adverse times. For that, the editor says thank you all for making this book real.

Contents

Editors and Contributors

About the Editor

Flavia Vischi Winck is a biologist, with a PhD in molecular biology. She has worked on protein chemistry, plant systems biology, functional genomics, molecular biology, and biochemistry with focus on understanding signaling mechanisms and complex gene regulatory networks using proteomics, transcriptomics, and bioinformatics approaches in a systems biology framework.

Contributors

Gabriela Arena Centro de Citricultura Sylvio Moreira, Laboratório de Biotecnologia, Instituto Agronômico, Cordeirópolis, São Paulo, Brasil

Luíza Lane de Barros Dantas John Innes Centre, Norwich Research Park, Norwich, UK

Marcelo Mendes Brandão Center for Molecular Biology and Genetic Engineering, State University of Campinas, Campinas, SP, Brazil

Camila Caldana Max Planck Institute of Molecular Plant Physiology, Potsdam, Germany

Maria Juliana Calderan-Rodrigues Max Planck Institute of Molecular Plant Physiology, Potsdam, Germany

J. Armando Casas-Mollano School of Biological Sciences and Engineering, Yachay Tech University, San Miguel de Urcuquí, Ecuador

The BioTechnology Institute, College of Biological Sciences, University of Minnesota, Saint Paul, MN, USA

Ian R. Castleden The Centre of Excellence in Plant Energy Biology, The University of Western Australia, Crawley, WA, Australia

Ronaldo J. D. Dalio Centro de Citricultura Sylvio Moreira, Laboratório de Biotecnologia, Instituto Agronômico, Cordeirópolis, SP, Brasil

IdeeLab Biotecnologia, Piracicaba, SP, Brasil

Hector Fabio Espitia-Navarro School of Biological Sciences, Georgia Institute of Technology, Atlanta, GA, USA

Sally V. Grasso The Centre of Excellence in Plant Energy Biology, The University of Western Australia, Crawley, WA, Australia

Cornelia M. Hooper The Centre of Excellence in Plant Energy Biology, The University of Western Australia, Crawley, WA, Australia

Celso Gaspar Litholdo Jr Centro de Citricultura Sylvio Moreira, Laboratório de Biotecnologia, Instituto Agronômico, Cordeirópolis, SP, Brasil

Marcos A. Machado Centro de Citricultura Sylvio Moreira, Laboratório de Biotecnologia, Instituto Agronômico, Cordeirópolis, SP, Brasil

Valeria Mafra Instituto Federal de Educação, Ciência e Tecnologia do Norte de Minas Gerais, Januária, Brazil

Diogo Magalhães Centro de Citricultura Sylvio Moreira, Laboratório de Biotecnologia, Instituto Agronômico, Cordeirópolis, SP, Brasil

IdeeLab Biotecnologia, Piracicaba, SP, Brasil

Gabriel Rodrigues Alves Margarido Department of Genetics, Luiz de Queiroz College of Agriculture, University of São Paulo, Piracicaba, Brazil

Marina C. M. Martins Departamento de Botânica, Instituto de Biociências, Universidade de São Paulo, São Paulo, Brazil

Danilo de Menezes Daloso Departamento de Bioquímica e Biologia Molecular, Universidade Federal do Ceará, Fortaleza, Brasil

A. Harvey Millar The Centre of Excellence in Plant Energy Biology, The University of Western Australia, Crawley, WA, Australia

Carolina C. Monte-Bello Universidade Estadual de Campinas, Campinas, Brazil

Laboratório Nacional de Ciência e Tecnologia do Bioetanol, Centro Nacional de Pesquisa em Energia e Materiais, Campinas, Brazil

Max Planck Institute of Molecular Plant Physiology, Potsdam, Germany

Lucca de F. R. Monteiro Institute of Chemistry, University of São Paulo, São Paulo, Brazil

Institute of Biosciences, University of São Paulo, São Paulo, Brazil

Natalia Faraj Murad Center for Molecular Biology and Genetic Engineering, State University of Campinas, Campinas, São Paulo, Brazil

Diego Mauricio Riaño-Pachón Laboratory of Computational, Evolutionary and Systems Biology, Center for Nuclear Energy in Agriculture, University of São Paulo, Piracicaba, Brazil

John Jaime Riascos Centro de Investigación de la Caña de Azúcar de Colombia, CENICAÑA, Cali, Valle del Cauca, Colombia

André Luis Wendt dos Santos Laboratory of Plant Cellular Biology, Instituto de Biociências, Universidade de São Paulo, São Paulo, Brazil

Glaucia M. Souza Institute of Chemistry, University of São Paulo, São Paulo, Brazil

Sandra K. Tanz The Centre of Excellence in Plant Energy Biology, The University of Western Australia, Crawley, WA, Australia

Thomas C. R. Williams Departamento de Botânica, Universidade de Brasília, Brasília, Brasil

Flavia Vischi Winck Institute of Chemistry, University of São Paulo, São Paulo, Brazil

Center for Nuclear Energy in Agriculture, University of São Paulo, Piracicaba, Brazil

Ericka Zacarias School of Biological Sciences and Engineering, Yachay Tech University, San Miguel de Urcuquí, Ecuador

Abbreviations

2-DE	Two-Dimensional Gel Electrophoresis
2D-DIGE	Two-Dimensional Difference Gel Electrophoresis
2D-PAGE	Two-Dimensional Polyacrylamide Gel Electrophoresis
ABA	Abscisic Acid
ABI1	ABA Insensitive 1
ABI2	ABA Insensitive 2
AD	Activation Domain
AdaBoost	Adaptive Boosting
ADP	Adenosine Diphosphate
AFB	Auxin Signaling F-Box
AGI	Arabidopsis Gene Identifier
AGL16	Agamous-Like16
AGO10	Protein argonaute 10
AI-1	Arabidopsis Interactome 1
AMP	Adenosine Monophosphate
AMPDB	Arabidopsis Mitochondrial Protein Database
API	Application Programming Interface
ARF	Auxin Response Factor
AS	Alternative Splicing
ASURE	Arabidopsis Subcellular Reference
ASV	Alternative Splice Variants
AT content	Adenine-Thymine Content
ATG	Autophagy Related
ATHENA	Arabidopsis THaliana ExpressioN Atlas
AtPIN, AtPINDB	Arabidopsis thaliana Protein Interaction Network Database
BaCelLo	Balanced Subcellular Localization Predictor
BANFF	Bayesian Network Feature Finder
BaNJO	Bayesian Network Inference with Java Objects
BDe	Bayesian Dirichlet Equivalence
BIC	Bayesian Information Criterion
bicor	Biweight Midcorrelation
BiFC	Bimolecular Fluorescence Complementation
Biogrid	Biological General Repository for Interaction Datasets
BLAST	Basic Local Alignment Search Tool

BN	Bayesian Networks
BNFinder	Bayes Net Finder
BRET	Bioluminescent Resonance Energy Transfer
BSR-Seq	Bulked Segregant RNA-seq
CAGE	Cap Analysis of Gene Expression
CAM	Crassulacean Acid Metabolism
cAMP	Cyclic Adenosine Monophosphate
Cas9	CRISPR-Associated Protein 9
CAT	Co-expression Adjacency Tool
CBP	CREB-Binding Protein
CCD	Charge-Coupled Device
CCS	Circular Consensus Sequence
CD distance	Czekanowski–Dice distance
cDNA	Complementary DNA
CE	Capillary Electrophoresis
CE-MS	Capillary Electrophoresis–Mass Spectrometry
CENP-B	Centromere Protein B
CFP	Cyan Fluorescent Protein
ChIP	Chromatin Immunoprecipitation
ChIP-chip	Chromatin Immunoprecipitation and Microarray Hybridization
ChIP-Seq	Chromatin Immunoprecipitation and High-Throughput Sequencing
CLF	Curly Leaf
CNV	Copy Number Variation
CO	Zinc Finger Protein Constans
Co-IP	Co-Immunoprecipitation
CREB	cAMP Response Element-Binding Protein
CRISPR	Clustered Regularly Interspaced Short Palindromic Repeats
cropPAL	Crop Proteins of Annotated Location Database
CUC1/2	Cup-Shaped Cotyledon 1/2
Da	Dalton
DAPG	2,4-Diacetyl Phloroglucinol
dATP	Deoxyadenosine Triphosphate
DB	DNA-Binding Domain
dCTP	Deoxycytidine Triphosphate
DDA	Data-Dependent Acquisition
DDB2	DNA Damage-Binding Protein 2
deoxy-UTP	Deoxyuridine Triphosphate
dGTP	Deoxyguanosine Triphosphate
DGTS	Diacylglyceryltrimethylhomoserine
DIA	Data-Independent Acquisition
DIP	Database of Interacting Proteins
DMR6	Downy Mildew Resistance 6
DNA	Deoxyribonucleic Acid
DREB2A	Dre-Binding Protein 2A
dscDNA	Double-Stranded Complementary DNA

dsDNA	Double-Stranded DNA
dsRNA	Double-Stranded RNA
dTTP	Deoxythymidine Triphosphate
EBE	Effector Binding Elements
EFR	Elongation Factor Thermo Unstable Receptor
ER	Endoplasmic Reticulum
ER model	Erdös–Rényi Model
ESI-MS	Electrospray Ionization Mass Spectrometry
EST	Expressed Sequence Tag
ETI	Effector-Triggered Immunity
FBA	Flux Balance Analysis
FDR	False Discovery Rate
FLAG-tag	DYKDDDDK Protein Tag
FP	Fluorescent Protein
FPKM	Fragments Per Kilobase Million
FRET	Fluorescent Resonance Energy Transfer, Förster Resonance Energy Transfer
FSS	First-Strand Synthesis
FSW	Functional Similarity Weight
GA	Gibberellin Acid
GABA	Gamma-Aminobutyric Acid
GABI	Genome Analysis of the Plant Biological System
GabiPD	GABI Primary Database of Arabidopsis
GC	Gas Chromatography
GC content	Guanine-Cytosine Content
GC-MS	Gas Chromatography–Mass Spectrometry
GC-NMR	Gas Chromatography–Nuclear Magnetic Resonance
GC–TOF-MS	Gas Chromatography–Time-of-Flight Mass Spectrometry
GCN	Gene Co-expression Network
GDA	Generalized Discriminant Analysis
GFP	Green Fluorescent Protein
GGM	Graphical Gaussian Model
GH	Glycoside Hydrolases
GM	Genetically Modified
GMO	Genetically Modified Organism
GO	Gene Ontology
GRID	Global Research Identifier Database
GSH	Glutathione
GST	Glutathione S-Transferase
GUI	Graphical User Interface
GWAS	Genome-Wide Association Studies
H1	Histone H One
H2A	Histone H Two A
H2B, HTB	Histone H Two B
H3	Histone H Three
H3K14ac	Histone H3 Lysine 14 Acetylation
H3K27me3	Histone H3 Lysine 27 Trimethylation

H3K36me2	Histone H3 Lysine 36 Dimethylation
H3K36me3	Histone H3 Lysine 36 Trimethylation
H3K4me1	Histone H3 Lysine 4 Methylation
H3K4me3	Histone H3 Lysine 4 Trimethylation
H3K56ac	Histone H3 Lysine 56 Acetylation
H3K9ac	Histone H3 Lysine 9 Acetylation
H3K9me1	Histone H3 Lysine 9 Methylation
H3K9me2	Histone H3 Lysine 9 Dimethylation
H3K9me3	Histone H3 Lysine 9 Trimethylation
H3S10ac	Histone H3 Serine 10 Acetylation
H3S10ph	Histone H3 Serine 10 Phosphorylation
H4	Histone H Four
HD-ZIPIII	Homeodomain-Leucine Zipper III
HEPES	4-(2-Hydroxyethyl)-1-Piperazineethanesulfonic Acid
HIGS	Host-Induced Gene Silencing
His	Histidine
HTS	High-Throughput Sequencing
HUPO	Human Proteome Organization
IAA	Indole-3-Acetic Acid
ICAT	Isotope-Coded Affinity Tag
IMAC	Immobilized Metal Affinity Chromatography
IMEx	International Molecular Exchange Consortium
indels	Insertions and Deletions
INPPO	International Plant Proteomics Organization
ISR	Induced Systemic Resistance
iTRAQ	Isobaric Tags for Relative and Absolute Quantitation
JAZ	Jasmonate-ZIM Domain
K	Lysine
kDa	Kilodalton
KEGG	Kyoto Encyclopedia of Genes and Genomes
KNN	*k*-Nearest Neighbor
KYP	Kryptonite
LASSO	Least Absolute Shrinkage and Selection Operator
LBD	Lateral Organ Boundaries Domain
LC	Liquid Chromatography
LC-MS	Liquid Chromatography–Mass Spectrometry
LC-MS/MS	Liquid Chromatography with Tandem Mass Spectrometry
LC-NMR	Liquid Chromatography–Nuclear Magnetic Resonance
LC/MRM-MS	Liquid Chromatography/Multiple Reaction Monitoring-Mass Spectrometry
LCR	Locus Control Region
LDA	Linear Discriminant Analysis
lncRNA	Long Non-Coding RNA
LOOCV	Leave One Out Cross Validation
LOPIT	Localization of Organelle Protein by Isotope Tagging

LTQ-FTICR	Linear Ion Trap-Fourier Transform Ion Cyclotron Resonance Trap
m7G	7-Methylguanylate
MALDI-MS	Matrix-Assisted Laser Desorption/Ionization Mass Spectrometry
MALDI-TOF/TOF	Matrix-Assisted Laser Desorption/Ionization Time-of-Flight
MAMP	Microbe-Associated Molecular Patterns
MAPK6	MAP Kinase 6
MASCP	Multinational Arabidopsis Steering Committee Proteomics Subcommittee
MCMC	Monte Carlo Markov Chain
MDa	Megadaltons
MDL	Minimal Description Length
MFA	Metabolic Flux Analysis
MGDG	Monogalactosyldiacylglycerol
MI	Mutual Information
MINT DB	Molecular Interaction Database
MIR	miRNA Gene
miRNA	Micro RNA
MIT	Mutual Information Test
ML	Machine Learning
mlDNA	Machine Learning-based Differential Network Analysis
Mlo	Mildew Resistance Locus O
MMAP	Multiple Marker Abundance Profiling
MOPS	3-(*N*-Morpholino)Propanesulfonic Acid
mQTL	Metabolite-based Quantitative Trait Loci
mRNA	Messenger RNA
MS	Mass Spectrometry
MS/MS	Tandem Mass Spectrometry
mW	Molecular Weight
MYB	MYELOBLASTOSIS Gene
NAD+	Nicotinamide Adenine Dinucleotide, oxidized form
NAF	Non-Aqueous Fractionation
NAT	Natural Antisense Transcripts
NCBI	National Center for Biotechnology Information
ncRNA	Non-Coding RNA
NGS	Next-Generation Sequencing
NHEJ	Non-Homologous End-Joining
NHS-propionate	Propionic Acid *N*-Hydroxysuccinimide Ester
NINJA	Novel Interactor of JAZ
NLR	Nucleotide-Binding Leucine-Rich Repeat
NMR	Nuclear Magnetic Resonance
NO	Nitric Oxide
NPAS	Normalized Protein Abundance Scores
NTA	Nitrilotriacetic Acid
O-GlcNAc	O-linked Beta-*N*-Acetylglucosamine

ORF	Open Reading Frame
OsGI	Oryza sativa GIGANTEA
OsPHO2	Oryza sativa Phosphate Over-Accumulator 2
OsSIRP2	Oryza sativa Salt-Induced Ring E3 Ligase 2
OsTKL1	Oryza sativa Transketolase 1
PAMP	Pathogen-Associated Molecular Patterns
PAT	PPI Adjacency Tool
PCA	Principal Component Analysis
PCA	Phenazine-1-Carboxylic Acid
PCC	Pearson Correlation Coefficient
PCR	Polymerase Chain Reaction
PEBL	Psychology Experiment Building Language
PGI	Phosphoglucose Isomerase
PGM	Probabilistic Graphical Models
PGPB	Plant Growth-Promoting Bacteria
PGPF	Plant Growth-Promoting Fungus
PGR	Plant Growth Regulator
pI	Isoelectric Point
PIPES	Piperazine-*N*,*N′*-bis(2-Ethanesulfonic Acid)
PlantCyC	Plant Metabolic Pathways Database
PlantPReS	Plant Stress Proteome Database
PMN	Plant Metabolic Network
poly-dT	Poly-Deoxythymidine
PPDB	Plastid Proteome DataBase
PPI	Protein–Protein Interaction
PredSL	Prediction of Subcellular Location from the N-terminal Sequence
PRM	Parallel Reaction Monitoring
PRR	Pattern Recognition Receptors
PSoL	Positive Sample only Learning algorithm
PTI	Pattern-Triggered Immunity
PTM	Post-Translational Modification
PTST	Protein Targeting to Starch
QTL	Quantitative Trait Loci
qToF, Q-TOF	Quadrupole Time of Flight
R genes	Resistance Genes
RBF	Radial Basis Function
RCC1	Regulator of Chromosome Condensation 1
RD29A	Responsive to Desiccation 29A
RD29B	Responsive to Desiccation 29B
RF	Random Forest
RISC	RNA-Induced Silencing Complexes
RNA	Ribonucleic Acid
RNA pol II	RNA Polymerase II
RNAi	RNA Interference
RNASeq, RNA-Seq	RNA Sequencing
RNS	Reactive Nitrogen Species
ROS	Reactive Oxygen Species

RP-HPLC	Reverse Phase-High Performance Liquid Chromatography
RPKM	Reads Per Kilobase Million
RPM1	Disease Resistance Protein RPM1
RPS2	Ribosomal Protein S2
rRNA	Ribosomal RNA
RSEM	RNA-Seq by Expectation Maximization
RT-qPCR, qRT-PCR	Quantitative Reverse Transcriptase Polymerase Chain Reaction
S	Serine
S genes	Susceptibility Genes
SAGE	Serial Analysis of Gene Expression
SAIL	Syngenta Arabidopsis Insertion Library
SAM	Shoot Apical Meristem
SCC	Spearman's Correlation Coefficient
SDS-PAGE	Sodium Dodecyl Sulfate–Polyacrylamide Gel Electrophoresis
SILAC	Stable Isotope Labeling by Amino Acids
siRNA	Small Interfering RNA
SJ	Splice Junctions
SLPFA	Subcellular Location Prediction with Frequency and Alignment
SMV-RFE	Support Vector Machines with Recursive Feature Elimination
snoRNA	Small Nucleolar RNA
SNP	Single Nucleotide Polymorphism
SNV	Single Nucleotide Variant
SOM	Self-Organized Maps
SP7	Rhizophagus Intraradices Secreted Protein 7
SPL	SQUAMOSA-Promoter Binding-Like Proteins
SPP	Sucrose-6-Phosphate Phosphatase
SPS	Sucrose-6-Phosphate Synthase
SQDG	Sulfoquinovosyldiacylglycerol
SRA	Short Read Archive
SRM	Selected Reaction Monitoring
SRM	Selective Reaction Monitoring
sRNA	Small Non-Coding-RNA
SS4	Starch Synthase 4
ssDNA	Single-Stranded DNA
SSI2	Suppressor of Salicylate Insensitivity of npr1-5
SSR	Simple Sequence Repeat
SSS	Second-Strand Synthesis
SUBA	Subcellular Location of Proteins in Arabidopsis Database
SUBAcon	SUBA Consensus Classifier
SUMO	Small Ubiquitin-Like Modifier
SVM	Support Vector Machines

SWATH-MS	Sequential Window Acquisition of all Theoretical Mass Spectra
T	Threonine
T-DNA	Transfer DNA
TAG	Triacylglycerol
TAIR	The Arabidopsis Information Resource
TAL	Transcription Activator-Like
TAP	Tandem Affinity Purification
TAP-MS	Tandem Affinity Purification–Mass Spectrometry
TCA	Tricarboxylic Acid
TDA	Target Data Acquisition
TEV	Tobacco Etch Virus
TGS	Third-Generation Sequencing
TMT	Tandem Mass Tags
TMV	Tobacco Mosaic Virus
ToF, TOF	Time-of-Flight
TPM	Transcripts Per Million
Tre6P	Trehalose 6-Phosphate
Tris	Tris(hydroxymethyl)aminomethane
tRNA	Transfer RNA
TRX	Thioredoxins
TSS	Transcription Start Site
Ubl	Ubiquitin-like Protein
UDG	Uracil-DNA-Glycosylase
UDP	Uridine Diphosphate
UHPLC-MS/MS	Ultra-High Performance Liquid Chromatography–Tandem Mass Spectrometry
UHPLC-QTOF-MS	Ultra-High Performance Liquid Chromatography–Quadrupole Time-of-Flight Mass Spectrometry
USDA	United States Department of Agriculture
UV	Ultraviolet
VSR	Vacuolar Sorting Receptor
WGS	Whole Genome Sequencing
WTA	Winner-Takes-All
Y	Tyrosine
Y2H	Yeast Two-Hybrid
YFP	Yellow Fluorescent Protein

Introduction: Advances in Plant Omics and Systems Biology

1

Flavia Vischi Winck, Lucca de F. R. Monteiro, and Glaucia M. Souza

Abstract

How the complexity of biological systems can be understood is currently limited by the amount of biological information we have available to be incorporate in the vastitude of possibilities that could represent how a biological organism function. This point of view is, of course, alive under the paradigm that describes a living thing as a whole that could never be interpreted as to the sole understanding of its separated parts.

If we are going to achieve the knowledge to understand all the complex relations between the molecules, pathways, organelles, cells, organs, phenotypes, and environments is unknown. However, that is exactly what moves us toward digging the most profound nature of relationships present in the living organisms.

During the last 20 years, a big workforce was dedicated to the development of techniques, instruments, and scientific approaches that guided a whole new generation of scientists into the universe of omics approaches. The implementation of technological advances in several omics applications, such as transcriptomics, proteomics, and metabolomics, has brought to light the information that nowadays reshape our previous thinking on specific aspects of plant sciences, including growth, development, organ communication, chromatin states, and metabolism, not to mention the underpinning role of regulatory mechanisms that in many cases are essentially the basis for the phenotypical expression of a biological phenomenon and plants adaptation to their environment.

In this chapter, some of the original concepts of complex systems theory were briefly discussed, and examples of omics approaches that are contributing to uncovering emergent characteristics of plants are presented and discussed. The combination of several experimental and computational or mathematical approaches indicated that there is room for improvements and novel discoveries. However, the level of complexity of biological systems seems to require and demand us to unify efforts toward the integration of the large omics datasets already available and the

F. V. Winck (✉)
Institute of Chemistry, University of São Paulo, São Paulo, Brazil

Center for Nuclear Energy in Agriculture, University of São Paulo, Piracicaba, Brazil
e-mail: winck@cena.usp.br

Lucca de F. R. Monteiro
Institute of Chemistry, University of São Paulo, São Paulo, Brazil

Institute of Biosciences, University of São Paulo, São Paulo, Brazil

G. M. Souza
Institute of Chemistry, University of São Paulo, São Paulo, Brazil

F. V. Winck (ed.), *Advances in Plant Omics and Systems Biology Approaches*, Advances in Experimental Medicine and Biology 1346, https://doi.org/10.1007/978-3-030-80352-0_1

ones to come. This unification may represent the necessary breakthrough to the achievement of the understanding of complex phenomena. To do so, the inclusion of systems biology thinking into the training of undergraduate and graduate students of plant sciences and related areas seems to be also a contribution that is necessary to be organized and implemented in a worldwide scale.

Keywords

Single cell · Networks · Integration · Regulation · Metabolism · Signaling

1.1 Overview of Systems Theory Applied to Plant Sciences

The extraordinary complexity of cellular responses of any living organism has long been an attractive topic to scientists, including the understanding of what complexity is and how it is structured in life.

In the endless endeavor of investigating all factors that act concurrently in the definition of cell fate, several new discoveries were made. In special, at the beginning of the twentieth century, several initiatives reemerged to foment the integrative (or holistic) investigation of living organisms, with a special dedication to transdisciplinary studies of animals and plants (Drack et al. 2007) that recapitulated the different views on system theory and thinking and a new modern time emerged in system theories applied to biology. The basic concepts that permeated the system thinking resided in the premise that the properties of the whole cannot be completely understood from the simple sum of the properties of its isolated components (Von Bertalanffy 1972).

Therefore, the systems property should emerge from the enormous number of dynamic interactions (direct and/or indirect) and relations between the components (e.g., molecules, organelles, organs, tissues) of the living organisms, which indeed represent the complexity of the biological systems and that also permits it to dynamically respond to environmental changes and internal perturbations, which is one of the most important characteristics of the living organisms.

One of the special messages that remained from the studies of the previous century and conceptual elements generated thereof is the need for transdisciplinary thinking and methodologies that could guide us into the understanding of complex phenomena.

In plant sciences, several efforts have been made to address complex phenomena, bringing novel insights into the structure, organization, regulation, and evolution of plant responses. The typical research on systems biology usually proposes the cyclic implementation of experimental and theoretical analysis of the biological systems with the investigation of dynamic biological responses. Experiments are performed in laboratories and computational and/or mathematical modeling and simulation of the biological systems are applied in an iterative manner (Kitano 2002). This approach will certainly enhance our knowledge of complex systems by gradual increments and additional breakthroughs that improve our understanding of underlying principles of the whole biological systems.

The advancement of molecular biology approaches and technologies, including the evolution of different types of OMICS analysis opened new venues for the analysis of the several layers of biological information (here referring to pools of different types of molecules) toward the reconstruction, generation, and validation of new proposed models of biological organisms. The progress and dissemination of OMIC approaches and high-throughput methods are intimately connected to the advances in plant systems biology (Provart and McCourt 2004), and several important initiatives have provided significant contributions of novel biological data, construction, and maintenance of biological databases that could be confidently applied in systems biology approaches (Falter-Braun et al. 2019). The underlying cellular mechanisms that define plant phenotypes are now being investigated through a systems approach in different plant models by researchers from several countries.

1.2 Advances in Omics and Systems Biology Applications in Plant Sciences

More than 10 years ago, proposed modeling approaches that integrated chemical and mechanical information have revealed important aspects of plant development, such as meristem development into shoots or flowers, and based on molecular modeling, methodologies have also revealed the challenging aspects of such studies, such as the need of time-resolved information and imaging of biological phenomena integration between molecular data into mechanical, phenotypic descriptions of plant growth (Chickarmane et al. 2010).

Essential processes such as chromatin remodeling and regulation of activation or deactivation of transcriptional units or modules were investigated through genome-wide analysis and meta-analysis of several datasets of *Arabidopsis* plants histone modifications. The analysis performed through a system view brought to light the high complexity of chromatin remodeling processes, evidencing the existence of nine possible different chromatin states. Computational analysis of biological data originated from published profiles of histone modifications, histone variants, nucleosome density, genomic G + C content, CG methylated residues, and chromatin immunoprecipitation (ChIP) data for histone acetylation was performed in an integrative fashion and revealed transcriptional active sites, repressed sites, elongation signatures, intergenic upstream promoter regions, Polycomb, intragenic regions associated to short and long transcript units and two heterochromatin profiles related to intergenic regions and pericentromeric regions. This discovery also revealed the correlation between these regions and gene expression activation or deactivation, which rendered knowledge with a higher confidence of topological organization of chromatin regions and their association or close proximity to each other (Sequeira-Mendes et al. 2014). This type of information thus suggests that chromatin remodeling processes may have a mechanism or structured biological information that influence a priori the position of the main epigenetic modifications, defining the chromatin topology in the plant cells.

These findings are intrinsically connected with the understanding of the multi-combinatorial nature of the control of gene expression. In *Arabidopsis*, for instance, it has been evidenced that most gene promoters (63%) are recognized by at least two Transcription Factor (TFs) proteins, while some promoters may be recognized by up to 18 different TFs, composing a highly interconnected hub of molecular interactions. Genes that are expressed in many different conditions are usually controlled by many different TFs (Brkljacic and Grotewold 2017).

The integration of spatial information of chromatin signatures and topology with the most recent findings of the distribution of cis-regulatory modules (CRMs) along the genomes may substantially reveal some principles of the global regulation of gene expression and causative structures that can be associated with phenotypes of interest, such as plant growth and stress responses. An exciting review on gene expression control can be consulted in the work from Brkljacic and Grotewold (2017). Some interesting questions arise from such studies on gene combinatorial expression, including the possible global and conserved preferences of groups or families of TFs for binding to correlated chromatin regions and states, the effects of TFs in the regulation of non-available CRMs and how the different groups of TFs are associated to work together in multiple different complexes, depending on the CRMs and chromatin they interact.

In this present book, you will find the presentation of interesting examples of epigenetic mechanisms and its principles, the basics of interaction networks and the strategies to generate models that can be implemented to describe the possible connections of the elements of the biological systems.

In a different but complementary perspective, the analysis of the transcripts and proteins expressed in a cell or tissue will be introduced and advances discussed. These omics data have rendered a massive amount of qualitative and quantitative biological information in the last two decades. The omics approaches revealed, in several instances, with a time resolution, the dynamics of the cellular responses with the indication of timely coordinated events, cyclic, inhibited, and induced responses of the gene complements (transcripts and proteins).

As a side note, the growing volume of multi-omics information, together with higher computational processing capabilities, states a duality that must be addressed: the enhancement of the integration of different layers of biological information from different datasets towards the improvement of the understanding of complex systems. This requires extensive communication, the development of databases and data sharing between researchers. The organization of multinational participative computational repositories and open access data analysis platforms guided to continuous data mining and data integrative analysis improvement could integrate the plant systems biology community around the problems that concern the major tasks of interpreting the complex systems. This would add a social benefit of contributing to expanding the access of non-developed countries to the advanced science in the field of plant biology and computational biology.

In such context, the *Crops in silico* initiative (https://cropsinsilico.org/) represents an interesting prospect for integrative modeling tools in plant sciences. In fact, there is a shortage of plant-based multi-scale simulations compared to the number of models of mammals, with the existing plant models being restricted to time-limited descriptions of several singular biological processes or phenotypic responses to environmental stimuli. There is indeed an urge for the development of a virtual physiological plant, such a model also integrating developmental timescales and environmental data to plant multi-omic networks and phenotypes to understand response complexities. The *initiative* tries to address such demands through constructing a plant community-centered platform, while also dealing with usual collective technical barriers: visualization, data imputation, coding standardizations, and accessibility issues (Marshall-Colon et al. 2017).

The massive scale of current transcriptome data analysis, in special for model plants such as *Arabidopsis thaliana*, resulted in the identification of many different molecular phenotypes, generating novel insights on how changes in the transcriptional state of the cells are associated with global patterns of gene expression control. For instance, time-resolved transcriptome analysis of the *Arabidopsis* root revealed that different nitrogen doses induced the modulation of 1153 genes in a pattern that fits a Michaelis–Menten kinetics, indicating the existence of a saturation trend of transcripts accumulation or depletion at upper levels of nutrient availability. This study also revealed that some early responsive TF genes are likely related to compound-dose-responsive transcription (Swift et al. 2020). Even though the existence of a gradual transcriptional response has been proposed for a while, the deep investigation of this type of global patterns is necessary, since the promoter architecture of several genes revealed the presence of repeated binding sites which suggested their capacity to interact with several protein complexes containing TFs (Brady et al. 2006).

These results raised some questions about how much of the gene expression control can directly be affected by the concentration of external compounds following a simple kinetics and if this type of control is happening through a direct or indirect way at the DNA level in different environmental conditions. In the same way, it is essential to know how these global patterns are established and conserved in the cells, in a multi-combinatorial regulatory network structure, where multiple TFs may bind to multiple *cis*-regulatory elements. The understanding of these mechanisms can contribute to the identification of the nature of the coordination of these modular transcriptional states and their integration into new cellular functions, opening a new path for identification and modeling of causative effects on cell functional regulation. In addition, the several transcriptome analyses performed in the past 15 years have also contributed to reveal several molecular phenotypes in detail that exposed the connection of many biological processes. For instance, the transcriptome depicted genome-wide oscillations of gene expression, with the identification of genes essential for circadian rhythm and photosynthesis, cell growth, and division, contributed to the identification of the promoter elements correlated to the genes unexpectedly modulated under circadian control (Harmer et al. 2000). More recently, the involvement of pre-mRNA processing, transcript stability, mRNA nuclear export, posttranslation, and non-protein coding RNAs (ncRNAs) in particular, long ncRNAs (lncRNAs), in the regulation of circadian rhythm in plants (Romanowski and Yanovsky 2015) exposed the complexity of this biological process compared to the first model of transcriptional–translational feedback loop described in plants with the participation of two *Arabidopsis* MYB transcription factors CIRCADIAN CLOCK ASSOCIATED 1 (CCA1) and LATE ELONGATED HYPOCOTYL (LHY) (Alabadi et al. 2001; Schaffer et al. 1998; Strayer et al. 2000; Wang and Tobin 1998).

In the light of evolution, the recent efforts on the analysis of more than one thousand plant transcriptomes opened a new path to a comprehensive understanding of evolutionary differences between plants and their cellular mechanisms. The integration of data on plant habitat and niches with molecular data may reveal novel sets of genes involved in specific adaptations and phenotypes. The broad phylogenetic analysis based on transcriptome data also indicated that most gene expansion events in plant lineage have occurred before the appearance of vascular plants (One Thousand Plant Transcriptomes 2019), intriguing by its great possibilities of future applications in plant biotechnology.

In addition, computational resources implemented for plant functional genomics data visualization and mining such as Bio-analytic Resource (BAR) for Plant Biology with eFP Browser (Winter et al. 2007) have been of great benefit to the plant community to interpret such vastitude of data. The same is true for other platforms such as Phytozome (Goodstein et al. 2012) that over the years have made available hundreds of genomic datasets and have expanded its applications into Phytomine (https://phytozome.jgi.doe.gov/phytomine/begin.do), integrating genomics and functional data and fostering data mining. Gramene is also an example bringing integration to the plant reactome (http://plantreactome.gramene.org/index.php?lang=en). The current expansion of such platforms to integrate other types of omics data would be beneficial to the future of plant systems biology applications.

Proteomics approaches have also revealed important aspects of plant phenotypes, especially the molecular description of metabolic pathways operating in cellular responses, and significantly increased the number of proteins identified that are related to crops productivity and stress responses (Salekdeh and Komatsu 2007). The last 15 years have witnessed an expansion of proteomics data ranging from land plants to unicellular algae, showing the particularities of these organisms in responses to variations of environmental conditions such as light (Mettler et al. 2014), CO_2 content (Santos and Balbuena 2017), metabolic regimes (Vidotti et al. 2020), among others. It is also noticeable that proteomics is opening venues for a broader understanding of the cell response regulation by revealing the identity and possible regulatory role of many proteins that undergo posttranslational modifications (Huang et al. 2019; Van Leene et al. 2019). The quantitative proteomic profiling of cellular responses has been applied to model plant species, and it is quickly expanding to cover non-model species, which is highly desirable and necessary to uncover the vast diversity of metabolic characteristics and nuances of the plant species naturally adapted to different environmental niches, such as desertic, tropical, semiarid, and rainy regions. This and other topics of plant proteomics are presented and discussed within this book.

The transferable application of omics knowledge into system-biology-based plant breeding is one important consequence of the development of large biological datasets of omics and plant systems biology data (Lavarenne et al. 2018). Natural breeding or genetic engineering of plants should now address great problems with a more holistic approach of plant systems contributing to the generation of novel stress-resistance crops (Zhang et al. 2018).

In a parallel trend, similar paradigmatic breakthroughs in plant systems biology have been recently achieved through metabolomics studies. Together with improved mass spectrometry (MS) techniques, advancements in computational biology and bioinformatics have been allowing for wider modeling efforts, permitting broader integration of metabolite data with other large amounts of transcriptome and proteome information, further bolstering the construction of interactomes. This has important consequences given the massive size of the plant metabolome and the fact that it is usually the first layer of biological information within the cell to be subjected to the effects of environmental changes since the response at transcriptional level can take longer to occur, if not to mention the direct influence of metabolites in transcriptional modulation and other possible interactions of the metabolome with transcripts and proteins.

An applied sample of this trend can be noted through the work of Veyel and collaborators with a study in *Arabidopsis* where they developed an improved proteomic-compatible, metabolomic-oriented method for system-wide analyses of protein-small molecule complexes. The protein-metabolite interactome is an often overlooked but functionally important regulatory feature to be dealt with when studying metabolite-diverse, metabolome data-rich systems such as plants—retroactively, this substantial amount of information also makes data collecting and processing troublesome. A simple, but innovative addressing to this issue is the proposed general-case, large-scale analysis-oriented co-fractionation method instead of canonical approaches such as cross linking or protein tagging for search of interaction sites. The approach relies on the hypothesis that proteins and protein-bound small molecules fractionate together when forming stable complexes: as such, the use of size separation techniques should concentrate protein-metabolite complexes in higher molecular weight fractions, the process could then be coupled with analytic techniques (i.e., mass spectrometry). Apart from the technological prospect, this proof-of-concept has also identified a plethora of novel stable protein-metabolite complexes from the *Arabidopsis* samples, suggesting emergent regulatory roles for some small molecules, with potential for extension to other biological systems (Veyel et al. 2017).

The further study of how the metabolic landscape of plants is shaped by varying environmental conditions can also unravel the metalinguistics of experimental design in plant sciences. A common consequence of the application of high-throughput analytical techniques to biological systems is the finding that often formerly overlooked properties of the environment can exert unexpected effects in an organism's metabolism, potentially biasing the reproducibility of some experiments. As a matter of fact, sensitivity to initial conditions is an inherent condition of complex systems, particularly in a multi-omics perspective.

While dealing with experimentation in plants, lighting conditions represent an essential, though often taken for granted environmental condition. Sunlight is characterized by sinusoidal changes in irradiance throughout the day cycle, with shading and clouds momentarily varying the amount of light absorbed by the plant. On the other hand, artificially lighted growth chambers usually offer constant light irradiance (square wave), abrupt light–dark shifts and different spectral quality when compared to naturally lighted environments. Ironically, although growth chambers are considered essential for experimental reproducibility for scientific approaches, the vastitude of differences from the phenotypes of plants grown in such environments with the phenotypes observed in plants in natural environments and in the field enhance the number of possible dynamic phenotypes that may populate the universe of metabolomics.

This issue is illustrated through metabolite analysis of samples obtained from *Arabidopsis* plants grown under greenhouse and growth chamber (with sinusoidal or square lighting patterns, fluorescent or sunlight spectra-simulating LED light) conditions. The combination of enzymatic assays with HPLC and LC-MS indicates major differences (fold change) in occurrence of components of central carbon and nitrogen pathways when data from greenhouse and growth chamber conditions are compared altogether. In the context of adaptation difficulties faced by the plant while dealing with unexpected changes in sunlight within and between days, emphasis should be placed on the importance of variations (for greenhouse versus growth chamber experiments) in photoassimilate partitioning optimization, amino acid synthesis (Ser, Gln, Gly), C:N ratio, and synthesis of components of sugar signaling networks such as sucrose and the sucrose status indicator, trehalose 6-phosphate, those finely linked with the sugar feedback-operated regulation of starch metabolism (turnover) and sucrose homeostasis through day–night cycles by the circadian clock in *Arabidopsis*: the circadian oscillator itself suffering adjustments through feedback from sunlight pattern changes (Seki et al. 2017). The metabolite analysis also reveals that milder differences were observed when comparing different conditions of artificial illumination (i.e., sinusoidal versus square patterns) and that LED lighting may not fully represent natural lighting conditions (Annunziata et al. 2017).

In this present book, different aspects of metabolomics basics and applications are introduced, discussing scientific and methodological approaches that are contributing to broadening the knowledge on the metabolic regulation of plant phenotypes. Nevertheless, the metabolomics analysis still faces a great challenge of identifying the cellular compartmentalization of the metabolites in the different cell responses, which may include an extraordinary level of complexity in the cellular responses, especially if other dual interactions of metabolites, such as metabolite–protein or metabolite–miRNAs, occur in a dimension of hundreds of thousands.

1.3 Challenges in Plant Systems Biology and Paths to Expand the Research Field

Besides their sensitivity to initial conditions, biological systems are also characterized by emergence. The interaction between parts of a system can generate emergent properties, sometimes loosely linked with such parts. As a remarkable example, the use of mass spectrometry (affinity purification-mass spectrometry, gas chromatography-mass spectrometry) has revealed that glycolytic enzymes can mediate mitochondria-chloroplast colocalization in *Arabidopsis*. This finding sums to an already extensive array of supposed properties of the glycolytic pathway in non-plant cells, such as the formation of multienzyme complexes, colocalization with ATP-demanding areas and enzyme chemotactic movements, not to mention the physical association of glycolytic enzymes with the mitochondria (Zhang et al. 2020). Along stating that even well-studied central pathways can perform unprecedented roles, one can also argue that the given groundbreaking work—in a similar fashion to other featured research works—asserts the insufficiency of reductionism to deal with complexity. This context has a special impact on undergraduate biochemistry disciplines.

The known biological complexity substantially impacts experimental design in plant molecular physiology studies. In cultivated crops where multiple copies of genomes are present, the computational challenges alongside those of mathematics and statistical genetics are indeed formidable. The investigation of emergent properties emphasizes to us the importance of the chosen approach to teach complexity in some life science courses. Despite its non-intuitive nature, emergence is an important component of the Dynamical Systems Theory, which constitutes one of the three systems theories, along with Bertalanffy's General Systems Theory and Cybernetics: the so-called systems thinking concept has regained recent relevance in primary and secondary education. Its crosscutting characteristic and multidisciplinary applicability are especially helpful in teaching skills to comprehend biological complexity (Verhoeff et al. 2018). Yet, there is comparatively less extensive research and application of systems thinking in STEM education, a substantial lack of integration with chemistry courses is particularly noticeable while most of existing peer-reviewed literature on systems thinking generally focuses on biology education.

The implementation of systems thinking approach in areas such as chemistry and biochemistry may enhance student's strength in taking complex decisions that are important for global issues, such as sustainability (York et al. 2019; York and Orgill 2020). Although not extensively, some life science courses have recently been incorporating systems theory concepts into their programs, a trend exemplified by attempts to apply ecological perspectives and systems modeling into redesigning an undergraduate botany major course (Zangori and Koontz 2017). This is specially targeted at meeting novel demands on teaching biological complexity and retroactively generates content-changing demands for the disciplines that service those courses: still, while paramount to most life science careers, biochemistry is such a complicated example.

The inclusion of space for the teaching of the basics of systems biology-related topics such as complex system theories that could be explored through the concepts of the regulatory landscape of biological systems (i.e., molecular binding to sites in proteins and nucleic acid sequences as a topic for structure–function classes, transcriptional regulation, posttranslational modifications), multi-omics networks, high-throughput techniques for data acquisition (mass spectrometry, sequencing), and computational analysis would contribute to change the oftentimes discrete, linear fashion for teaching the structure, components and behavior of certain well-studied, canonical cellular mechanisms and pathways. Although time and resource constraints indeed interfere in which contents (and to what extent they can be deepened) should be taught, one could argue that an aged linear, pathway-focused attempt to teach basic biochemistry stretches away from thoroughly portraying complexity by snatching its integrative, network-based, and emergent attributes.

Systems biology and omics approaches can also be wisely applied to explore more mundane, unforeseen problems in teaching scenarios. The implementation of basic bioinformatic skills in early years of the undergraduate courses in biological and biochemistry sciences is essential to prepare the students for the future of populated biological databases and data mining schemes that are going to permeate their future academic and professional lives. Examples on how to connect the systems biology and omics knowledge into the curriculum and lectures for undergraduate students can cross the successful examples of omics applications in the several fields of research, including plant sciences and the more recent synthetic biology approaches for bioproducts production. Therefore, we envisage that some examples of studies presented and discussed in the present book may in the future be applied in systems biology-based graduate and undergraduate schools worldwide.

References

Alabadi D et al (2001) Reciprocal regulation between TOC1 and LHY/CCA1 within the Arabidopsis circadian clock. Science 293(5531):880–883

Annunziata MG et al (2017) Getting back to nature: a reality check for experiments in controlled environments. J Exp Bot 68(16):4463–4477

Brady SM, Long TA, Benfey PN (2006) Unraveling the dynamic transcriptome. Plant Cell 18(9):2101–2111

Brkljacic J, Grotewold E (2017) Combinatorial control of plant gene expression. Biochim Biophys Acta Gene Regul Mech 1860(1):31–40

Chickarmane V et al (2010) Computational morphodynamics: a modeling framework to understand plant growth. Annu Rev Plant Biol 61:65–87

Drack M, Apfalter W, Pouvreau D (2007) On the making of a system theory of life: Paul A Weiss and Ludwig von Bertalanffy's conceptual connection. Q Rev Biol 82(4):349–373

Falter-Braun P et al (2019) iPlant Systems Biology (iPSB): an international network hub in the plant community. Mol Plant 12(6):727–730

Goodstein DM et al (2012) Phytozome: a comparative platform for green plant genomics. Nucleic Acids Res 40(Database issue):D1178–D1186

Harmer SL et al (2000) Orchestrated transcription of key pathways in Arabidopsis by the circadian clock. Science 290(5499):2110–2113

Huang D et al (2019) Protein S-nitrosylation in programmed cell death in plants. Cell Mol Life Sci 76(10):1877–1887

Kitano H (2002) Systems biology: a brief overview. Science 295(5560):1662–1664

Lavarenne J et al (2018) The spring of systems biology-driven breeding. Trends Plant Sci 23(8):706–720

Marshall-Colon A et al (2017) Crops in silico: generating virtual crops using an integrative and multi-scale modeling platform. Front Plant Sci 8:786
Mettler T et al (2014) Systems analysis of the response of photosynthesis, metabolism, and growth to an increase in irradiance in the photosynthetic model organism chlamydomonas reinhardtii. Plant Cell 26(6):2310–2350
One Thousand Plant Transcriptomes, I (2019) One thousand plant transcriptomes and the phylogenomics of green plants. Nature 574(7780):679–685
Provart NJ, McCourt P (2004) Systems approaches to understanding cell signaling and gene regulation. Curr Opin Plant Biol 7(5):605–609
Romanowski A, Yanovsky MJ (2015) Circadian rhythms and post-transcriptional regulation in higher plants. Front Plant Sci 6:437
Salekdeh GH, Komatsu S (2007) Crop proteomics: aim at sustainable agriculture of tomorrow. Proteomics 7(16):2976–2996
Santos BM, Balbuena TS (2017) Carbon assimilation in Eucalyptus urophylla grown under high atmospheric CO2 concentrations: a proteomics perspective. J Proteome 150:252–257
Schaffer R et al (1998) The late elongated hypocotyl mutation of arabidopsis disrupts circadian rhythms and the photoperiodic control of flowering. Cell 93(7):1219–1229
Seki M et al (2017) Adjustment of the Arabidopsis circadian oscillator by sugar signalling dictates the regulation of starch metabolism. Sci Rep 7(1):8305
Sequeira-Mendes J et al (2014) The functional topography of the arabidopsis genome is organized in a reduced number of linear motifs of chromatin states. Plant Cell 26(6):2351–2366
Strayer C et al (2000) Cloning of the Arabidopsis clock gene TOC1, an autoregulatory response regulator homolog. Science 289(5480):768–771
Swift J et al (2020) Nutrient dose-responsive transcriptome changes driven by Michaelis-Menten kinetics underlie plant growth rates. Proc Natl Acad Sci U S A 117(23):12531–12540
Van Leene J et al (2019) Capturing the phosphorylation and protein interaction landscape of the plant TOR kinase. Nat Plants 5(3):316–327
Verhoeff RP et al (2018) The theoretical nature of systems thinking. Perspectives on systems thinking in biology education. Front Educ 3:40
Veyel D et al (2017) System-wide detection of protein-small molecule complexes suggests extensive metabolite regulation in plants. Sci Rep 7:42387
Vidotti ADS et al (2020) Analysis of autotrophic, mixotrophic and heterotrophic phenotypes in the microalgae Chlorella vulgaris using time-resolved proteomics and transcriptomics approaches. Algal Res 51:102060
Von Bertalanffy L (1972) The history and status of general systems theory. Acad Manag J 15(4):407–426
Wang Z-Y, Tobin EM (1998) Constitutive expression of the CIRCADIAN CLOCK ASSOCIATED 1 (CCA1) gene disrupts circadian rhythms and suppresses its own expression. Cell 93(7):1207–1217
Winter D et al (2007) An "Electronic Fluorescent Pictograph" browser for exploring and analyzing large-scale biological data sets. PLoS One 2(8):e718
York S, Orgill M (2020) ChEMIST Table: a tool for designing or modifying instruction for a systems thinking approach in chemistry education. J Chem Educ 97(8):2114–2129
York S et al (2019) Applications of systems thinking in STEM education. J Chem Educ 96(12):2742–2751
Zangori L, Koontz JA (2017) Supporting upper-level undergraduate students in building a systems perspective in a botany course. J Biol Educ 51(4):399–411
Zhang H, Li Y, Zhu JK (2018) Developing naturally stress-resistant crops for a sustainable agriculture. Nat Plants 4(12):989–996
Zhang Y et al (2020) A moonlighting role for enzymes of glycolysis in the co-localization of mitochondria and chloroplasts. Nat Commun 11(1):4509

Modern Approaches for Transcriptome Analyses in Plants

2

Diego Mauricio Riaño-Pachón, Hector Fabio Espitia-Navarro, John Jaime Riascos, and Gabriel Rodrigues Alves Margarido

Abstract

The collection of all transcripts in a cell, a tissue, or an organism is called the transcriptome, or meta-transcriptome when dealing with the transcripts of a community of different organisms. Nowadays, we have a vast array of technologies that allow us to assess the (meta-)transcriptome regarding its composition (which transcripts are produced) and the abundance of its components (what are the expression levels of each transcript), and we can do this across several samples, conditions, and time-points, at costs that are decreasing year after year, allowing experimental designs with ever-increasing complexity. Here we will present the current state of the art regarding the technologies that can be applied to the study of plant transcriptomes and their applications, including differential gene expression and coexpression analyses, identification of sequence polymorphisms, the application of machine learning for the identification of alternative splicing and ncRNAs, and the ranking of candidate genes for downstream studies. We continue with a collection of examples of these approaches in a diverse array of plant species to generate gene/transcript catalogs/atlases, population mapping, identification of genes related to stress phenotypes, and phylogenomics. We finalize the chapter with some of our ideas about the future of this dynamic field in plant physiology.

Keywords

RNA-Seq · Crops · Transcription · Gene expression · Polyploidy · Next-generation sequencing · Long reads · Short reads · Assembly

D. M. Riaño-Pachón (✉)
Laboratory of Computational, Evolutionary and Systems Biology, Center for Nuclear Energy in Agriculture, University of São Paulo, Piracicaba, Brazil
e-mail: diego.riano@cena.usp.br

H. F. Espitia-Navarro
School of Biological Sciences, Georgia Institute of Technology, Atlanta, GA, USA
e-mail: hspitia@gatech.edu

J. J. Riascos
Centro de Investigación de la Caña de Azúcar de Colombia, CENICAÑA,
Cali, Valle del Cauca, Colombia
e-mail: jjriascos@cenicana.org

G. R. A. Margarido
Department of Genetics, Luiz de Queiroz College of Agriculture, University of São Paulo, Piracicaba, Brazil
e-mail: gramarga@usp.br

F. V. Winck (ed.), *Advances in Plant Omics and Systems Biology Approaches*, Advances in Experimental Medicine and Biology 1346, https://doi.org/10.1007/978-3-030-80352-0_2

2.1 Introduction

The transcriptome is the collection of all RNA molecules found at a given time in an organism, in a tissue, or in a cell. Researchers today can study the full transcriptome, or a targeted transcriptome (a defined subset of transcripts under a certain condition) using an array of different technologies, like microarrays, reverse transcription quantitative PCR (RT-qPCR), and nucleic acid sequencing. In most approaches, the population of RNA molecules should be first converted into the more stable cDNA, but recent advances and the development of new sequencing platforms are allowing the direct sequencing of the RNA molecules, removing biases that could be introduced by the synthesis of cDNA (Garalde et al. 2018; Keller et al. 2018). Assessing the transcriptome offers an overview of the functional component of a genome and of the genes that must be active in order to achieve a given transcriptional state. Transcriptomics studies have been employed to develop catalogs of expressed sequences, by the identification of mRNAs, small-RNAs (e.g., miRNA, snoRNAs), long-non-coding RNAs (lncRNAs) among others. Also, to aid in the annotation of newly sequenced genomes, improving the inference and definition of gene structure, like start and end sites of the transcription, position of introns and exons, and alternative splicing patterns. Perhaps the most prevalent use of transcriptomics is the quantification of gene expression levels under different conditions aiming at revealing the molecular mechanisms underlying the establishment of phenotypes and responses to stresses. Transcriptomics is increasingly being used to infer the function of genes, by exploiting co-expression, under the assumption of "guilt-by-association," and for the identification of coordinated expression modules. The rapidly decreasing costs and wide availability of the diverse transcriptomics technologies are allowing studies in diverse groups of plants and addressing evolutionary questions about the evolution of expression patterns, gene expression and regulation networks, at a scale without precedent.

The earliest approaches that can be called transcriptomics studies relied on sequencing expressed sequence tags (ESTs) using the low-throughput Sanger chain-termination sequencing technology and started in the 1980s (see Fig. 2.1). EST sequencing projects were expensive and laborious but allowed assessing the functional fraction of a genome sequence at a fraction of the effort and cost. The wealth of sequence information generated in these projects could be leveraged with the development of array-based hybridization technologies (macroarrays used nylon membranes and microarrays used glass slides), which offered higher throughput and had lower application costs than EST projects, once the development of the membranes/slides had been deduced. The first use of the words microarray or macroarray in the scientific literature dates back to 1996, but their use really takes off in the 2000s (Fig. 2.1). The use of ESTs and array-based technologies was superseded by high-throughput sequencing-based methods, first exploiting small transcript signatures (tags) and later the sequencing of complete or close to complete transcripts.

In this chapter, we will introduce you to the basics of transcriptome studies, applications, and some examples in non-model plants.

2.2 Transcriptomics Approaches

2.2.1 Array-Based Approaches

Large-scale characterization of transcriptomes was made possible with the use of microarrays. In this technology, an array of oligonucleotide probes that are complementary to known transcripts is immobilized on a glass slide. Next, cDNA molecules synthesized from RNA are hybridized with the probes, and signal intensities are assessed to provide a measure of transcript abundance. This provides an economical way of analyzing transcriptomes on a genome-wide scale. Microarrays are used nowadays for model species and economically important crops, primarily due to low cost and laboratory routine.

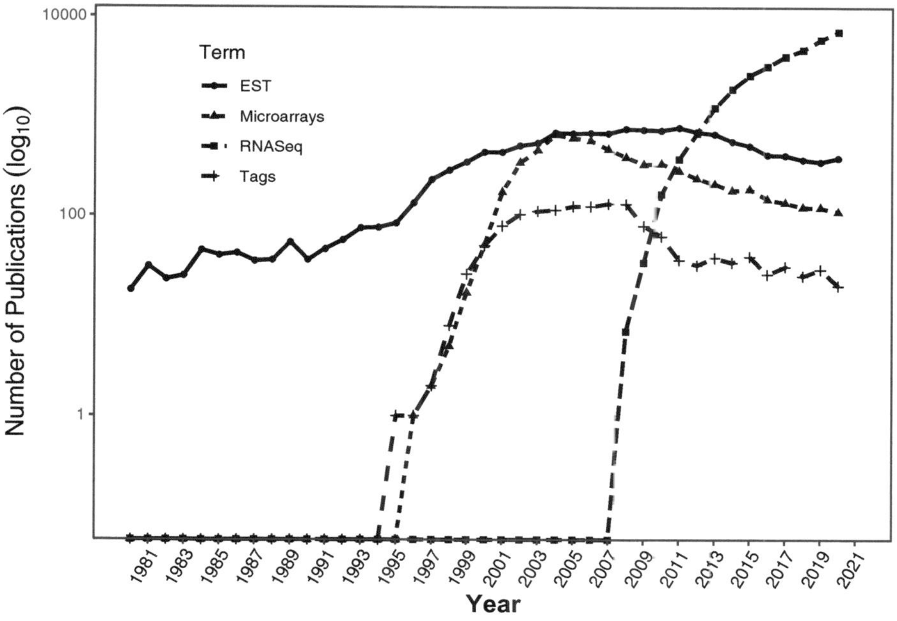

Fig. 2.1 Number of publications in the last four decades for different transcriptome technologies

However, this approach presents a number of disadvantages that have relevant practical implications. First, previous knowledge about the transcripts of interest is required for designing the array chip, which hinders application for non-model species. This may introduce bias toward the specific sequences used to obtain the probes, which is particularly important for genes with multiple isoforms. Second, transcript abundance estimation is not accurate for lowly expressed genes, owing to background noise from nonspecific hybridization, or for very highly expressed genes, due to probe saturation. The dynamic range of detection is thus limited. Third, cross-hybridization of transcripts with similar sequence can adversely affect expression estimates. Finally, intrinsic differences in hybridization exist between probes because of their sequence content (Marioni et al. 2008; Wang et al. 2009; Zhao et al. 2014).

Sequencing-based approaches resolve many of these issues and are now the method of choice for large-scale transcriptome profiling in a variety of scenarios. From now on, we will focus on these more recent strategies.

2.2.2 Sequencing-Based Approaches

In-depth knowledge and understanding of a plant genome, or any organism for that matter, involves the elaboration of a catalog of the genes present in the genome and information about the expression levels of the transcripts derived from these genes under a wide array of conditions. In both cases, one requires sequence data.

The most widely used technology in early genome projects was Expressed Sequence Tag (EST) sequencing (reviewed by Parkinson and Blaxter 2009). EST sequencing was employed to generate gene catalogs, both in model plants (Delseny et al. 1997; Weng et al. 2005; Asamizu et al. 1999; Banks et al. 2011) and in crops (e.g., Yamamoto and Sasaki 1997; Vettore et al. 2003; Ma et al. 2004; Pavy et al. 2005). In many cases, ESTs also served as a basis for the development of cDNA microarrays to query gene expression under different plant conditions or developmental stages (Lembke et al. 2012; Pavy et al. 2008). In most projects, ESTs were derived from normalized libraries, which meant that all transcripts have approximately the same probability to be sequenced. This readily reduces costs for gene discovery, but the gene expression levels and the dynamics of transcription regulation cannot be assessed.

With the creation and advance of high-throughput sequencing (HTS) technologies toward the end of the 1990s and in the early 2000s, new approaches were applied to discover plant genes and transcripts and to assess the dynamics of transcription, and its regulation, like alternative transcription starting sites (TSS) and alternative splicing form usage. Among these approaches, one could mention Cap Analysis of Gene Expression—CAGE (de Hoon and Hayashizaki 2008) and Serial Analysis of Gene Expression—SAGE (Velculescu et al. 1995; Matsumura et al. 2005), to name just a few, which are collectively known as tag sequencing approaches (Harbers and Carninci 2005) (see "Tags" in Fig. 2.1). These technologies started by exploiting the traditional Sanger DNA sequencing method to assess transcription, but moved soon to exploit the newer, highly parallel and HTS technologies, and thus gained suffixes like –deep or –seq and prefixes like ultra–, to differentiate them from their older lower throughput versions. Briefly, tag sequencing approaches aim to generate short sequence tags from the transcript ends, either the 5′ or the 3′ end. These short tags should unequivocally identify each transcript or genomic region, although it was not uncommon that a single tag could be mapped to more than one transcript/gene, particularly in cases of large gene families which are common in plants. In addition, the number of tags sequenced for each transcript is directly related to the transcript abundance in the original sample. Being based on short sequence tags from the transcript ends, these approaches were better suited for organisms whose genomes were already sequenced.

On the one hand, one of the main advantages of either EST or tag-sequencing approaches is the generation of a digital measure of gene expression, the number, or count, of a certain event, i.e., the sequencing of a complete, or part of a, RNA molecule. In contrast to an analogous measure, such as that offered by cDNA microarrays which is subject to probe saturation and thus has a low dynamic range, this digital measure is not saturated in the case of highly abundant transcripts. For the case of lowly expressed transcripts, the trivial alternative is to continue counting events until a certain number of rare events (lowly expressed transcripts) have been achieved, although this could have an important impact on the overall cost of the experiment. If lowly expressed transcripts are the focus of the study, then alternative approaches can be employed, such as targeted sequencing and reverse transcription quantitative PCR (RT-qPCR). On the other hand, the main drawback of both approaches (ESTs and tag-sequencing) is that neither of them provides the full representation of the underlying transcripts. Additionally, tag-sequencing and microarray approaches require preexisting knowledge about the transcript space of the species of interest, which impose serious limitations to its application in non-model organisms.

2.2.2.1 RNA-Seq

The sequencing of transcriptomes employing HTS technologies, without focus on any particular region of the mRNA, in contrast to CAGE or SAGE, is known as RNA-Seq. The first publications using the word RNA-Seq appeared between 2006 and 2008 applied to few organisms (Mortazavi et al. 2008; Nagalakshmi et al. 2008; Bainbridge et al. 2006; Wilhelm et al. 2008; Cloonan et al. 2008), also including *Arabidopsis thaliana*, the model land plant (Lister et al. 2008) (see "RNA-Seq" in Fig. 2.1).

The synthesis and maturation of transcripts is a finely regulated process that allows the plant cell to produce the required gene products in the proper quantities and at the proper times and places. Within a single experiment, RNA-Seq allows the discovery of expression levels, splicing events (Marquez et al. 2012; Shang et al. 2017; Brown et al. 2017), RNA editing (Hackett and Lu 2017), and mutations (Peng et al. 2016; Serin et al. 2017). RNA-Seq paves the way for the understanding of the rules governing RNA regulation and the underlying regulatory networks, thus generating new insights on plant development and the response to biotic and abiotic (Imadi et al. 2015) stresses at the cellular and molecular levels.

The main steps in any RNA-Seq project are (1) sample preparation, (2) library preparation and (3) sample sequencing.

(1) Sample preparation consists on the isolation of RNA from the biological samples of interest. Plant cells have different types of RNA molecules, like messenger RNA (mRNA), ribosomal RNA (rRNA), transfer RNA (tRNA), and other types of non-coding RNA (ncRNA). Over 95% of the transcript population in a cell consists of rRNA and tRNA species (Rosenow et al. 2001). Thus, to assess, via HTS technologies, the other transcript species, samples must be processed in special ways. For instance, if the objective of the project is to assess mRNAs transcribed by RNA pol II (which are mostly genes that will eventually undergo translation), one can exploit the fact that these eukaryotic mRNAs are polyadenylated, by fishing for these transcripts using poly-dT oligonucleotides, effectively excluding the large fraction of rRNA and other ncRNAs. On the other hand, if one is interested in evaluating the whole transcriptome (mRNA + all types of ncRNAs, only excluding rRNA), then there are approaches to specifically remove rRNA from the sample, usually employing hybridization techniques, methods that are usually referred to as ribo-depletion (O'Neil et al. 2013). Additionally, the goal of the study could be to focus on small ncRNAs, in that case one would perform a size fractionation and selection step.

As part of (2) library preparation, for short-read HTS technologies (see below for long-read HTS technologies), the isolated RNA must be converted into double-stranded cDNA and fragmented. Fragments should be ligated to adapters to allow amplification and sequencing. At this point, it is important to remember that a given message in the genome is encoded in one of the two strands of the DNA double helix, and thus it is important in most cases to keep the information of which strand was transcribed. In general, one can divide the library preparation methods in two groups, those that keep the strand information (strand-specific protocols) and those that do not (often called unstranded protocols). Today, most RNA-Seq datasets are still being generated using library preparation protocols that do not keep the strand information. For instance, from 219,832 green plant datasets using RNA as source in RNA-Seq experiments in the Short Read Archive (SRA; https://www.ncbi.nlm.nih.gov/sra/; July 2020), only 5995 have 'strand-specific' in their description.

(3) Sample sequencing is carried out in massively parallel sequencing instruments, paying attention to the dependence between library preparation method and sequencing instrument. The most widely available technologies for RNA-Seq are those released by Illumina Inc, i.e., using reversible-terminators sequencing-by-synthesis technology (Bentley et al. 2008; Illumina 2010), within their sequencing instruments MiSeq, HiSeq, NextSeq, or NovaSeq. Samples prepared with Illumina library construction methods are compatible with any of their instruments, the only difference being on the throughput obtained, e.g., number of sequenced fragments and number of samples that can be analyzed simultaneously.

Before you start your RNA-Seq project, you must develop the experimental design that will allow you to answer biologically relevant questions with a predefined level of certainty. Here we will only highlight two factors among the many that must be taken into account during the experimental design phase: (1) number of biological replicates and (2) number of sequenced fragments per sample. The number of replicates depends on your final goal. On the one hand, if your goal is to make a catalog of genes present in an organism's genome, typical when sequencing a new genome and preparing for annotating it, then preparing a single, or few, library from a pool of tissues and/or conditions might be enough. On the other hand, if you plan to evaluate the statistically significant differences in gene expression values between different conditions, then a higher number of replicates is required. Depending on the size of the effects that are desired to be detected, if only changes around two to threefold are sought, then a number of biological replicates around five should suffice in most cases; a higher number of replicates would be required to detect smaller changes in expression values (Schurch et al. 2016). Regarding the number of sequenced fragments, you should keep in mind that RNA-Seq is basically a random sampling process. If your goal is to assess statistical differences among conditions, you must check whether your sampling is deep enough to support your conclusions. A few approaches have been proposed to check for this, all of them are based on resampling your reads, and counting a feature of interest for each subsample for increasingly large subsamples. If the sequencing depth is high enough, you would expect that the number of a given feature is close to saturation with increasing number of resampled reads. There are a few approaches to achieve this. First you could count the number of transcripts that are detected at different fractions of the original datasets, e.g., 5%, 10%, 20, of the original reads; if sampling is deep enough, you would expect to find a plateau (Garcia-Ortega and Martinez 2015). Similarly, instead of looking at the number of transcripts, you can look at the number of exon–exon junctions detected with increasingly large samples of the reads; again you expect to achieve a plateau if your sequencing depth was saturated. This can be achieved with the junction-saturation.py script part of RSeQC (Wang et al. 2012). It is important to note that, despite sequencing depth being important, especially for lowly expressed genes, the number of biological replicates is much more important, and if you have to choose between more depth or more biological replicates, you should always choose the latter (Liu et al. 2014; Lamarre et al. 2018; Baccarella et al. 2018).

Regarding the sequencing depth, it is important to keep in mind that under several conditions, a large fraction of the reads would originate from one or a few transcripts. For instance, when doing sequencing of total RNA, you will have a large fraction of sequencing reads originating from rRNA transcripts, which can be up to 90% of the total RNA in the cell (Conesa et al. 2016). In these cases, you should try to deplete your sample from rRNA transcripts, for which several options are available in the market (Conesa et al. 2016; Hrdlickova et al. 2017; NuGen n.d.; siTOOLsBiotech 2018). However, not only rRNA transcripts exhibit such high abundance. A recent study of the *A. thaliana* transcriptome identified over 4000 ubiquitously and extremely highly expressed transcripts (Sun et al. 2014). If your specific project aims at assessing the expression of lowly expressed and rare transcripts, it might be important to deplete these ubiquitous and highly expressed transcripts, for such case, some alternatives for library preparation are available, as the AnyDeplete or riboPools technologies (NuGen n.d.; siTOOLsBiotech 2018).

2.2.2.2 Strand-Specific RNA-Seq

The existence of overlapping genes (genes whose transcripts are encoded—completely, or most frequently partially—in opposite strands of the same genomic region) in plants has been known for some years (Quesada et al. 1999; Xiao et al. 2005). Natural antisense transcripts (NATs) are RNA molecules that can have regions of sequence complementary to other RNAs and that can regulate the expression level of their target genes. Particularly, cis-NATs are pairs of transcripts that overlap on the genome. Disambiguating the expression levels of the two overlapping transcripts requires data that keep the information about which strand was transcribed (see for example, Britto-Kido Sde et al. 2013; Li et al. 2013a; Jin et al. 2008; Riano-Pachon et al. 2016). Between 7% and 8% of genes in rice (Osato et al. 2003) and *Arabidopsis* (Wang et al. 2005; Jen et al. 2005), respectively, are cis-NATs, recent studies suggesting even higher rates of cis-NATs (Oono et al. 2017; Zhao et al. 2018). Figure 2.2 illustrates the importance to have strand information for transcriptome analyses.

Currently, three technologies are widely available that can maintain strand information: Illumina's TruSeq Stranded library preparation kits, Pacific Biosciences's IsoSeq, and Oxford Nanopore Technologies's direct rRNA sequencing. Perhaps the most pervasive of the three in the market is the one commercialized by Illumina in their TruSeq Stranded library preparation kits, which use the deoxy-UTP strand-marking strategy. The Illumina instruments are capable of

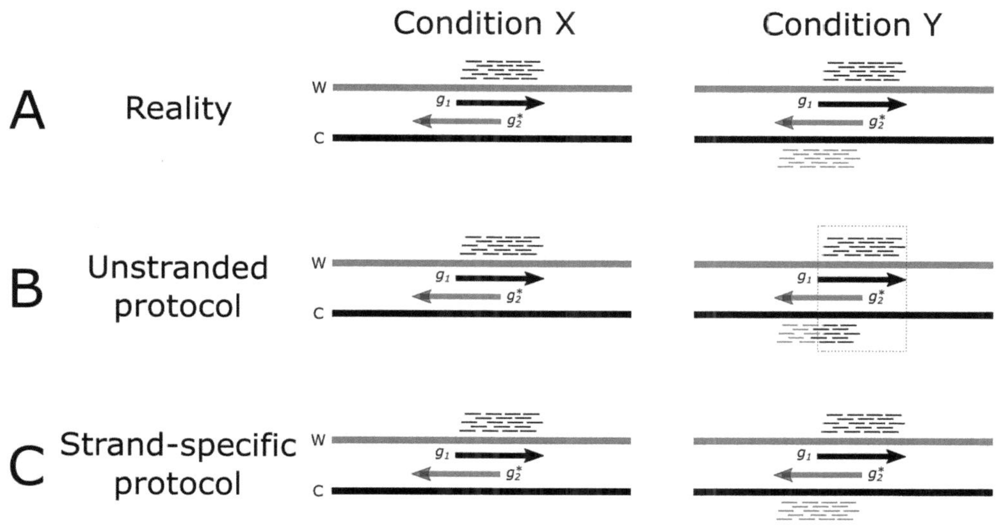

Fig. 2.2 Use of strand-specific information to disambiguate the expression of overlapping genes. Two overlapping genes g_1 in the Watson strand and g_2 in the Crick strand shown in two different experimental conditions, *X* and *Y*. The symbol * indicates that g_2 is an unknown (unannotated gene). Short sequencing reads appear either above or below the DNA strands as short line, each line representing a sequencing fragment. (**a**). The real case: g_1 is expressed in both conditions *X* and *Y*, with similar or identical abundances, while g_2 is only expressed in condition *Y*. (**b**) Sequencing results obtained with a protocol that ignores (or loses during library preparation) the information about which strand generated the reads. Only reads that overlap with annotated features are counted (dashed line in condition *Y*). In condition *Y*, many of the reads originated from the gene g_2 will counted as if they were from gene g_1 (reads shown in black). This will lead to the wrong conclusion that the expression of g_1 in condition *Y* is higher than in condition *X*. (**c**) Using a protocol that keeps strand information (strand-specific), in condition *Y* only the reads in black will be assigned to g_1, and the additional reads in gray will hint toward the existence of an additional gene in the same locus that is only expressed in condition *Y*. The abundances of g_1 in condition *X* and *Y* will be similar and will not lead to a differential expression call, as in (**b**)

sequencing double-stranded DNA molecules (dsDNA), but not single-stranded RNA molecules (ssRNA), so transcript sequences, which are made of ssRNA, must be transformed into dsDNA molecules by a process called cDNA synthesis. Briefly, the RNA molecules are fragmented, and each resulting fragment will be used for the synthesis of dscDNA in a two-step process. The first step, called First-Strand Synthesis (FSS), uses random primers, reverse transcriptase, and all the four deoxy nucleotides (dATP, dTTP, dCTP, and dGTP), resulting in a hybrid double-stranded RNA-DNA molecule. After FSS, the RNA molecule is degraded. In the second step, called Second-Strand Synthesis, the dTTP is replaced with dUTP. At the end of SSS, there is a dsDNA molecule, in which the strand with dTTP is the reverse complement of the sequence that was transcribed, and the strand with dUTP corresponds to the transcribed sequence. At this stage, the information about which strand was transcribed is already encoded in the chemistry of the created dscDNA. In the following step, the typical asymmetric Illumina Y-adapters are ligated to the dscDNA fragments. The incorporation of dUTP will quench the synthesis of the second strand during downstream amplification steps (Illumina 2017) or could be selectively degraded by Uracil-DNA-Glycosylase (UDG) (Borodina et al. 2011). Deciding whether an RNA-Seq dataset is stranded or not is quite easy and can be achieved by visual inspection of the reads mapped to either the genome or the transcriptome. However, some packages can aid inferring this, and are very useful when dealing with tens or hundreds of samples, some examples are the infer_experiment.py module part of RSeQC (Wang et al. 2012), or the option --lib-

Type A in Salmon (Patro et al. 2017), to name just a couple.

Data obtained from sequencing libraries prepared in such a way can be exploited either to map directly to a reference genome or transcriptome or build a de novo transcriptome assembly, in both cases exploiting the strand information and leading to correct directionality of the identified transcripts, with the potential for the identification of novel transcripts.

2.2.2.3 Long Read RNA Sequencing

Next-generation sequencing (NGS) technologies afforded the most widely used tools for transcriptome analysis in the recent years and are likely to remain pervasively used for many years to come. Still, RNA-Seq is not devoid of biases and limitations, notably about transcript identification and isoform disambiguation, as well as expression-level estimation. Short reads can be ambiguous, map to multiple locations, and originate from low-complexity sequences that hamper alignment.

The ability to sequence full-length transcripts, from the 5′ end to the poly-A tail, in principle allows complete differentiation of isoforms, with no ambiguity in assigning fragments to transcripts. It also eliminates the need for (de novo) transcript assembly. Third-generation sequencing (TGS) technologies already provide the means for achieving this goal, at least for a large fraction of the transcripts, with long reads that completely cover molecules with lengths upwards of 10 kbp. Besides facilitating transcript identification, long reads boost transcriptome analyses through the discovery of novel genes, novel isoforms, and detection of fusion transcripts (Rhoads and Au 2015; Shi et al. 2016). Even previously annotated sequences can be enhanced with these technologies, through correction of existing gene models (Liu et al. 2017). Furthermore, PCR-free protocols get rid of amplification biases that affect expression quantification.

One such technology is the Iso-Seq method (Rhoads and Au 2015) from Pacific Biosciences (PacBio). This isoform sequencing strategy has shown power to discriminate transcript isoforms in some important species (Abdel-Ghany et al. 2016; Li et al. 2018), including some with very complex genomes, such as cotton (Wang et al. 2018b), coffee (Cheng et al. 2017), and even the highly polyploid sugarcane (Hoang et al. 2017; Piriyapongsa et al. 2018). These studies collectively show that RNA-Seq based exclusively on short reads renders a limited view of the transcriptome, because of partial isoform identification and inaccuracies in expression quantification.

Long reads can also be obtained with the Oxford Nanopore technology. In addition to sequencing cDNA molecules, this approach allows direct RNA sequencing (Garalde et al. 2018), an alternative that removes reverse transcription biases and helps in identifying other types of RNA molecules, such as long non-coding and antisense RNAs (Jenjaroenpun et al. 2018). These technologies can also be applied for characterizing transcriptomes of individual cells (Byrne et al. 2017).

Despite these benefits, a series of practical concerns still limit the widespread application of third-generation sequencing technologies. Even though success in sequencing full-length transcripts is highly advantageous for cataloging the transcriptome of cells, quantitation is a different matter. Although potentially less biased for transcript abundance estimation (Byrne et al. 2017), the current lower throughput of these approaches prevents accurate quantification of transcripts in the wide dynamic range of expression levels, with more pronounced effects on lowly expressed transcripts. Increasing sequencing depth can circumvent this issue, but this is presently limited by the higher cost of long reads, such that efforts in improving throughput and lowering costs are vital.

Another obstacle is that sequencing errors rates are substantially higher for state-of-the-art long read technologies (Jenjaroenpun et al. 2018). Error rates in Iso-Seq reads can be greatly reduced by the so-called circular consensus sequence (CCS), in which the same molecule is repeatedly sequenced (Rhoads and Au 2015; Liu et al. 2017). However, this is not yet feasible for long, single-pass transcripts, which still suffer from lower sequencing accuracy. Hybrid strategies that combine the transcript identification power of TGS with the massive read volume of

NGS enable error correction and abundance estimation for a more complete and trustworthy transcriptome characterization (Li et al. 2018; Jenjaroenpun et al. 2018).

2.2.3 Transcriptome Assembly

2.2.3.1 Genome-Guided Transcriptome Assembly

When the genome sequence of the species under study is available, one can choose to try assembling the transcriptome from raw data (short reads) using the genome as a guide. This procedure consists of mapping the RNA-Seq reads onto the reference genome sequence and then looking for clusters of sequencing reads representing putative isoform transcripts that should be assembled. During the mapping step, the read mapper employed must be aware of spliced-reads, that is reads that span exon–exon borders, like HiSAT2 (Kim et al. 2015, 2019), STAR (Dobin et al. 2013), or GSNAP (Wu and Nacu 2010), among others. After reads have been mapped and clustered along the genome sequence, these clusters of reads are usually represented as a graph (Florea and Salzberg 2013). The graph model could be a splice graph, where exons or parts of exons are represented as nodes and edges represent possible splice variants, implemented in the software Stringtie (Pertea et al. 2015), or an overlap graph, where nodes represent sequence fragments or reads (*k*-mers) and edges connect sequence fragments if they overlap and have a compatible splice pattern, implemented in software such as Cufflinks, Scripture, and Trinity (Trapnell et al. 2010; Haas et al. 2013; Guttman et al. 2010). Alternatively the genome sequence could be just used to cluster reads together to be then de novo assembled, using software such as Trinity (Haas et al. 2013; Grabherr et al. 2011).

Genome-guided transcriptome assembly is usually more precise than de novo transcriptome assembly (see below), as it is less sensitive to sequencing errors, polymorphisms, and paralogous loci (Ungaro et al. 2017; Zhao et al. 2011). It is important to note, though, that it could only help in recovering/assembling the transcripts that are present in the sequence used as reference, so variation between individual, ecotypes, cultivars, etc. would be missed. This has been highlighted in recent studies about the pan-transcriptome and pan-genome of diverse plant species (Gao et al. 2019; Ma et al. 2019). Also, if the genome sequence used as reference is fragmented, exons or whole transcripts could be located in sequencing gaps. An alternative to overcome these limitations would be the generation of a comprehensive, or non-redundant, transcriptome, that leverages the information of the genome-guided transcript assembly and of de novo transcript assemblies (Visser et al. 2015; Jain et al. 2013). The PASA pipeline (Haas et al. 2003) and CD-HIT-EST (Fu et al. 2012) can generate such non-redundant transcriptome representations, by controlling the minimum fraction identity, and length aligned to create transcript clusters. Clustering at 100% identity would be the most basic level of clustering, and lower values, like 99% or 95% identity, could be useful to cluster transcripts originating from the same locus via alternative splicing, allelic versions, or closely related paralogous genes. GET-HOMOLOGUES-EST could enhance the generation of a comprehensive transcriptome, while taking into account coding potential, the presence of conserved protein domains, and information from closely related species or individuals within a polymorphic species (Contreras-Moreira et al. 2017).

2.2.3.2 De Novo Transcriptome Assembly

The availability of an annotated reference genome sequence eases the analysis of RNA-Seq data, by dividing the problem of transcript assembly and quantification into substantially smaller subsets. In this situation, sets of reads aligning against a particular genomic region can be analyzed independently of the remainder of the sequencing data.

It is nevertheless possible to carry out a thorough transcriptome analysis for non-model plant species lacking a reference genome (Collins et al. 2008). When available, the genome sequence of a closely related species can be used as a reference.

Alternatively, instead of aligning the reads against genomic sequences, a transcriptome reference can be assembled de novo based on the RNA-Seq reads alone. This provides a cost-effective means of applying functional genomics tools to less well-studied organisms. It can also shorten the path to biological insight because any species can potentially be studied without the need for previous genomic knowledge. However, de novo transcript assembly is one of the most difficult tasks in bioinformatics (Garg and Jain 2013).

The most widely used de novo transcriptome assemblers are based on a *de Bruijn* graph, a data structure that compactly represents the sequences of hundreds of millions of short sequencing reads. Construction of a *de Bruijn* graph involves parsing the collection of reads and extracting k-mers of a certain size. A k-mer is a subsequence of length k contained in any biological sequence segment, such as a read, a transcript, or even an entire chromosome. In a standard *de Bruijn* graph, each existing k-mer is represented by a node, or vertex. If a suffix of length $k - 1$ of a given node matches the $k - 1$ prefix of another node, an edge connecting these vertices is used to represent this overlap. After obtaining this graph, assembly software packages usually perform several (combinations of) steps of error correction, graph simplification and collapsing, scaffolding, and gap closure. Finally, graph traversal based on sequencing read information can be used to reconstruct contigs representing transcripts.

Contig assembly algorithms based on *de Bruijn* graphs were initially devised for genome assembly based on high depth sequencing data. Indeed, many of the currently available transcriptome assemblers were built relying on previously existing genome assemblers. For example, Oases (Schulz et al. 2012) is a pipeline built on top the Velvet genome assembler (Zerbino and Birney 2008). Similarly, Trans-ABySS (Robertson et al. 2010) is based on ABySS (Simpson et al. 2009), and SOAPdenovo-Trans (Xie et al. 2014) uses the *de Bruijn* graph from SOAPdenovo2 (Luo et al. 2012) as a starting point. Following a more widespread adoption of RNA sequencing studies, proper de novo assemblers such as Trinity (Grabherr et al. 2011; Haas et al. 2013), were also developed from scratch to tackle the challenges posed by these datasets.

Despite using an underlying data structure similar to genome assemblers, these software packages take into account unique features of the RNA-Seq data to drive the assembly strategy and address several particular issues. While the goal in genome assembly is to produce a few large (chromosome-sized) sequences, transcriptome assembly aims to reconstruct tens of thousands of sequences, each representing a different transcript. Also, coverage depth in RNA sequencing is heavily dependent on gene expression levels, such that approaches for assembling lowly or highly expressed genes can differ.

These de novo assembly methods can naturally handle alternative splicing arising from RNA processing after transcription. Ideally, a transcriptome assembly should contain full-length transcripts accurately representing different isoforms, while also separating paralogs from large gene families. For polyploid species, the presence of multiple alleles and homeologs adds another layer of complexity that makes assembly an even harder exercise. In this context, it is noteworthy that long-range information from paired-end and/or longer sequencing reads provide a valuable resource that can greatly enhance assembly quality by simplifying the recovery of full-length transcripts.

Even though the current transcriptome assemblers are based on similar basic concepts and share many features, they differ widely in running time and required memory. They also stand apart in their ability to recover full-length transcripts from datasets with varying sequencing depth, obtained from species with distinct transcriptome complexity. Comparisons among assemblers can reveal scenarios in which particular combinations of software and parameters show superior performance (Zhao et al. 2011).

Finally, functional annotation of the assembled transcripts is commonly done to provide meaningful biological information about each resulting sequence. This usually entails adding gene ontology terms (Ashburner et al. 2000; Gene Ontology Consortium 2017) and pathway information from KEGG (Kyoto Encyclopedia of Genes and Genomes) (Kanehisa and Goto 2000; Kanehisa et al. 2016), to the transcripts, as well as searching for protein domains. Pipelines for performing such annotation include Blast2GO (Conesa et al. 2005) and Trinotate (Bryant et al. 2017).

2.2.3.3 Assessment of Transcript Assemblies

The goal of transcriptome assembly, either genome-guided or de novo, is to generate a truly complete collection of all the transcripts produced by an organism. However, attaining that goal is in most real cases unlikely, some of the reasons for this include: (1) Sequencing depth is limiting, and lowly abundant transcripts are not represented in the sequencing data. (2) Biases of the sequencing depth limit the observation of certain transcripts, e.g., problems with high GC content sequences. (3) Not all possible transcripts are expressed at a given moment, a good transcriptome coverage should include a survey of samples from different developmental stages, growing conditions, tissues, and organs. Thus, we need tools to assess the quality and completeness of a generated transcriptome assembly (Honaas et al. 2016; Moreton et al. 2015; Li et al. 2014; Smith-Unna et al. 2016). In the following, we describe some of the most important metrics to evaluate a transcriptome assembly.

Evaluation of Sequencing Depth

There are two related questions that are often asked at the beginning of any transcriptome study using NGS. (1) How many reads should be generated to capture most/all of the transcripts? (2) Are the reads generated enough to make statistical inferences or to get a complete overview of the transcriptome? In order to answer these, one can evaluate the degree of read saturation present in the assembly as a function of sampling effort, using an approach analogous to that of species accumulation curves (rarefaction curves) in biodiversity studies. This approach will allow to decide whether sequencing depth has been enough to capture all transcripts in the sample (Hale et al. 2009). At the beginning of a study, before generating the data, one could carry out a pilot study with shallow sequencing depths, that could help estimating the depth required to capture all or most of the transcripts. Alternatively, and if a genome reference is available, one could evaluate the saturation of orthogonal features, for instance the number of exon–exon junctions supported by the sequencing reads at different levels of sequencing effort, this approach has been implemented in the tool junction_saturation.py in the package RSeQC (Wang et al. 2012).

Percent Reads Mapped

The proportion of reads that map back to the assembly is also a measure of assembly and data quality. In principle one wants most of the original read data (after quality trimming) mapping to the transcriptome assembly. However, when using a genome as a reference (or the transcriptome derived from the genome sequence), a low percent of reads mapping could also be indicative of large diversity between the reference and the sample, or of contamination, and further analyses would be required.

Identification of Sets of Conserved Genes

Genes that appear in all of the best-known genomes can be exploited to evaluate the completeness of a transcriptome assembly. The tool Benchmarking Universal Single-Copy Ortholog (BUSCO) has sets of conserved single-copy orthologous genes present at diverse taxonomic levels, e.g., Viridiplantae (green plants), Embryophyta (land plants) (Waterhouse et al. 2017). A transcriptome that was assembled from samples representing different developmental stages, growth conditions, tissues and organs, should have a good representation of these conserved single-copy gene sets. On the other hand, a transcriptome representing a single condition could have a low value for this metric, corroborating its specificity. Alternatively one could also compare the assembled transcripts to the transcripts (or proteins) of a related species, these are usually called reference-based or comparative metrics and are implemented in tools such as TransRate (Smith-Unna et al. 2016) or Detonate's REF-EVAL (Li et al. 2014).

Contamination Screening and Filtration
NGS data can easily be contaminated, but it is important to note that there are different sources of contaminants. There can be internal contaminants, for instance, mitochondrial and plastid sequences, or ribosomal RNA sequences. Or there could be external contaminants, genetic material from other organism present in the sample, e.g., symbionts, pests, fungi, or bacteria. In general, contamination should be removed as early as possible, in order to reduce computational costs, fragmentation of the assembly and the chance to generate chimeric transcripts (Zhou et al. 2018). For example, BBDuk (https://jgi.doe.gov/data-and-tools/bbtools/) can be used to efficiently remove rRNA reads by comparing them against the SILVA database (Quast et al. 2013). A similar approach could be followed to eliminate reads from other contaminants if they have been previously identified. The presence of rRNA could be exploited to identify which contaminants (if any) are present in the sample.

2.2.4 Transcript Quantification

2.2.4.1 Alignment/Mapping-Based Approaches

Transcriptome characterization via RNA-Seq not only provides a catalog of transcripts present in a particular sample of cells, but also yields quantitative information that allows expression levels to be assessed. This is true both for species with and without a reference genome. A major step for obtaining expression estimates is to assign sequencing reads to genes or transcripts, which is commonly accomplished by first aligning them to a reference genome or transcriptome sequence.

Development and application of alignment algorithms has been one of the most active research areas in bioinformatics, and consequently, there is a wide range of tools available for various purposes. The majority of alignment algorithms tailored for short reads use indexing strategies that can be categorized into two main approaches: a seed-and-extend strategy based on hash tables or alignment based on a Burrows-Wheeler transform (Flicek and Birney 2009; Trapnell and Salzberg 2009; Li and Homer 2010).

Short read sequence aligners were initially developed for aligning genomic reads against a reference genome. In this situation, reads are expected to align contiguously against the reference, except for minor gaps which may stem from small indels or sequencing errors. Reads from RNA-Seq libraries, on the other hand, originate from cDNA molecules synthesized from mature mRNA templates, from which introns have been stripped off. Aligning RNA-Seq reads against a reference genome then requires splice-aware aligners, which appropriately handle reads that span exon junctions, without penalizing long gaps corresponding to introns. This class of aligners includes TopHat2 (Trapnell et al. 2009; Kim et al. 2013), which has been superseded by HISAT2 (Kim et al. 2015) and STAR (Dobin et al. 2013). An interesting quality of these aligners is that they can not only use previously annotated splice junctions, but also discover novel junctions and isoforms.

Following alignment, mapped reads can be assigned to annotated features in the genome. A simple and widely used way to measure expression levels is to count the number of reads overlapping a feature of interest. This is the approach implemented in programs such as HTSeq (Anders et al. 2015) and the featureCounts (Liao et al. 2014) component of the Subread package (Liao et al. 2013).

Reflecting the nature of gene expression, feature annotation follows a hierarchy of terms, with a gene frequently corresponding to the highest-level term. Any given gene may originate one or more transcripts, which in turn may contain one or more exons and compose one or more coding sequences. Read counts can be obtained for features at any level desired, but it is frequent to count reads overlapping exons. Depending on the goals of the study, features may then be grouped to obtain expression levels for meta-features. For instance, counts for all exons of a given transcript may be combined to get a transcript-level expression estimate, or all exons of all transcripts of a gene may be used to yield a gene-level read count. It is important to realize that, when working with paired-end read information, both reads of a pair come from a single molecule fragment, such that they should contribute only once to the expression count.

It is not always possible to uniquely assign a read to a feature or meta-feature. In some cases, there are overlapping features in an annotated genome reference, as a consequence of the structural organization of genes in the species of interest. Reads that align to a genomic region covered by two or more genes may not unequivocally be assigned to any one of them. Much of this ambiguity can be worked out by using stranded RNA-Seq library preparation, because overlapping genes may be transcribed in opposing directions.

Additionally, different gene isoforms can share a common exon, such that reads overlapping this exon are ambiguous. Lastly, the aligner may report multiple possible mappings for some reads, due to sequence similarity between members of a gene family, conserved protein domains and sequencing errors. The researcher can decide whether to simply discard multimapping or ambiguous reads, count them for all overlapping features or assign them heuristically. It should be noted that ambiguities at a given annotation level may not represent ambiguities at a higher level (e.g., a read mapping to an exon shared by multiple isoforms is ambiguous at the transcript level, but not at the gene level).

When using a de novo assembled transcriptome, introns are virtually absent from the reference, and therefore, one may use standard sequence aligners, such as BWA-MEM (Li and Durbin 2009; Li 2013) and Bowtie2 (Langmead et al. 2009; Langmead and Salzberg 2012). Splice-aware aligners also have modes for aligning reads against a splice junction-free reference sequence. For expression level quantification, in this case each contig can independently be treated as a feature. In fact, some assemblers such as Trans-ABySS may internally leverage the alignment of reads to contigs and automatically provide a measure of the per-contig expression level. The simplicity of the feature annotation in an assembled transcriptome does not mean that alignment and quantification are an easier endeavor. In fact, the issue of multiply aligned reads can be even more challenging in this situation, as it can be hard to distinguish between paralogs of the same gene.

These ambiguity issues have prompted alternative approaches for obtaining expression estimates to be devised. Because of the uncertainty in determining the transcript of origin of sequencing reads, one such possibility is to use mixture-model procedures that probabilistically assign reads to features, instead of simply counting overlapped fragments. As an example, the RSEM method (Li et al. 2010; Li and Dewey 2011) generates maximum likelihood or Bayesian expression estimates based on several variables of the annotated feature set and of the aligned reads, such as length, orientation, and quality scores. The main underlying principle is that uniquely aligned reads can also provide information for the (probabilistic) assignment of ambiguous reads. For example, suppose that two isoforms of a gene share one common exon, but also contain one exclusive exon each. If a large number of fragments align to one of the exclusive exons, while the other shows no overlapping reads, it is likely that fragments overlapping the common exon also originate from the isoform with a higher expression level based on the uniquely aligned reads.

Similarly, the Stringtie package formulates the simultaneous estimation of isoform assembly and abundance as a maximum network flow problem (Kovaka et al. 2019; Pertea et al. 2015). This maximum flow approach has been shown to be as accurate as the maximum likelihood approach in cufflinks (Trapnell et al. 2010), but it is able to recover a larger fraction of bona fide transcripts (Kovaka et al. 2019). In the maximum flow approach, a path in the splice graph with the heaviest coverage is used to build a flow network, this path represents a transcript, which is then removed from the splice graph, and a new path with the heaviest coverage is sought, until no more transcripts are assembled. The coverage for each assembled transcript is used to represent expression values as FPKM (fragments per kilobase million) and TPM (transcripts per million).

These difficulties in estimating expression levels are substantial enough for diploid model species. The situation may be considerably harder for researchers dealing with polyploid organisms, because of the added complexity from homeologs and multiple alleles. It is reasonable to assume that probabilistic strategies for read assignment may provide more accurate estimates of transcript abundance in this case.

Finally, a brief comment on expression-level normalization is needed. Transcript read counts are influenced by the length of the transcript and the size of the sequenced library, i.e., the number of fragments obtained from a given sample. Read counts are expected to be higher for longer transcripts and larger libraries. Many downstream application packages directly handle raw read counts, but it is not always straightforward to interpret raw values. For reporting expression levels, for instance, it is useful to use normalized values, such as the TPM (*transcripts per million*) value (Li et al. 2010; Wagner et al. 2012). It represents the number of transcripts of a certain type present in a total of one million sequenced transcripts from a given sample and thus estimates the fraction of that transcript in a pool of RNA molecules. The TPM is normalized by the length of the transcript, the sequencing depth, and the mean transcript length in the sample. Relative expression levels represented by TPM values do not depend on the expression levels of other genes in the transcriptome and appropriately measure the fraction of fragments from a given gene or isoform. Other measure of gene expression includes the RPKM (reads per kilobase million) and FPKM (fragments per kilobase million), but they have been largely superseded by the TPM.

2.2.4.2 Alignment-Free Approaches

Recent methods have tried to let go of the traditional strategy of mapping reads to a reference and then count, to arrive at estimates of gene expression levels, approach described above. The main reason for this is that these traditional approaches require large computational resources, and do not scale well with the amount of available data. These newer approaches implement what they call as pseudo-alignment, lightweight mapping, or quasi-mapping (Patro et al. 2017, 2014; Bray et al. 2016) and are known as alignment-free methods. Another important difference to the traditional approach is that instead of using reference genomes, these approaches use reference and well-annotated transcriptomes, including transcript isoforms, allowing the accurate estimation of isoform expression levels. Expression-level estimates at the level of isoforms are important given that most plant genes are interrupted (i.e., they have introns), and the removal of introns is a regulated process that can generate alternative splicing forms, which can have different, even antagonistic functions (Shang et al. 2017). In order to estimate isoform expression levels, tools like Kallisto or Salmon, let go of the idea of knowing where a read aligns in a given transcript, with base-to-base correspondence, and instead try to identify a transcript, or a set of transcripts, that could have originated such read, without keeping track of base-to-base correspondences. Such approaches have been shown to be extremely fast and accurate (Zhang et al. 2017). Some of these methods, besides their speed, can model different sources of sample-specific biases that can affect transcript quantification, like sequence-specific, fragment GC-content and positional biases (Patro et al. 2017; Bray et al. 2016). Refinement of the initial lightweight mapping of reads to the transcriptome, using Selective Alignment, allows the elimination of most mapping errors, by providing alignment scores that allow to distinguish alternative mapping locations that otherwise would appear the same (Srivastava et al. 2019).

2.3 Applications

Figure 2.3 shows some of the paths that can be followed in RNA-Seq studies. Table 2.1 lists some of the main software packages to carry out the operations shown in Fig. 2.3.

2.3.1 Differential Gene Expression

RNA sequencing is frequently done with the goal of detecting differences in expression levels between two or more contrasting groups of samples. One may be interested in evaluating the effect of different experimental treatments, genotypes, or stress conditions, for instance, on the

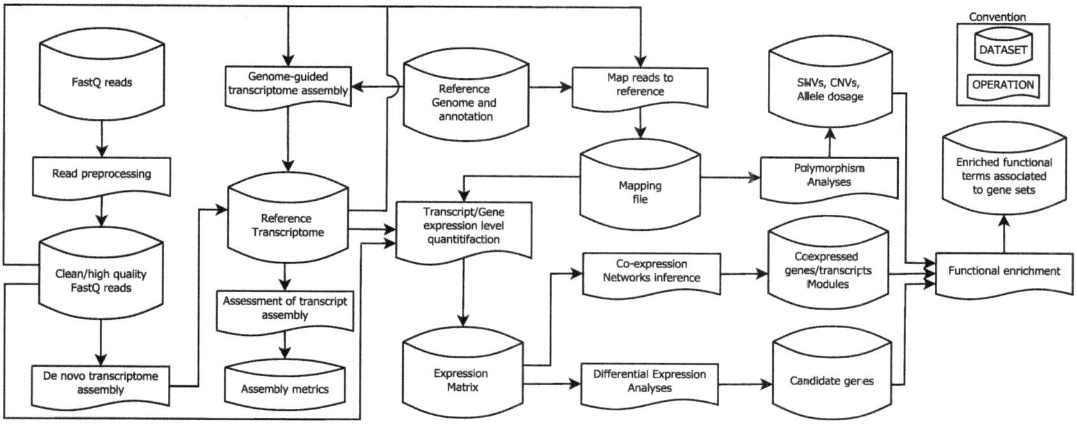

Fig. 2.3 General steps in an RNA-Seq analysis pipeline. Not all steps/paths are taken in a given study

transcriptome of particular cells. Gene or isoform expression measures are thus often used for identifying transcripts that are significantly up- or downregulated in a condition of interest, in comparison to a distinct condition.

Differences in the expression levels of two (groups of) samples can be represented by the fold change, which is simply the ratio of the expression levels estimated for both cases. Usually the expression estimate of a control or reference condition is used in the denominator, whereas the expression level of the treatment group is used in the numerator. As a result, genes that are upregulated in the treatment samples show a fold change greater than one (with no upper boundary), while downregulated genes display a fold change between zero and one. This discrepancy in scale led to the representation of these ratios in the log2 scale, such that fold changes in both directions are symmetric around zero.

Several methodologies are available for testing whether an observed fold change is statistically significant. Many of these methods use read count data directly, which calls for modeling of the expression levels with discrete distributions. The first statistical approaches proposed for such tests used the Poisson distribution to model read counts, assuming that the variance in the estimates was directly proportional to the mean expression level (Wang et al. 2010). This proved to be appropriate for technical replicates of the same sample (Marioni et al. 2008), but variance for biological replicates was shown to be higher than expected based on the mean alone (Robinson and Smyth 2008).

An alternative to the Poisson distribution is the negative binomial, which adds a second parameter (often denoted dispersion), allowing the sample variance to be different from the mean; hence, it corresponds to a Poisson distribution with overdispersion. This is the approach taken by most of the modern differential expression analysis packages (Wang et al. 2010; Robinson et al. 2010; Trapnell et al. 2013; Love et al. 2014).

The need to estimate sample variances makes it clear that biological replication is necessary in RNA-Seq experiments. Appropriate design planning is required, and all treatment combinations should be replicated, as alternatives devised for data without replicates are far from ideal. Yet, despite continual reduction in sequencing costs, RNA-Seq for large numbers of samples may still be impractical for many research goals. In order to increase reliability of variance estimates obtained from small numbers of replicates, techniques that share information between genes were proposed and implemented (Robinson and Smyth 2007).

Software packages edgeR (Robinson et al. 2010), DESeq (Anders and Huber 2010; Love et al. 2014), and Cuffdiff (Trapnell et al. 2010, 2013) are among the most extensively used tools

Table 2.1 Some of the software packages for different steps in RNA-Seq analysis pipelines

Activity		Software	Reference
Read pre-processing	Sequencing diagnostics	FastQC	bioinformatics.babraham.ac.uk/projects/fastqc
		RSeQC	Wang et al. (2012)
		RNA-SeQC	DeLuca et al. (2012)
	Removal of adapters and low-quality bases	Trimmomatic	Bolger et al. (2014)
		Atropos	Didion et al. (2017)
		BBDuk	sourceforge.net/projects/bbmap/
	Removal of ribosomal RNA	SortMeRNA	Kopylova et al. (2012)
		BBDuk	sourceforge.net/projects/bbmap/
	Identification of duplication artifacts	dupRadar	Sayols et al. (2016)
De novo transcriptome assembly		Trinity	Grabherr et al. (2011)
		Trans-ABySS	Robertson et al. (2010)
		Velvet/Oases	Schulz et al. (2012)
Genome-guided transcriptome assembly		Trinity	Grabherr et al. (2011)
		Stringtie	Kovaka et al. (2019)
		PASA	Haas et al. (2003)
Assessment of transcriptome assembly		BUSCO	Waterhouse et al. (2017)
		DETONATE	Li et al. (2014)
		Transrate	Smith-Unna et al. (2016)
Functional annotation		Trinotate	Bryant et al. (2017)
		Blast2GO	Conesa et al. (2005)
Read mapping		STAR	Dobin et al. (2013)
		GSNAP	Wu and Nacu (2010)
		HISAT2	Kim et al. (2015), Kim et al. (2019)
Transcript/gene expression-level quantitation		Stringtie	Kovaka et al. (2019)
		featureCounts	Liao et al. (2014)
		kallisto	Bray et al. (2016)
		Salmon	Patro et al. (2017)
Differential expression analyses		Limma	Ritchie et al. (2015)
		edgeR	Robinson et al. (2010)
		Ballgown	Frazee et al. (2015)
		Sleuth	Pimentel et al. (2017)
		DESeq2	Love et al. (2014)
Co-expression network inference		WGCNA	Langfelder and Horvath (2008)
		HRR	Liesecke et al. (2018)
		HCCA	Mutwil et al. (2010)
Polymorphism analyses		GATK	DePristo et al. (2011), McKenna et al. (2010)
		NGSEP V4.0	sourceforge.net/p/ngsep/
Functional enrichment		goseq	Young et al. (2010)
		topGO	Alexa et al. (2006)
		Blast2GO	Conesa et al. (2005)

for differential expression analyses. In more detail, edgeR uses raw read counts and models sample variation in terms of the biological coefficient of variation, which corresponds to the square root of the dispersion. It allows estimating a common dispersion for all genes, or a trended dispersion via a locally weighted adjusted profile likelihood for genes with similar average read count. It further allows moderated gene-wise dispersion estimates to be obtained by a weighted likelihood method combining individual and trended or common estimates (McCarthy et al. 2012). Normalization is carried out with a trimmed mean of log2 fold changes (Robinson and Oshlack 2010).

Similarly, the DESeq2 package uses size factors estimated based on the median of ratios of observed read counts to normalize expression levels. It empirically estimates the relationship between mean and variance of the negative binomial distribution, fitting a smooth curve of the dispersion as a function of the average expression of genes with similar means. Finally, it employs empirical Bayes approaches to shrink gene-wise dispersion estimates and also the fold changes, which is particularly relevant for lowly expressed genes and/or those with highly variable expression levels. Both edgeR and DESeq were initially designed for performing differential expression analyses of simple experiments, commonly involving pairwise comparisons of contrasting conditions. More recent implementations of edgeR and DESeq2 allow fitting generalized linear models for analysis of more complex designs, with the inclusion of experimental blocking factors and modeling of interactions, for example.

Cuffdiff 2 was developed for testing differential expression at both the isoform and the gene levels. Instead of using raw read counts, it models variability across replicated expression estimates by jointly considering overdispersion and uncertainty in the assignment of reads to their possible originating transcripts. Because of differences in the normalization procedures and model assumptions, these methods differ in their statistical power to detect differential expression over the range of expression values, as well as in the occurrence of false positives. Note also that conducting differential expression analyses at the transcript level may have important implications for statistical power. Greater uncertainty in expression estimates, because of more ambiguously mapped reads, negatively influences statistical power. Differential isoform expression analyses may require higher coverage depth, as more reads are needed to provide accurate estimates of individual isoform expression levels, especially for genes with many isoform variants and many shared exons. On the other hand, failure to adequately model uncertainty in read to transcript assignment can result in higher rates of false positives, even at the gene level.

RNA-Seq is a high-throughput screen that yields quantitative information for tens of thousands of genes (or hundreds of thousands of transcripts). Consequently, statistical tests are applied for multiple comparisons, which can result in many false positives if liberal significance levels are used for individual tests. Multiple testing correction is generally used to control for the occurrence of such false positives. One of most well-known corrections is the Benjamini and Hochberg (Benjamini and Hochberg 1995) false discovery rate (FDR) correction, aimed at controlling the proportion of false discoveries among the rejected hypotheses, while minimizing the drop in statistical power.

The output of these analyses is a list of significantly differentially expressed genes. Because of the large number of genes studied, this list may be quite long, which complicates summarization and reporting of the results. More easily interpretable biological meaning can be extracted from such lists through functional enrichment analyses, that look for overrepresented groups of genes among the statistically significant ones. Groupings of interest are usually obtained by categorizing genes according to their functional annotation, including gene ontology terms and/or biological pathways. Each functional group is tested for overrepresentation in the gene list against a background set, which includes all (expressed) genes in the transcriptome.

2.3.2 Co-expression Networks

Networks have recently emerged as a robust and holistic approach to understand complex cellular processes that comprise multiple and parallel interactions between cellular constituents such as DNA, RNA, and proteins. The network approach allows analyzing components and interactions as a system instead of analyzing them as separate entities. In a general way, a network, or graph, is defined as a set of elements called nodes, which are related through connections called edges. When edges have a direction, that is, they have source and target nodes, the network is called directed; otherwise, the network is undirected. These simple definitions are used to create biological networks that model cellular processes by taking nodes to represent molecules such as genes, proteins, or metabolites, and edges to represent physical, functional, or chemical interactions (Barabasi and Oltvai 2004). Depending on the molecules and interactions used, biological networks can be gene co-expression networks (GCN), genetic interaction networks, gene regulatory networks, protein–protein interaction (PPI) networks, metabolic networks, and signaling networks (Serin et al. 2016; Vital-Lopez et al. 2012). This section will focus on gene co-expression networks, in which each node corresponds to a gene, and edges represent co-expression relationships.

An advantageous feature of GCNs is the ability to reduce data complexity drastically. Nodes in a GCN, rather than solely representing a gene per se, represent its whole expression profile when the studied organism is under a condition, such as a treatment or biotic/abiotic stress. Edges in a GCN represent associations between gene expression profiles and can be interpreted as the simultaneous and coordinated expression of two or more genes under the studied perturbations. Thus, GCNs reduce the complexity of expression data of multiple samples from one or multiple experiments.

GCNs can be constructed from expression data derived from DNA microarrays and RNA-Seq. Traditionally DNA microarrays were the primary source of data expression for constructing GCNs, as this technology has been used intensively for almost two decades in gene expression studies. Recently, with the advent of next-generation sequencing (NGS) technologies, RNA-Seq has turned in a natural source for constructing GCNs. Among the advantages that microarrays had over RNA-Seq for the reconstruction of GCNs, we can name the considerable amount of information available in public databases, the well-established and mature data normalization approaches, and data homogeneity. Although RNA-Seq was shown as a promising source of data for GCNs (Iancu et al. 2012), some limitations related to normalization methods used for this technology were also demonstrated (Giorgi et al. 2013). However, with the increased number of RNA-Seq samples publicly available, more recent studies have shown that bigger datasets can overcome those caveats (Ballouz et al. 2015; Huang et al. 2017a) and highlight multiple advantages of RNA-Seq over microarrays for GCNs.

GCN inference comprises three main steps: similarity calculation, filtering, and edges construction (Serin et al. 2016). In the first step, a measure of similarity (or relatedness) is computed for each pair of genes. Multiple measures can be used in this step, such as mutual information (MI) (Meyer et al. 2008, 2007), or the prevalent correlation coefficients. The latter category includes the Pearson correlation coefficient (PCC), Spearman's correlation coefficient (SCC), and biweight midcorrelation (bicor) (Langfelder and Horvath 2008). Although MI is useful for finding nonlinear relationships between genes (Langfelder and Horvath 2008), it has been shown that it has several caveats and can be outperformed in many situations by correlation measures (Liesecke et al. 2018; Song et al. 2012). In the second step, the pairs of genes (edges) are either filtered based on a relatedness threshold that specifies the minimum level of similarity between expression profiles to define if a pair of genes is connected, or weighted. When using a threshold, it can be defined as a simple cutoff (hard threshold) (Tsaparas et al. 2006; Qiao et al. 2017), or as a result of more elaborated approaches. Some of these approaches include selecting a subset of the most positive/negative correlations (Lee et al. 2004), relying on topological features of co-expression networks like the clustering coefficient (Elo et al. 2007) or a power law distribution of the number of edges per node (Zhang and Horvath 2005), or applying models such as the Random Matrix Theory (Luo et al. 2007). Finally, in the third step, edges of the GCN are defined based on the resultant list of genes after filtering.

Depending on the type of connection between nodes, GCNs can be unweighted and weighted.

In the unweighted networks, edges indicate whether there is an association between a pair of nodes. They are derived from applying a hard threshold, i.e., an edge is present if the similarity measure between nodes is above the cutoff value. In the weighted networks, the degree of association between nodes is quantified by an attribute called weight, which commonly corresponds to a value in the range [0, 1]. This weight can result from applying a soft similarity threshold (Langfelder and Horvath 2008; Zhang and Horvath 2005) or from assigning a value derived from correlations such as the coefficient rank (Ballouz et al. 2015).

After constructing a GCN, a wide repertoire of analyses from graph theory, computer science, and engineering can be applied for elucidating valuable information hidden in the expression data. For example, by applying clustering algorithms like the hierarchical clustering (Langfelder and Horvath 2008), or the Markov Cluster (Zhang et al. 2012), it is possible to identify groups of highly coexpressed nodes (modules) with similar functions or involved in common biological processes. Modules are annotated with functional and metabolic information publicly available in databases such as Gene Ontology (GO, http://www.geneontology.org), Reactome (https://reactome.org/), and the Kyoto Encyclopedia of Genes and Genomes (KEGG, http://www.genome.jp/kegg/).

Another example of methods applied to GCNs are the topological analyses that examine the structural properties of networks. One of the most used topological properties is the node degree, which indicates how many connections each node has. It has been suggested that some biological networks are scale-free, which means that their degree distribution $P(k)$ approximates a power law $P(k) \sim k^{-\gamma}$ (Barabasi and Oltvai 2004). However, in many cases, proper statistical tests have revealed otherwise (Arita 2005; Broido and Clauset 2019; Lima-Mendez and van Helden 2009; Khanin and Wit 2006; Stumpf and Ingram 2005), and methods that strongly rely on the power-law distribution of the node degree must be assessed critically. In general, biological networks, including co-expression networks, exhibit many nodes poorly connected (low degree) and a relatively small number of nodes with many connections. Highly connected nodes (hubs) are usually representative of the biological function associated with a module and also have been associated with interesting processes like regulation (Hollender et al. 2014), and evolution (Masalia et al. 2017). Another biologically relevant topological property is the betweenness centrality that indicates the level to which a node works as a bridge between other nodes and allows to detect bottlenecks (genes with high centrality). Since high connectivity and betweenness centrality tend to be related to essentiality in functional processes (Carlson et al. 2006), they can be used to identify key genes with biological relevance. Other topological properties with biological relevance, including clustering coefficient, density, centralization, and heterogeneity, have also been explored (Dong and Horvath 2007; Horvath and Dong 2008).

GCNs have been used mainly for two purposes, gene function prediction, and the selection and prioritization of genes associated with specific phenotypes like diseases or traits. The first application is derived from module identification and annotation, which infer functions for uncharacterized genes following the "guilt by association" principle (Oliver 2000). For instance, functions for unknown genes have been predicted in yeast (Luo et al. 2007) and grapevine (Liang et al. 2014) using GCNs. The second application is perhaps the most popular of GCNs, and it is derived from exploiting network centrality properties (e.g., degree and betweenness) combined with module information. For example, several studies have used GCNs to identify genes associated with traits of interest in plants, such as heat shock recovery in grapevine (Liang et al. 2014), aluminum stress response in soybean (Das et al. 2017), sugar/acid ratio in sweet orange (Qiao et al. 2017), regulation of cell wall biosynthesis in sugarcane and bamboo (Ferreira et al. 2016; Ma et al. 2018), wood formation in *Populus trichocarpa* (Shi et al. 2017), the regulation of catechins, theanine, and caffeine metabolism in the tea plant *Camellia sinensis* (Tai et al. 2018), and plant height in maize (Wang et al. 2018a).

GCNs also have some caveats that are worth mentioning. GCNs provide only direct information for co-expression and not of direct interactions between its components like in PPIs. Additional information such as functional relationships or the essentiality of genes is elucidated by applying analyses that can be prone to biases, for example, clustering or annotation methods. Biologically meaningful conclusions are only supported by reliable networks that sometimes are difficult to obtain due to multiple factors in the construction like the amount and quality of the expression data, or the appropriate selection of similarity measures, parametrization (e.g., thresholds), and clustering methods.

Despite the caveats and difficulties in their inference, it has been shown that GCNs remain useful tools in gene expression analysis. They allow to reduce the complexity of the currently growing expression data, suggest functions of unknown genes, and identify essential genes involved in biological processes of interest.

2.3.3 Polymorphisms

Sequencing reads from RNA-Seq studies are often used for identifying polymorphisms in the expressed regions of the genome. The principles of variant identification from transcriptomic data are similar to those involved in variant calling from DNA sequencing and many important applications are possible. Briefly, software such as GATK (McKenna et al. 2010; DePristo et al. 2011) and BCFtools (Li et al. 2009; Li 2011) traverse genomic positions from a reference sequence and compare the aligned reads to identify single-nucleotide polymorphisms (SNPs) and insertions and deletions (indels). However, there are important particularities when working with RNA-Seq data and care must be taken when interpreting the results.

If these aligned reads are originated from transcriptomics datasets, polymorphic sites can only be identified between expressed transcripts. This is useful, for instance, if the goal is to search for imbalance of expression levels among different alleles of the same gene, or allele-specific expression (Pham et al. 2017; Shao et al. 2019). Accuracy for detecting polymorphisms and estimating allele expression ratios depends on the depth of coverage. This can be improved by increasing the sequencing depth but also depends on the expression level of each gene (Castel et al. 2015). Highly expressed genes naturally draw on a larger proportion of the sequencing data and thus offer more power to identify variants and higher accuracy of allelic expression estimates. On the other hand, lowly expressed genes are more prone to false negatives and require deeper sequencing to accurately identify polymorphisms.

Also, the fact that identified variants are constrained to expressed exons can limit the scope of the study. Polymorphic sites in introns, regulatory and intergenic sequences, which can be more numerous and may have key biological significance, cannot be identified from RNA-Seq data alone (Cubillos et al. 2012; Magalhaes et al. 2007). Genomic variants located in alleles that are not expressed in a given transcriptome will also be missed. Finally, many possible posttranscriptional modifications may negatively impact variant calling results and lead to flawed conclusions (Lee et al. 2013).

Variant calling efforts and studies of allelic imbalance are even more complicated in polyploid organisms, where more than two different alleles can be found (Cai et al. 2020). First, for allopolyploids, it can be difficult to differentiate between true alleles and homeologous sequences, which may not be polymorphic within each subgenome (Yang et al. 2018a). Additionally, it is important to note that allele ratio information from RNA-Seq data is not appropriate for quantitative genotyping (estimating genomic dosage) in autopolyploids, because of differences in the expression levels of different alleles. In other words, while the variation in allelic expression levels does provide valuable biological information, these ratios are affected by expression control mechanisms and do not necessarily reflect allele dosage at the DNA level (Pham et al. 2017).

Considering these complications and limitations, in most scenarios a combination of variant calling with other strategies is more valuable, such as identifying polymorphisms from both RNA-Seq and whole-genome sequencing (WGS) data, for instance.

2.3.4 Machine Learning Technologies for Transcriptomics

The advent of high-throughput technologies like microarrays and next-generation sequencing has led researchers in biosciences to face the challenges of analyzing large amounts of data. These challenges include heterogeneity, high dimensionality, noisiness, incompleteness, and computational expensiveness, among others. Machine learning (ML) has emerged as a suitable solution for analyzing massive data while dealing well with its challenges. ML has been extensively applied for large-scale data analysis in fields such as genetics (Libbrecht and Noble 2015), biomedicine (Mamoshina et al. 2016; Leung et al. 2016), genomics, transcriptomics, proteomics, and systems biology (Larranaga et al. 2006; Min et al. 2017). This section presents an overview of ML that includes basic concepts and applications on transcriptomics in plants.

ML can be defined as the computational process of automatically learning from experience to make predictions on new data (Murphy 2012). The process of learning is carried out by extracting knowledge from exemplary data by identifying hidden patterns. ML methods are classified into two main groups, supervised and unsupervised learning. Supervised learning is a predictive approach that comprises data examples with inputs and outputs. This approach uses evidence from the example data to make a model that generates reasonable predictions for new unseen datasets. More formally, the example data corresponds to a set of input–output pairs D called training set and defined as,

$$D = \left\{ \left(x_i, y_i \right) \right\}_{i=1}^{N},$$

where x_i is a training input of the set x, y_i is the response variable that represents an output from the set y, and N is the number of training examples. Hence, the model is trained to learn how to map each x_i to a corresponding output y_i.

Supervised learning methods can be subdivided into two categories according to the nature of predictions. When the response variable is discrete or categorical, e.g., male or female, healthy or diseased, the method falls into the classification category. General applications of classification algorithms are voice and handwriting recognition, and document and image classification. Common algorithms of this category include support vector machines (SVM) Support Vector Regression (SVR), *k*-nearest neighbor (KNN), decision trees, logistic regression, and neural networks. When the response variable is continuous, e.g., the height of a person, or a temperature, the method corresponds to the regression category. Regression algorithms include linear and nonlinear models, neural networks, and regularization. A variation of the late category is the ordinal regression, which comprises methods whose response variable has a natural ordering.

The second main group of ML, unsupervised learning, uses data examples with just inputs, i.e., the set

$$D = \left\{ x_i \right\}_{i=1}^{N}.$$

This type of ML tries to elucidate hidden patterns in data, which can be considered "interesting" to the researcher. In this case, there is no information about the kind of patterns that are expected to be found in the data. Unsupervised learning, also called knowledge discovery, is more commonly used than unsupervised techniques. Two notorious categories within unsupervised learning are clustering and dimensionality reduction. Clustering algorithms are intended to group data by looking for similarities among the features of each element from the input. Standard clustering algorithms include *k*-means, self-organized maps (SOM), hierarchical clustering, and hidden Markov models. Dimensionality reduction algorithms try to extract the "essence" of data (Murphy 2012) by selecting a subset of features that represents better the dataset (feature selection) or by transforming the high-dimensional space of the original data into a lower one (feature extraction). Usual algorithms for dimensionality reduction are principal component analysis (PCA), linear discriminant analysis (LDA), and generalized discriminant analysis (GDA).

Supervised ML techniques have been applied in transcriptomics-related tasks such as assembly, identification, and abundance estimation of transcripts, splicing sites/events detection, non-coding

RNA identification, and gene selection. Transcriptome assembly is one of the essential tasks in RNA-Seq-based studies that is followed by analyses such as, the estimation of gene expression levels or differential gene/trnascript expression. IsoLasso is a reference-based RNA-Seq transcriptome assembler that uses an ML regression algorithm called Least Absolute Shrinkage and Selection Operator (LASSO) and has the interesting feature of identifying and quantifying novel isoforms (Li et al. 2011b). Another ML-based tool for transcript identification and abundance estimation is SLIDE, which uses a linear model that models the sampling probability of RNA-Seq reads from mRNA isoforms, and a modified LASSO algorithm for estimating parameters (Li et al. 2011a). Unlike IsoLasso, SLIDE requires the coordinates of transcripts and exons previously assembled with other tools.

Identifying splicing sites and splicing events is crucial for determining isoforms and, thus, for estimating the abundance of transcripts. TrueSight is a tool developed for detecting splice junctions (SJs) based on an iterative regression algorithm that uses RNA-Seq mapping information and splicing signals from the DNA sequence of a reference genome (Li et al. 2013b). TrueSight was tested using simulated and real datasets from humans, *D. melanogaster*, *C. elegans*, and *A. thaliana*, and showed better specificity and sensitivity compared to other SJs detection applications. A recently developed tool called DeepBound also uses alignment information to determine SJs and infer boundaries of expressed transcripts from RNA-Seq data (Shao et al. 2017). DeepBound utilizes deep convolutional neural fields (DeepCNF), a technique that belongs to an emerging ML branch referred to as deep learning (Mamoshina et al. 2016; Min et al. 2017; Angermueller et al. 2016). All the described applications for transcript abundance and SJ detection can be used in plants. However, except for SLIDE, these tools are not suitable for being applied directly to non-model species, as they depend on a reference genome.

In plants, supervised learning methods have also been used for detecting alternative splicing (AS) events. SVM classifiers were employed to detect two types of AS events, exon skips and intron retentions, in *A. thaliana* from tiling arrays data (Eichner et al. 2011). EST and cDNA data were used for training with two SVM layers: one for classifying sequence segments as introns or exons, assigning probabilities of being included in mature mRNA, and a second layer to predict AS events by using the probabilities from the first layer. In addition to SVM, Random Forest (RF) has been used to detect intron retention in *A. thaliana*, the most common type of alternative splicing in this species. These RF were created using a hybrid approach that combines essential features (i.e., length, nucleotide occurrence probabilities, AT and GC content) with additional features (i.e., common motifs, splice sites, and flanking sequences) to differentiate retained introns from constitutively spliced introns. These RFs had a better classification performance than SVM (Mao et al. 2014).

Noncoding RNAs (ncRNAs) are determinant in cellular processes like regulation and alternative splicing. Several ML methods have been applied to discover ncRNAs, including micro RNAs (miRNA) and long non-coding RNAs (lncRNA), using NGS datasets. In the case of miRNAs, decision trees (based on the C4.5 algorithm) combined with genetic algorithms, allowed the prediction of miRNA targets in humans from datasets that comprise genomic and transcriptomic information (Rabiee-Ghahfarrokhi et al. 2015). miRNAs were predicted in 18 different plant species from data extracted from RNA-Seq, chromosome sequences, or ESTs, exploiting decision trees (C5.0 algorithm) (Williams et al. 2012). An SVM approach was employed to identify miRNAs associated with cold stress in *A. thaliana* (Zhou et al. 2008). Multiple Kernel Learning has been applied to the identification of circularRNA, a type of lncRNA, in humans, which can identify them with high accuracy in de novo assembled transcriptomes (Pan and Xiong 2015).

Gene selection from expression data is a problem in which ML methods can be used naturally. Given an expression dataset that usually comprises thousands of genes, the goal here is to select a handful of relevant genes associated with a specific condition of interest, e.g., a disease or a treatment. A common ML-based approach for gene selection from expression datasets is variable ranking, in which genes (variables) are prioritized according to a value derived from the applied classification algorithm. This value is a proxy for the importance or relevance of each gene among the whole dataset. In this way, genes at the top of the rank are more relevant to the condition of interest, e.g., healthy/diseased tissue, treated/untreated tissue, and genes at the lower positions are redundant and less relevant. Following this approach, ML algorithms such as RFs, SVMs, and decision trees have been used with microarray data to select subsets of cancer-related genes which can be used as markers in diagnosis (Diaz-Uriarte and Alvarez de Andres 2006; Horng et al. 2009; Guyon et al. 2002).

Although most of the proposed ML-based gene selection methods are tested in cancer expression datasets, some studies have applied similar approaches to plants using gene expression data from microarrays. An SVM with Recursive Feature Elimination (SVM-RFE) and a Radial Basis Function (RBF) was used to identify four genes related to resistance to tungro disease in rice (Ren et al. 2010). This was a modification of the application of the same technique to cancer (Guyon et al. 2002). A caveat in this study is the small dataset used (21 samples), as the amount of data for training is a decisive factor to get revealing results in ML. A further study refined the same SVM-RFE approach to identify genes related to drought resistance in *A. thaliana* (Liang et al. 2011). Although authors of this study used a dataset with only 22 samples, they mitigated the small sample size effect by implementing a Leave One Out Cross Validation (LOOCV) scheme to select the training dataset and bootstrapping strategy to iterate the variable ranking process. In such a way, a subset of ten genes were identified, seven of which have previous biological information that links them to processes involved in drought resistance. ML and GCN were combined into the R package "machine learning-based differential network analysis" (mlDNA), which implements a two-phase ML method for selecting genes from expression data. In the first phase, the method identifies and discards irrelevant genes from the dataset using an RF classifier with the Positive Sample only Learning algorithm (PSoL), a technique that discriminates positive from negative data after using only positive samples for training. The second phase involves the construction of GCNs from the filtered genes, the extraction of topological features from the GCNs, and an RF algorithm to select the candidate genes based on the extracted features. This approach proved to successfully select candidate genes in *A. thaliana* responding to drought, cold, heat, wound, and genotoxic stress conditions (Huang et al. 2011).

2.4 Case/Examples of Transcriptomics in Non-model Plants

Perhaps the most notable quality of transcriptomics is the possibility of producing robust amounts of data for a reduced representation of the genome, which is of importance in non-model plant species and species with complex genomes. This quality allows for a diverse series of biological questions to be asked and for which answers can be obtained. In this section we will exemplify the most relevant uses of recent transcriptomics studies.

2.4.1 Construction of Improved Transcripts Catalogs

Although, in principle, transcriptomic studies derived from RNA-Seq do not require any prior genetic information, it is true that having a high-quality reference transcriptome undoubtedly favors high-quality research. Current assembly tools and sequencing technologies have advanced our capacity to produce de novo assemblies. In constructing high-quality transcriptomes for polyploid (allopolyploid) species, where two or more sub-genomes are present, one particular challenge is the identification of homeologous copies of the same genes which tend to be highly similar and difficult to separate in a de novo assembly. Classical assemblers such as SOAPdenovo-Trans Trinity and TransABySS have been tested for this task. This is exemplified in the study by (Chopra et al. 2014) aiming at reconstructing the transcriptome of tetraploid and diploid peanut species, using RNA-Seq data. After examining several variables including contig length and number, results showed that Trinity and TransAByss performed in a similar way for the diploid species, while Trinity performed better for the tetraploid genotype. In addition, the transcriptome produced for the tetraploid genotype almost doubled in number of contigs, total size and transcript N50 compared to the existing resources. It also produced at least 40% more full-length sequences.

Others have searched to develop specific software to tackle the problem. Such is the case of the software HomeoSplitter which takes into consideration the elevated rates of heterozygosity of certain contigs (alleles) to target possible homeoalleles. Once identified, the software uses a likelihood model-based method to disentangle the mixed alleles taking into consideration their expression levels. For durum wheat (*Triticum turgidum*) HomeoSplitter showed capacity to separate homeologous sequences, as assessed by comparison to the diploid progenitors, and allowed to recover a greater number of SNPs for the population genotyped (Ranwez et al. 2013).

From the sequencing-and-assembly point of view, this issue has been approached through the use of normalized libraries, which increases the likelihood of seeing rare or less abundant transcript, and the use of single-molecule long read sequencing technologies, which can produce near complete transcript sequences represented in a single-sequencing read. The protocol called Iso-Seq has been applied to several crop species, including sorghum (Abdel-Ghany et al. 2016), maize (Wang et al. 2016), cotton (Wang et al. 2018b), coffee (Cheng et al. 2017), *Salvia miltiorrhiza* (Xu et al. 2015), grape wine (Minio et al. 2019), the Chinese herb *Astragalus membranaceus* (Li et al. 2017a), *Arabidopsis pumila* (Yang et al. 2018b), the shrub *Zanthoxylum bungeanum* (Tian et al. 2018), the giant timber bamboo native to China (Zhang et al. 2018), wild strawberry (Li et al. 2017b), and the highly complex sugarcane (Hoang et al. 2017). Iso-Seq has been shown to recover full-length isoforms, which was not possible with short-read technologies, but also it has allowed the detection of alternative start sites, alternative splicing and alternative polyadenylation (Zhao et al. 2019). In the case of sugarcane, Iso-Seq was further complemented with short RNA-Seq reads in order to correct errors present in long reads. The same dataset also served to compare the transcriptomes created by the hybrid approach and a de novo approach based solely on RNA-Seq reads. The hybrid transcriptome recovered more full-length transcripts, with a longer N50, more ORFs and predicted transcripts, and higher average length of the largest 1000 proteins, compared to the de novo contigs. Importantly, RNA-Seq covered more gene content, and more RNA classes than Iso-Seq, which was attributed to the greater sequencing depth (Hoang et al. 2017).

Oxford Nanopore Technologies (ONT) have a platform option that allows for the direct sequencing of RNA molecules, which in addition to producing full-length transcript sequences, study of alternative polyadenylation and splice and start sites, reveals the status of RNA modifications, and could revolutionize the transcriptomics field (Hussain 2018). This approach is still very recent and has not yet been applied to many plant species. Direct RNA sequencing was performed on seeds of soybean to quantify transcript degradation as a proxy of seed viability (Fleming et al. 2018). Eukaryotic transcripts are usually modified on their 5′-end by the addition of a 7-methylguanylate (m^7G) cap which protects mRNA from decay and has several implications in mRNA-downstream processes. However, a recent study, using direct RNA sequencing, showed that in *A. thaliana*, up to 5% of the transcripts of several thousand genes have instead a NAD+ cap (Zhang et al. 2019a), an RNA modification that had been reported before in bacteria (Chen et al. 2009), yeast (Walters et al. 2017), and humans (Jiao et al. 2017).

Overall, despite current advances in the construction of de novo transcriptomes, there is still room for improvement in assemblers tailored to polyploid genomes. Also, given the current rate of innovation in high-throughput sequencing, and provided a decrease in costs, the construction of novel transcriptomes through the use of long RNA molecules are expected to increase rapidly.

2.4.2 Populations Mapping

Transcriptomics can also be used to identify polymorphisms to map populations of interest. Two alternative strategies are often followed: In the first, the genetic variants are identified from transcriptomic data, from a diverse group of individuals. The variants identified are then used to design probes to test DNA samples from the same or an alternative, bigger, population. Contrary to the classic DNA mapping studies, this strategy increases the probability of identifying causal mutations given that the majority of the selected variants will be located within coding sequences. This is specially the case of species with big genomes and a high percentage of repetitive sequences which, for mapping studies, require a considerable number of markers to increase the probability of having a significant association. Markers, particularly SNP and SSR, derived from transcriptomic data have been produced for different crops including, but not limited to soybean (Guo et al. 2018), sugarcane (Bundock et al. 2009), grasspea (Hao et al. 2017), peanut (Chopra et al. 2015), and oilseed rape (Trick et al. 2009). More recently, and through the implementation of the Bulk Segregant RNA-Seq analyses (BSR-Seq) principle, which requires the formation of pooled samples contrasting for the phenotype of interest, markers linked to traits of interest have been mapped in crop species such as wheat (Wang et al. 2017; Ramirez-Gonzalez et al. 2015; Wu et al. 2018) and Chinese cabbage (Huang et al. 2017b).

In the second strategy, transcriptomics data is produced for a biparental population, and the markers identified (SNP markers) are directly used for construction of genetic maps. The value of these maps lies in the fact that "unlike sequence assembly, linkage analysis is essentially unaffected by allopolyploidy and repeated sequences as long as homeologous recombination is rare and genome-specific alleles can be identified" (reviewed in McKay and Leach 2011). This strategy, to the best of our knowledge, has been only used in the tetraploid *Brassica napus* (oilseed rape) (Bancroft et al. 2011). In this case, twin genetic maps were constructed for the two progenitor species (*B. oleracea* and *B. rapa*) of the modern *B. napus* genotypes, which also served as parents for the population tested. These genetic maps were next aligned to the existing genome of *B. napus* and that of *A. thaliana*. The whole strategy allowed to identify genome rearrangements between *B. oleracea* and *B. rapa* and therefore helped to refine the existing assemblies for these species. Likewise, it helped to pinpoint genomic regions involved in the recent breeding history of the crop. Considering these implications and the urgent necessity of genomic tools to tackle polyploid genomes, it is expected that linkage maps derived from transcriptomic data will be on the rise.

2.4.3 Stress-Related Studies

As sessile organisms, plants must deal with a variety of environmental conditions that can impact on their potential for growth and reproduction. In order to study the molecular mechanisms underlying the response to such conditions plant transcriptomics is being widely used. The most common approach consists of comparing gene expression levels of a specific genotype under a control and a stress-induced treatment. Oftentimes, contrasting genotypes (tolerant and susceptible) for the trait of interest are used. By identifying the changes in gene expression between control and treatment conditions, it is possible to determine the mRNAs activated by the stress under consideration. This in turn allows for exploring the mRNAs that are differentially expressed among the genotypes selected (tolerant vs. susceptible). Following this approach, it has been possible to study the molecular regulation of salt stress tolerance in cotton (Zhang et al. 2016a), the roles of the photosynthetic system during drought in upland rice (Zhang et al. 2016b), the molecular mechanisms driving copper stress tolerance in grapevine (Leng et al. 2015), the mechanisms for lipid accumulation in response to nitrogen deprivation in the green algae *Chlamydomonas reinhardtii* (Park et al. 2015), the molecular responses underlying drought tolerance in sugarcane (Pereira-Santana et al. 2017; Belesini et al. 2017), just to mention a few.

Perhaps, one of the most studied traits through comparative transcriptomic is drought. When "drought" and "RNA-Seq" are used as keywords in PubMed, 217 different titles, excluding reviews, show up as a result. Studies have been performed on nearly every major crop (Zhang et al. 2014; Chen et al. 2016; Divya Bhanu et al. 2016; Mofatto et al. 2016), but also on non-major crops and other plants whose original habitat are water-deprived locations and thus can contribute to better understanding of the physiological bases of this condition (Gross et al. 2013; Yang et al. 2015; Li et al. 2015). In polyploids, the challenge resides on having a high-quality reference transcriptome that allows to distinguish among isoforms derived from different sub-genomes. In fact, in hexaploid wheat, where different genomic resources have been recently developed (Pearce et al. 2015), it has been found that a large proportion of wheat homeologs exhibited expression partitioning under normal and abiotic stresses, indicating a specialized gene expression coordination among genomes.

2.4.4 Phylogenomics

Phylogenomics is a new biological discipline focusing on the resolution of relationships among taxa and the reconstruction of evolutionary histories through the use of genomic data. It involves the analysis of entire genomes, transcriptomes, or specific sequences that can be targeted (Yu et al. 2018) through the mining of already published information (Washburn et al. 2017).

In order to resolve relationships among species, phylogenomics relies heavily on the identification of single-copy genes to reduce the possibility of paralogy and thus limiting to conclusions based solely on orthologous genes. However, information on single-copy genes is difficult to obtain especially for non-model, polyploid species, where the entire genome is expected to be duplicated. Chloroplast genes are often targeted for phylogenomics; however, this part of the plant genome has its own problems such as a low recombinant nature, and thus low polymorphism levels, exclusive maternal inheritance, and these genes are subject to processes such as chloroplast capture and hybrid speciation which reduce its resolution capacity. Still, due to its high-throughput nature, transcriptomics offers the possibility to mine for nuclear single-copy markers in a rich set of genic sources. This is even possible in the case of polyploids and despite their repetitive nature. Due to evolutionary mechanisms such as gene conversion and loss, the number of retained duplicates in polyploids decreases over the time, allowing single-copy signals (coding and non-coding sequences) to arise (Wen et al. 2015). In the case of ferns, for example, which have a long history of polyploidy, 20 new nuclear regions spanning ten coding sequences have been identified by comparative transcriptomics which has increased significantly the taxonomic resolution across these group of plants (Rothfels et al. 2013).

Comparative transcriptomics can also contribute to detect and characterize polyploidy speciation. Although ancient polyploidy could be reconstructed through the comparison of high-quality, chromosome-level genomes, the lack of high-quality assemblies for the vast majority of polyploid species has positioned transcriptomics as a viable alternative. For this purpose, the rate of synonymous substitution (Ks), in coding sequences, derived from transcriptomics is widely used. This is possible because whole-genome duplications produce peaks in the cumulative distributions of pairwise Ks between paralogs within a genome. By evaluating the distribution of Ks among evolutionary lineages, it has been possible to better understand polyploidy speciation in the flax genus (Sveinsson et al. 2014), the evolution of gene families like CYP75 after the events of whole-genome duplication

(Zhang et al. 2019b), the redistribution of the seed plants in phylogenetic trees explaining the origin of angiosperms (Ran et al. 2018), the evolutionary patterns of agricultural traits in strawberry (Qiao et al. 2016), or the origin and early diversification of green (One Thousand Plant Transcriptomes Initiative 2019) and land plants (Wickett et al. 2014), among others.

2.5 Future Directions in the Field

Over the past decades, transcriptomics has seen a revolution. The technologies employed to produce expression data are nowadays much more efficient and with their regular decrease in costs, they are a realistic possibility even for small labs, and so it has become practical to be applied to non-model exotic plant species, and to perform more complex experimental designs. Nonetheless, the cost of sequencing is still not at reach for projects in which hundreds to thousands of samples need to be sequenced. This level of sequencing capacity is a reality for consortiums and greater collaborative efforts but not for smaller groups, which commonly have the possibility of greater access to genetically diverse samples but smaller budgets. Further decrease in library preparation and sequencing costs will ameliorate this though.

Technical advances have made it possible to directly sequence RNA molecules, and together with PCR-free protocols, they aid in eliminating potential sources of bias that could be introduced during library preparation. In addition to building comprehensive transcript catalogs, these advances will allow more reliable estimation of transcript abundances when it becomes affordable to sequence at higher depths of coverage. Recently published genome assemblies are increasingly resolving the different sequence haplotypes in organims with ploidy levels greater than one in these cases long-read RNA sequencing will allow the study of allele-specific expression with unprecedented levels of detail.

Along with this new technological capacity to produce data, the questions that may be answered with transcriptomics-based strategies have also matured. However, for many of these questions, their answers are limited by the available bioinformatic software. For example, all the efforts that have been made to confidently identify orthologous genes and in general to filter out the noise caused by polyploidy are encouraging because, among other reasons, this has increased our understanding of complex genomes. Nonetheless, only a handful of genes or a small portion of the transcriptomes are used for these purposes. It is then reasonable to believe that further efforts in software development are necessary to truly take advantage of the level of information being produced in transcriptomics studies. A similar situation happens with all the studies aiming at better understanding of specific phenomena (e.g., stress-related studies) that after producing high-quality, robust data are still left with lists of hundreds to thousands of differentially expressed genes, from which it is difficult to define the key players for the process under study. Perhaps this type of studie could benefit from the integration of different OMICs approaches to the same problems, with a more integrative approach which requires further advances in tool development, for instance including machine learning algorithms, necessary to mine for the most relevant transcripts.

Overall, we can confidently say that the last decade has been a defining one for plant transcriptomics thanks to the greater access to sequencing data. However, the same breakthrough has yet to impact data analyses and storage. Our data processing capabilities are being surpassed by our capacity to produce data, and it is imperative to face this challenge if we want to further increase our ability to address the challenges posed by climate change, speed up the efforts to breed crop plants, and deepen our understanding of the history of evolution of plants.

References

Abdel-Ghany SE, Hamilton M, Jacobi JL, Ngam P, Devitt N, Schilkey F, Ben-Hur A, Reddy AS (2016) A survey of the sorghum transcriptome using single-molecule long reads. Nat Commun 7:11706. https://doi.org/10.1038/ncomms11706

Alexa A, Rahnenfuhrer J, Lengauer T (2006) Improved scoring of functional groups from gene expression data by decorrelating GO graph structure. Bioinformatics 22(13):1600–1607. https://doi.org/10.1093/bioinformatics/btl140

Anders S, Huber W (2010) Differential expression analysis for sequence count data. Genome Biol 11(10):R106. https://doi.org/10.1186/gb-2010-11-10-r106

Anders S, Pyl PT, Huber W (2015) HTSeq--a Python framework to work with high-throughput sequencing data. Bioinformatics 31(2):166–169. https://doi.org/10.1093/bioinformatics/btu638

Angermueller C, Parnamaa T, Parts L, Stegle O (2016) Deep learning for computational biology. Mol Syst Biol 12(7):878. https://doi.org/10.15252/msb.20156651

Arita M (2005) Scale-freeness and biological networks. J Biochem 138(1):1–4. https://doi.org/10.1093/jb/mvi094

Asamizu E, Nakamura Y, Sato S, Fukuzawa H, Tabata S (1999) A large scale structural analysis of cDNAs in a unicellular green alga, Chlamydomonas reinhardtii. I. Generation of 3433 non-redundant expressed sequence tags. DNA Res 6(6):369–373

Ashburner M, Ball CA, Blake JA, Botstein D, Butler H, Cherry JM, Davis AP, Dolinski K, Dwight SS, Eppig JT, Harris MA, Hill DP, Issel-Tarver L, Kasarskis A, Lewis S, Matese JC, Richardson JE, Ringwald M, Rubin GM, Sherlock G (2000) Gene ontology: tool for the unification of biology. The Gene Ontology Consortium. Nat Genet 25(1):25–29. https://doi.org/10.1038/75556

Baccarella A, Williams CR, Parrish JZ, Kim CC (2018) Empirical assessment of the impact of sample number and read depth on RNA-Seq analysis workflow performance. BMC Bioinformatics 19(1):423. https://doi.org/10.1186/s12859-018-2445-2

Bainbridge MN, Warren RL, Hirst M, Romanuik T, Zeng T, Go A, Delaney A, Griffith M, Hickenbotham M, Magrini V, Mardis ER, Sadar MD, Siddiqui AS, Marra MA, Jones SJ (2006) Analysis of the prostate cancer cell line LNCaP transcriptome using a sequencing-by-synthesis approach. BMC Genomics 7:246. https://doi.org/10.1186/1471-2164-7-246

Ballouz S, Verleyen W, Gillis J (2015) Guidance for RNA-seq co-expression network construction and analysis: safety in numbers. Bioinformatics 31(13):2123–2130. https://doi.org/10.1093/bioinformatics/btv118

Bancroft I, Morgan C, Fraser F, Higgins J, Wells R, Clissold L, Baker D, Long Y, Meng J, Wang X, Liu S, Trick M (2011) Dissecting the genome of the polyploid crop oilseed rape by transcriptome sequencing. Nat Biotechnol 29(8):762–766. https://doi.org/10.1038/nbt.1926

Banks JA, Nishiyama T, Hasebe M, Bowman JL, Gribskov M, dePamphilis C, Albert VA, Aono N, Aoyama T, Ambrose BA, Ashton NW, Axtell MJ, Barker E, Barker MS, Bennetzen JL, Bonawitz ND, Chapple C, Cheng C, Correa LG, Dacre M, DeBarry J, Dreyer I, Elias M, Engstrom EM, Estelle M, Feng L, Finet C, Floyd SK, Frommer WB, Fujita T, Gramzow L, Gutensohn M, Harholt J, Hattori M, Heyl A, Hirai T, Hiwatashi Y, Ishikawa M, Iwata M, Karol KG, Koehler B, Kolukisaoglu U, Kubo M, Kurata T, Lalonde S, Li K, Li Y, Litt A, Lyons E, Manning G, Maruyama T, Michael TP, Mikami K, Miyazaki S, Morinaga S, Murata T, Mueller-Roeber B, Nelson DR, Obara M, Oguri Y, Olmstead RG, Onodera N, Petersen BL, Pils B, Prigge M, Rensing SA, Riano-Pachon DM, Roberts AW, Sato Y, Scheller HV, Schulz B, Schulz C, Shakirov EV, Shibagaki N, Shinohara N, Shippen DE, Sorensen I, Sotooka R, Sugimoto N, Sugita M, Sumikawa N, Tanurdzic M, Theissen G, Ulvskov P, Wakazuki S, Weng JK, Willats WW, Wipf D, Wolf PG, Yang L, Zimmer AD, Zhu Q, Mitros T, Hellsten U, Loque D, Otillar R, Salamov A, Schmutz J, Shapiro H, Lindquist E, Lucas S, Rokhsar D, Grigoriev IV (2011) The Selaginella genome identifies genetic changes associated with the evolution of vascular plants. Science 332(6032):960–963. https://doi.org/10.1126/science.1203810

Barabasi AL, Oltvai ZN (2004) Network biology: understanding the cell's functional organization. Nat Rev Genet 5(2):101–113. https://doi.org/10.1038/nrg1272

Belesini AA, Carvalho FMS, Telles BR, de Castro GM, Giachetto PF, Vantini JS, Carlin SD, Cazetta JO, Pinheiro DG, Ferro MIT (2017) De novo transcriptome assembly of sugarcane leaves submitted to prolonged water-deficit stress. Genet Mol Res 16(2):gmr16028845. https://doi.org/10.4238/gmr16028845

Benjamini Y, Hochberg Y (1995) Controlling the false discovery rate: a practical and powerful approach to multiple testing. J R Stat Soc Ser B Methodol 57(1):289–300

Bentley DR, Balasubramanian S, Swerdlow HP, Smith GP, Milton J, Brown CG, Hall KP, Evers DJ, Barnes CL, Bignell HR, Boutell JM, Bryant J, Carter RJ, Keira Cheetham R, Cox AJ, Ellis DJ, Flatbush MR, Gormley NA, Humphray SJ, Irving LJ, Karbelashvili MS, Kirk SM, Li H, Liu X, Maisinger KS, Murray LJ, Obradovic B, Ost T, Parkinson ML, Pratt MR, Rasolonjatovo IM, Reed MT, Rigatti R, Rodighiero C, Ross MT, Sabot A, Sankar SV, Scally A, Schroth GP, Smith ME, Smith VP, Spiridou A, Torrance PE, Tzonev SS, Vermaas EH, Walter K, Wu X, Zhang L, Alam MD, Anastasi C, Aniebo IC, Bailey DM, Bancarz IR, Banerjee S, Barbour SG, Baybayan PA, Benoit VA, Benson KF, Bevis C, Black PJ, Boodhun A, Brennan JS, Bridgham JA, Brown RC, Brown AA, Buermann DH, Bundu AA, Burrows JC, Carter NP, Castillo N, Chiara ECM, Chang S, Neil Cooley R, Crake NR, Dada OO, Diakoumakos KD, Dominguez-Fernandez B, Earnshaw DJ, Egbujor UC, Elmore DW, Etchin SS, Ewan MR, Fedurco M, Fraser LJ, Fuentes Fajardo KV, Scott Furey W, George D, Gietzen KJ, Goddard CP, Golda GS, Granieri PA, Green DE, Gustafson DL, Hansen NF, Harnish K, Haudenschild CD, Heyer NI, Hims MM, Ho JT, Horgan AM, Hoschler K, Hurwitz S, Ivanov DV, Johnson MQ, James T, Huw Jones TA,

Kang GD, Kerelska TH, Kersey AD, Khrebtukova I, Kindwall AP, Kingsbury Z, Kokko-Gonzales PI, Kumar A, Laurent MA, Lawley CT, Lee SE, Lee X, Liao AK, Loch JA, Lok M, Luo S, Mammen RM, Martin JW, McCauley PG, McNitt P, Mehta P, Moon KW, Mullens JW, Newington T, Ning Z, Ling Ng B, Novo SM, O'Neill MJ, Osborne MA, Osnowski A, Ostadan O, Paraschos LL, Pickering L, Pike AC, Chris Pinkard D, Pliskin DP, Podhasky J, Quijano VJ, Raczy C, Rae VH, Rawlings SR, Chiva Rodriguez A, Roe PM, Rogers J, Rogert Bacigalupo MC, Romanov N, Romieu A, Roth RK, Rourke NJ, Ruediger ST, Rusman E, Sanches-Kuiper RM, Schenker MR, Seoane JM, Shaw RJ, Shiver MK, Short SW, Sizto NL, Sluis JP, Smith MA, Ernest Sohna Sohna J, Spence EJ, Stevens K, Sutton N, Szajkowski L, Tregidgo CL, Turcatti G, Vandevondele S, Verhovsky Y, Virk SM, Wakelin S, Walcott GC, Wang J, Worsley GJ, Yan J, Yau L, Zuerlein M, Mullikin JC, Hurles ME, McCooke NJ, West JS, Oaks FL, Lundberg PL, Klenerman D, Durbin R, Smith AJ (2008) Accurate whole human genome sequencing using reversible terminator chemistry. Nature 456(7218):53–59. https://doi.org/10.1038/nature07517

Bolger AM, Lohse M, Usadel B (2014) Trimmomatic: a flexible trimmer for Illumina sequence data. Bioinformatics 30(15):2114–2120. https://doi.org/10.1093/bioinformatics/btu170

Borodina T, Adjaye J, Sultan M (2011) A strand-specific library preparation protocol for RNA sequencing. Methods Enzymol 500:79–98. https://doi.org/10.1016/B978-0-12-385118-5.00005-0

Bray NL, Pimentel H, Melsted P, Pachter L (2016) Near-optimal probabilistic RNA-seq quantification. Nat Biotechnol 34(5):525–527. https://doi.org/10.1038/nbt.3519

Britto-Kido Sde A, Ferreira Neto JR, Pandolfi V, Marcelino-Guimaraes FC, Nepomuceno AL, Vilela Abdelnoor R, Benko-Iseppon AM, Kido EA (2013) Natural antisense transcripts in plants: a review and identification in soybean infected with Phakopsora pachyrhizi SuperSAGE library. ScientificWorldJournal 2013:219798. https://doi.org/10.1155/2013/219798

Broido AD, Clauset A (2019) Scale-free networks are rare. Nat Commun 10(1):1017. https://doi.org/10.1038/s41467-019-08746-5

Brown JW, Calixto CP, Zhang R (2017) High-quality reference transcript datasets hold the key to transcript-specific RNA-sequencing analysis in plants. New Phytol 213(2):525–530. https://doi.org/10.1111/nph.14208

Bryant DM, Johnson K, DiTommaso T, Tickle T, Couger MB, Payzin-Dogru D, Lee TJ, Leigh ND, Kuo TH, Davis FG, Bateman J, Bryant S, Guzikowski AR, Tsai SL, Coyne S, Ye WW, Freeman RM Jr, Peshkin L, Tabin CJ, Regev A, Haas BJ, Whited JL (2017) A tissue-mapped axolotl de novo transcriptome enables identification of limb regeneration factors. Cell Rep 18(3):762–776. https://doi.org/10.1016/j.celrep.2016.12.063

Bundock PC, Eliott FG, Ablett G, Benson AD, Casu RE, Aitken KS, Henry RJ (2009) Targeted single nucleotide polymorphism (SNP) discovery in a highly polyploid plant species using 454 sequencing. Plant Biotechnol J 7(4):347–354. https://doi.org/10.1111/j.1467-7652.2009.00401.x

Byrne A, Beaudin AE, Olsen HE, Jain M, Cole C, Palmer T, DuBois RM, Forsberg EC, Akeson M, Vollmers C (2017) Nanopore long-read RNAseq reveals widespread transcriptional variation among the surface receptors of individual B cells. Nat Commun 8:16027. https://doi.org/10.1038/ncomms16027

Cai M, Lin J, Li Z, Lin Z, Ma Y, Wang Y, Ming R (2020) Allele specific expression of Dof genes responding to hormones and abiotic stresses in sugarcane. PLoS One 15(1):e0227716. https://doi.org/10.1371/journal.pone.0227716

Carlson MR, Zhang B, Fang Z, Mischel PS, Horvath S, Nelson SF (2006) Gene connectivity, function, and sequence conservation: predictions from modular yeast co-expression networks. BMC Genomics 7:40. https://doi.org/10.1186/1471-2164-7-40

Castel SE, Levy-Moonshine A, Mohammadi P, Banks E, Lappalainen T (2015) Tools and best practices for data processing in allelic expression analysis. Genome Biol 16:195. https://doi.org/10.1186/s13059-015-0762-6

Chen YG, Kowtoniuk WE, Agarwal I, Shen Y, Liu DR (2009) LC/MS analysis of cellular RNA reveals NAD-linked RNA. Nat Chem Biol 5(12):879–881. https://doi.org/10.1038/nchembio.235

Chen W, Yao Q, Patil GB, Agarwal G, Deshmukh RK, Lin L, Wang B, Wang Y, Prince SJ, Song L, Xu D, An YC, Valliyodan B, Varshney RK, Nguyen HT (2016) Identification and comparative analysis of differential gene expression in soybean leaf tissue under drought and flooding stress revealed by RNA-Seq. Front Plant Sci 7:1044. https://doi.org/10.3389/fpls.2016.01044

Cheng B, Furtado A, Henry RJ (2017) Long-read sequencing of the coffee bean transcriptome reveals the diversity of full-length transcripts. GigaScience 6(11):1–13. https://doi.org/10.1093/gigascience/gix086

Chopra R, Burow G, Farmer A, Mudge J, Simpson CE, Burow MD (2014) Comparisons of de novo transcriptome assemblers in diploid and polyploid species using peanut (*Arachis* spp.) RNA-Seq data. PLoS One 9(12):e115055. https://doi.org/10.1371/journal.pone.0115055

Chopra R, Burow G, Farmer A, Mudge J, Simpson CE, Wilkins TA, Baring MR, Puppala N, Chamberlin KD, Burow MD (2015) Next-generation transcriptome sequencing, SNP discovery and validation in four market classes of peanut, *Arachis hypogaea* L. Mol Gen Genomics 290(3):1169–1180. https://doi.org/10.1007/s00438-014-0976-4

Cloonan N, Forrest AR, Kolle G, Gardiner BB, Faulkner GJ, Brown MK, Taylor DF, Steptoe AL, Wani S, Bethel G, Robertson AJ, Perkins AC, Bruce SJ, Lee CC, Ranade SS, Peckham HE, Manning JM, McKernan KJ, Grimmond SM (2008) Stem cell transcriptome

profiling via massive-scale mRNA sequencing. Nat Methods 5(7):613–619. https://doi.org/10.1038/nmeth.1223
Collins LJ, Biggs PJ, Voelckel C, Joly S (2008) An approach to transcriptome analysis of non-model organisms using short-read sequences. Genome Inform 21:3–14
Conesa A, Gotz S, Garcia-Gomez JM, Terol J, Talon M, Robles M (2005) Blast2GO: a universal tool for annotation, visualization and analysis in functional genomics research. Bioinformatics 21(18):3674–3676. https://doi.org/10.1093/bioinformatics/bti610
Conesa A, Madrigal P, Tarazona S, Gomez-Cabrero D, Cervera A, McPherson A, Szczesniak MW, Gaffney DJ, Elo LL, Zhang X, Mortazavi A (2016) A survey of best practices for RNA-seq data analysis. Genome Biol 17:13. https://doi.org/10.1186/s13059-016-0881-8
Contreras-Moreira B, Cantalapiedra CP, Garcia-Pereira MJ, Gordon SP, Vogel JP, Igartua E, Casas AM, Vinuesa P (2017) Analysis of plant pan-genomes and transcriptomes with GET_HOMOLOGUES-EST, a clustering solution for sequences of the same species. Front Plant Sci 8:184. https://doi.org/10.3389/fpls.2017.00184
Cubillos FA, Coustham V, Loudet O (2012) Lessons from eQTL mapping studies: non-coding regions and their role behind natural phenotypic variation in plants. Curr Opin Plant Biol 15(2):192–198. https://doi.org/10.1016/j.pbi.2012.01.005
Das S, Meher PK, Rai A, Bhar LM, Mandal BN (2017) Statistical approaches for gene selection, hub gene identification and module interaction in gene co-expression network analysis: an application to aluminum stress in soybean (Glycine max L.). PLoS One 12(1):e0169605. https://doi.org/10.1371/journal.pone.0169605
Delseny M, Cooke R, Raynal M, Grellet F (1997) The *Arabidopsis thaliana* cDNA sequencing projects. FEBS Lett 405(2):129–132
DeLuca DS, Levin JZ, Sivachenko A, Fennell T, Nazaire MD, Williams C, Reich M, Winckler W, Getz G (2012) RNA-SeQC: RNA-seq metrics for quality control and process optimization. Bioinformatics 28(11):1530–1532. https://doi.org/10.1093/bioinformatics/bts196
DePristo MA, Banks E, Poplin R, Garimella KV, Maguire JR, Hartl C, Philippakis AA, del Angel G, Rivas MA, Hanna M, McKenna A, Fennell TJ, Kernytsky AM, Sivachenko AY, Cibulskis K, Gabriel SB, Altshuler D, Daly MJ (2011) A framework for variation discovery and genotyping using next-generation DNA sequencing data. Nat Genet 43(5):491–498. https://doi.org/10.1038/ng.806
Diaz-Uriarte R, Alvarez de Andres S (2006) Gene selection and classification of microarray data using random forest. BMC Bioinformatics 7:3. https://doi.org/10.1186/1471-2105-7-3
Didion JP, Martin M, Collins FS (2017) Atropos: specific, sensitive, and speedy trimming of sequencing reads. PeerJ 5:e3720. https://doi.org/10.7717/peerj.3720
Divya Bhanu B, Ulaganathan K, Shanker AK, Desai S (2016) RNA-seq analysis of irrigated vs. water stressed transcriptomes of Zea mays Cultivar Z59. Front Plant Sci 7:239. https://doi.org/10.3389/fpls.2016.00239
Dobin A, Davis CA, Schlesinger F, Drenkow J, Zaleski C, Jha S, Batut P, Chaisson M, Gingeras TR (2013) STAR: ultrafast universal RNA-seq aligner. Bioinformatics 29(1):15–21. https://doi.org/10.1093/bioinformatics/bts635
Dong J, Horvath S (2007) Understanding network concepts in modules. BMC Syst Biol 1:24. https://doi.org/10.1186/1752-0509-1-24
Eichner J, Zeller G, Laubinger S, Ratsch G (2011) Support vector machines-based identification of alternative splicing in *Arabidopsis thaliana* from whole-genome tiling arrays. BMC Bioinformatics 12:55. https://doi.org/10.1186/1471-2105-12-55
Elo LL, Jarvenpaa H, Oresic M, Lahesmaa R, Aittokallio T (2007) Systematic construction of gene coexpression networks with applications to human T helper cell differentiation process. Bioinformatics 23(16):2096–2103. https://doi.org/10.1093/bioinformatics/btm309
Ferreira SS, Hotta CT, Poelking VG, Leite DC, Buckeridge MS, Loureiro ME, Barbosa MH, Carneiro MS, Souza GM (2016) Co-expression network analysis reveals transcription factors associated to cell wall biosynthesis in sugarcane. Plant Mol Biol 91(1–2):15–35. https://doi.org/10.1007/s11103-016-0434-2
Fleming MB, Patterson EL, Reeves PA, Richards CM, Gaines TA, Walters C (2018) Exploring the fate of mRNA in aging seeds: protection, destruction, or slow decay? J Exp Bot 69(18):4309–4321. https://doi.org/10.1093/jxb/ery215
Flicek P, Birney E (2009) Sense from sequence reads: methods for alignment and assembly. Nat Methods 6(11 Suppl):S6–S12. https://doi.org/10.1038/nmeth.1376
Florea LD, Salzberg SL (2013) Genome-guided transcriptome assembly in the age of next-generation sequencing. IEEE/ACM Trans Comput Biol Bioinformatics 10(5):1234–1240
Frazee AC, Pertea G, Jaffe AE, Langmead B, Salzberg SL, Leek JT (2015) Ballgown bridges the gap between transcriptome assembly and expression analysis. Nat Biotechnol 33(3):243–246. https://doi.org/10.1038/nbt.3172
Fu L, Niu B, Zhu Z, Wu S, Li W (2012) CD-HIT: accelerated for clustering the next-generation sequencing data. Bioinformatics 28(23):3150–3152. https://doi.org/10.1093/bioinformatics/bts565
Gao L, Gonda I, Sun H, Ma Q, Bao K, Tieman DM, Burzynski-Chang EA, Fish TL, Stromberg KA, Sacks GL, Thannhauser TW, Foolad MR, Diez MJ, Blanca J, Canizares J, Xu Y, van der Knaap E, Huang S, Klee HJ, Giovannoni JJ, Fei Z (2019) The tomato pan-genome uncovers new genes and a rare allele regulating fruit flavor. Nat Genet 51:1044. https://doi.org/10.1038/s41588-019-0410-2
Garalde DR, Snell EA, Jachimowicz D, Sipos B, Lloyd JH, Bruce M, Pantic N, Admassu T, James P, Warland A, Jordan M, Ciccone J, Serra S, Keenan J, Martin S, McNeill L, Wallace EJ, Jayasinghe L, Wright C, Blasco J, Young S, Brocklebank D, Juul S, Clarke J, Heron AJ,

Turner DJ (2018) Highly parallel direct RNA sequencing on an array of nanopores. Nat Methods 15(3):201–206. https://doi.org/10.1038/nmeth.4577
Garcia-Ortega LF, Martinez O (2015) How many genes are expressed in a transcriptome? Estimation and results for RNA-Seq. PLoS One 10(6):e0130262. https://doi.org/10.1371/journal.pone.0130262
Garg R, Jain M (2013) RNA-Seq for transcriptome analysis in non-model plants. Methods Mol Biol 1069:43–58. https://doi.org/10.1007/978-1-62703-613-9_4
Gene Ontology Consortium T (2017) Expansion of the gene ontology knowledgebase and resources. Nucleic Acids Res 45(D1):D331–D338. https://doi.org/10.1093/nar/gkw1108
Giorgi FM, Del Fabbro C, Licausi F (2013) Comparative study of RNA-seq- and microarray-derived coexpression networks in *Arabidopsis thaliana*. Bioinformatics 29(6):717–724. https://doi.org/10.1093/bioinformatics/btt053
Grabherr MG, Haas BJ, Yassour M, Levin JZ, Thompson DA, Amit I, Adiconis X, Fan L, Raychowdhury R, Zeng Q, Chen Z, Mauceli E, Hacohen N, Gnirke A, Rhind N, di Palma F, Birren BW, Nusbaum C, Lindblad-Toh K, Friedman N, Regev A (2011) Full-length transcriptome assembly from RNA-Seq data without a reference genome. Nat Biotechnol 29(7):644–652. https://doi.org/10.1038/nbt.1883
Gross SM, Martin JA, Simpson J, Abraham-Juarez MJ, Wang Z, Visel A (2013) De novo transcriptome assembly of drought tolerant CAM plants, Agave deserti and Agave tequilana. BMC Genomics 14:563. https://doi.org/10.1186/1471-2164-14-563
Guo Y, Su B, Tang J, Zhou F, Qiu LJ (2018) Gene-based SNP identification and validation in soybean using next-generation transcriptome sequencing. Mol Gen Genomics 293(3):623–633. https://doi.org/10.1007/s00438-017-1410-5
Guttman M, Garber M, Levin JZ, Donaghey J, Robinson J, Adiconis X, Fan L, Koziol MJ, Gnirke A, Nusbaum C, Rinn JL, Lander ES, Regev A (2010) Ab initio reconstruction of cell type-specific transcriptomes in mouse reveals the conserved multi-exonic structure of lincRNAs. Nat Biotechnol 28(5):503–510. https://doi.org/10.1038/nbt.1633
Guyon I, Weston J, Barnhill S, Vapnik V (2002) Gene selection for cancer classification using support vector machines. Mach Learn 46(1–3):389–422. https://doi.org/10.1023/A:1012487302797
Haas BJ, Delcher AL, Mount SM, Wortman JR, Smith RK Jr, Hannick LI, Maiti R, Ronning CM, Rusch DB, Town CD, Salzberg SL, White O (2003) Improving the Arabidopsis genome annotation using maximal transcript alignment assemblies. Nucleic Acids Res 31(19):5654–5666. https://doi.org/10.1093/nar/gkg770
Haas BJ, Papanicolaou A, Yassour M, Grabherr M, Blood PD, Bowden J, Couger MB, Eccles D, Li B, Lieber M, MacManes MD, Ott M, Orvis J, Pochet N, Strozzi F, Weeks N, Westerman R, William T, Dewey CN, Henschel R, LeDuc RD, Friedman N, Regev A (2013) De novo transcript sequence reconstruction from RNA-seq using the Trinity platform for reference generation and analysis. Nat Protoc 8(8):1494–1512. https://doi.org/10.1038/nprot.2013.084
Hackett JB, Lu Y (2017) Whole-transcriptome RNA-seq, gene set enrichment pathway analysis, and exon coverage analysis of two plastid RNA editing mutants. Plant Signal Behav 12(5):e1312242. https://doi.org/10.1080/15592324.2017.1312242
Hale MC, McCormick CR, Jackson JR, Dewoody JA (2009) Next-generation pyrosequencing of gonad transcriptomes in the polyploid lake sturgeon (Acipenser fulvescens): the relative merits of normalization and rarefaction in gene discovery. BMC Genomics 10:203. https://doi.org/10.1186/1471-2164-10-203
Hao X, Yang T, Liu R, Hu J, Yao Y, Burlyaeva M, Wang Y, Ren G, Zhang H, Wang D, Chang J, Zong X (2017) An RNA sequencing transcriptome analysis of grasspea (Lathyrus sativus L.) and development of SSR and KASP markers. Front Plant Sci 8:1873. https://doi.org/10.3389/fpls.2017.01873
Harbers M, Carninci P (2005) Tag-based approaches for transcriptome research and genome annotation. Nat Methods 2(7):495–502. https://doi.org/10.1038/nmeth768
Hoang NV, Furtado A, Mason PJ, Marquardt A, Kasirajan L, Thirugnanasambandam PP, Botha FC, Henry RJ (2017) A survey of the complex transcriptome from the highly polyploid sugarcane genome using full-length isoform sequencing and de novo assembly from short read sequencing. BMC Genomics 18(1):395. https://doi.org/10.1186/s12864-017-3757-8
Hollender CA, Kang C, Darwish O, Geretz A, Matthews BF, Slovin J, Alkharouf N, Liu Z (2014) Floral transcriptomes in woodland strawberry uncover developing receptacle and anther gene networks. Plant Physiol 165(3):1062–1075. https://doi.org/10.1104/pp.114.237529
Honaas LA, Wafula EK, Wickett NJ, Der JP, Zhang Y, Edger PP, Altman NS, Pires JC, Leebens-Mack JH, dePamphilis CW (2016) Selecting superior de novo transcriptome assemblies: lessons learned by leveraging the best plant genome. PLoS One 11(1):e0146062. https://doi.org/10.1371/journal.pone.0146062
de Hoon M, Hayashizaki Y (2008) Deep cap analysis gene expression (CAGE): genome-wide identification of promoters, quantification of their expression, and network inference. BioTechniques 44(5):627–628., 630, 632. https://doi.org/10.2144/000112802
Horng JT, Wu LC, Liu BJ, Kuo JL, Kuo WH, Zhang JJ (2009) An expert system to classify microarray gene expression data using gene selection by decision tree. Expert Syst Appl 36(5):9072–9081. https://doi.org/10.1016/j.eswa.2008.12.037
Horvath S, Dong J (2008) Geometric interpretation of gene coexpression network analysis. PLoS Comput Biol 4(8):e1000117. https://doi.org/10.1371/journal.pcbi.1000117
Hrdlickova R, Toloue M, Tian B (2017) RNA-Seq methods for transcriptome analysis. Wiley Interdiscip Rev RNA 8(1):e1364. https://doi.org/10.1002/wrna.1364

Huang Q, Lin B, Liu H, Ma X, Mo F, Yu W, Li L, Li H, Tian T, Wu D, Shen F, Xing J, Chen ZN (2011) RNA-Seq analyses generate comprehensive transcriptomic landscape and reveal complex transcript patterns in hepatocellular carcinoma. PLoS One 6(10):e26168. https://doi.org/10.1371/journal.pone.0026168

Huang J, Vendramin S, Shi L, McGinnis KM (2017a) Construction and optimization of a large gene coexpression network in maize using RNA-Seq data. Plant Physiol 175(1):568–583. https://doi.org/10.1104/pp.17.00825

Huang Z, Peng G, Liu X, Deora A, Falk KC, Gossen BD, McDonald MR, Yu F (2017b) Fine mapping of a clubroot resistance gene in Chinese cabbage using SNP markers identified from bulked segregant RNA sequencing. Front Plant Sci 8:1448. https://doi.org/10.3389/fpls.2017.01448

Hussain S (2018) Native RNA-sequencing throws its hat into the transcriptomics ring. Trends Biochem Sci 43(4):225–227. https://doi.org/10.1016/j.tibs.2018.02.007

Iancu OD, Kawane S, Bottomly D, Searles R, Hitzemann R, McWeeney S (2012) Utilizing RNA-Seq data for de novo coexpression network inference. Bioinformatics 28(12):1592–1597. https://doi.org/10.1093/bioinformatics/bts245

Illumina (2010) Illumina sequencing technology

Illumina (2017) TruSeq stranded total RNA - reference guide

Imadi SR, Kazi AG, Ahanger MA, Gucel S, Ahmad P (2015) Plant transcriptomics and responses to environmental stress: an overview. J Genet 94(3):525–537

Jain P, Krishnan NM, Panda B (2013) Augmenting transcriptome assembly by combining de novo and genome-guided tools. PeerJ 1:e133. https://doi.org/10.7717/peerj.133

Jen CH, Michalopoulos I, Westhead DR, Meyer P (2005) Natural antisense transcripts with coding capacity in Arabidopsis may have a regulatory role that is not linked to double-stranded RNA degradation. Genome Biol 6(6):R51. https://doi.org/10.1186/gb-2005-6-6-r51

Jenjaroenpun P, Wongsurawat T, Pereira R, Patumcharoenpol P, Ussery DW, Nielsen J, Nookaew I (2018) Complete genomic and transcriptional landscape analysis using third-generation sequencing: a case study of Saccharomyces cerevisiae CEN.PK113-7D. Nucleic Acids Res 46(7):e38. https://doi.org/10.1093/nar/gky014

Jiao X, Doamekpor SK, Bird JG, Nickels BE, Tong L, Hart RP, Kiledjian M (2017) 5' end nicotinamide adenine dinucleotide cap in human cells promotes RNA decay through DXO-mediated deNADding. Cell 168(6):1015–1027.e1010. https://doi.org/10.1016/j.cell.2017.02.019

Jin H, Vacic V, Girke T, Lonardi S, Zhu JK (2008) Small RNAs and the regulation of cis-natural antisense transcripts in *Arabidopsis*. BMC Mol Biol 9:6. https://doi.org/10.1186/1471-2199-9-6

Kanehisa M, Goto S (2000) KEGG: kyoto encyclopedia of genes and genomes. Nucleic Acids Res 28(1):27–30

Kanehisa M, Sato Y, Kawashima M, Furumichi M, Tanabe M (2016) KEGG as a reference resource for gene and protein annotation. Nucleic Acids Res 44(D1):D457–D462. https://doi.org/10.1093/nar/gkv1070

Keller MW, Rambo-Martin BL, Wilson MM, Ridenour CA, Shepard SS, Stark TJ, Neuhaus EB, Dugan VG, Wentworth DE, Barnes JR (2018) Direct RNA sequencing of the coding complete influenza A virus genome. Sci Rep 8(1):14408. https://doi.org/10.1038/s41598-018-32615-8

Khanin R, Wit E (2006) How scale-free are biological networks. J Comput Biol 13(3):810–818. https://doi.org/10.1089/cmb.2006.13.810

Kim D, Pertea G, Trapnell C, Pimentel H, Kelley R, Salzberg SL (2013) TopHat2: accurate alignment of transcriptomes in the presence of insertions, deletions and gene fusions. Genome Biol 14(4):R36. https://doi.org/10.1186/gb-2013-14-4-r36

Kim D, Langmead B, Salzberg SL (2015) HISAT: a fast spliced aligner with low memory requirements. Nat Methods 12(4):357–360. https://doi.org/10.1038/nmeth.3317

Kim D, Paggi JM, Park C, Bennett C, Salzberg SL (2019) Graph-based genome alignment and genotyping with HISAT2 and HISAT-genotype. Nat Biotechnol 37(8):907–915. https://doi.org/10.1038/s41587-019-0201-4

Kopylova E, Noe L, Touzet H (2012) SortMeRNA: fast and accurate filtering of ribosomal RNAs in metatranscriptomic data. Bioinformatics 28(24):3211–3217. https://doi.org/10.1093/bioinformatics/bts611

Kovaka S, Zimin AV, Pertea GM, Razaghi R, Salzberg SL, Pertea M (2019) Transcriptome assembly from long-read RNA-seq alignments with StringTie2. Genome Biol 20(1):278. https://doi.org/10.1186/s13059-019-1910-1

Lamarre S, Frasse P, Zouine M, Labourdette D, Sainderichin E, Hu G, Le Berre-Anton V, Bouzayen M, Maza E (2018) Optimization of an RNA-Seq differential gene expression analysis depending on biological replicate number and library size. Front Plant Sci 9:108. https://doi.org/10.3389/fpls.2018.00108

Langfelder P, Horvath S (2008) WGCNA: an R package for weighted correlation network analysis. BMC Bioinformatics 9:559. https://doi.org/10.1186/1471-2105-9-559

Langmead B, Salzberg SL (2012) Fast gapped-read alignment with Bowtie 2. Nat Methods 9(4):357–359. https://doi.org/10.1038/nmeth.1923

Langmead B, Trapnell C, Pop M, Salzberg SL (2009) Ultrafast and memory-efficient alignment of short DNA sequences to the human genome. Genome Biol 10(3):R25. https://doi.org/10.1186/gb-2009-10-3-r25

Larranaga P, Calvo B, Santana R, Bielza C, Galdiano J, Inza I, Lozano JA, Armananzas R, Santafe G, Perez A, Robles V (2006) Machine learning in bioinformatics. Brief Bioinform 7(1):86–112. https://doi.org/10.1093/bib/bbk007

Lee HK, Hsu AK, Sajdak J, Qin J, Pavlidis P (2004) Coexpression analysis of human genes across many microarray data sets. Genome Res 14(6):1085–1094. https://doi.org/10.1101/gr.1910904

Lee JH, Ang JK, Xiao X (2013) Analysis and design of RNA sequencing experiments for identifying RNA editing and other single-nucleotide variants. RNA 19(6):725–732. https://doi.org/10.1261/rna.037903.112

Lembke CG, Nishiyama MY Jr, Sato PM, de Andrade RF, Souza GM (2012) Identification of sense and antisense transcripts regulated by drought in sugarcane. Plant Mol Biol 79(4–5):461–477. https://doi.org/10.1007/s11103-012-9922-1

Leng X, Jia H, Sun X, Shangguan L, Mu Q, Wang B, Fang J (2015) Comparative transcriptome analysis of grapevine in response to copper stress. Sci Rep 5:17749. https://doi.org/10.1038/srep17749

Leung MKK, Delong A, Alipanahi B, Frey BJ (2016) Machine learning in genomic medicine: a review of computational problems and data sets. Proc IEEE 104(1):176–197. https://doi.org/10.1109/Jproc.2015.2494198

Li H (2011) A statistical framework for SNP calling, mutation discovery, association mapping and population genetical parameter estimation from sequencing data. Bioinformatics 27(21):2987–2993. https://doi.org/10.1093/bioinformatics/btr509

Li H (2013) Aligning sequence reads, clone sequences and assembly contigs with BWA-MEM. ArXiv

Li B, Dewey CN (2011) RSEM: accurate transcript quantification from RNA-Seq data with or without a reference genome. BMC Bioinformatics 12:323. https://doi.org/10.1186/1471-2105-12-323

Li H, Durbin R (2009) Fast and accurate short read alignment with Burrows-Wheeler transform. Bioinformatics 25(14):1754–1760. https://doi.org/10.1093/bioinformatics/btp324

Li H, Homer N (2010) A survey of sequence alignment algorithms for next-generation sequencing. Brief Bioinform 11(5):473–483. https://doi.org/10.1093/bib/bbq015

Li H, Handsaker B, Wysoker A, Fennell T, Ruan J, Homer N, Marth G, Abecasis G, Durbin R (2009) The sequence alignment/map format and SAMtools. Bioinformatics 25(16):2078–2079. https://doi.org/10.1093/bioinformatics/btp352

Li B, Ruotti V, Stewart RM, Thomson JA, Dewey CN (2010) RNA-Seq gene expression estimation with read mapping uncertainty. Bioinformatics 26(4):493–500. https://doi.org/10.1093/bioinformatics/btp692

Li JJ, Jiang CR, Brown JB, Huang H, Bickel PJ (2011a) Sparse linear modeling of next-generation mRNA sequencing (RNA-Seq) data for isoform discovery and abundance estimation. Proc Natl Acad Sci U S A 108(50):19867–19872. https://doi.org/10.1073/pnas.1113972108

Li W, Feng J, Jiang T (2011b) IsoLasso: a LASSO regression approach to RNA-Seq based transcriptome assembly. J Comput Biol 18(11):1693–1707. https://doi.org/10.1089/cmb.2011.0171

Li S, Liberman LM, Mukherjee N, Benfey PN, Ohler U (2013a) Integrated detection of natural antisense transcripts using strand-specific RNA sequencing data. Genome Res 23(10):1730–1739. https://doi.org/10.1101/gr.149310.112

Li Y, Li-Byarlay H, Burns P, Borodovsky M, Robinson GE, Ma J (2013b) TrueSight: a new algorithm for splice junction detection using RNA-seq. Nucleic Acids Res 41(4):e51. https://doi.org/10.1093/nar/gks1311

Li B, Fillmore N, Bai Y, Collins M, Thomson JA, Stewart R, Dewey CN (2014) Evaluation of de novo transcriptome assemblies from RNA-Seq data. Genome Biol 15(12):553. https://doi.org/10.1186/s13059-014-0553-5

Li H, Yao W, Fu Y, Li S, Guo Q (2015) De novo assembly and discovery of genes that are involved in drought tolerance in Tibetan Sophora moorcroftiana. PLoS One 10(1):e111054. https://doi.org/10.1371/journal.pone.0111054

Li J, Harata-Lee Y, Denton MD, Feng Q, Rathjen JR, Qu Z, Adelson DL (2017a) Long read reference genome-free reconstruction of a full-length transcriptome from Astragalus membranaceus reveals transcript variants involved in bioactive compound biosynthesis. Cell Discov 3:17031. https://doi.org/10.1038/celldisc.2017.31

Li Y, Dai C, Hu C, Liu Z, Kang C (2017b) Global identification of alternative splicing via comparative analysis of SMRT- and Illumina-based RNA-seq in strawberry. Plant J 90(1):164–176. https://doi.org/10.1111/tpj.13462

Li Y, Wei W, Feng J, Luo H, Pi M, Liu Z, Kang C (2018) Genome re-annotation of the wild strawberry Fragaria vesca using extensive Illumina- and SMRT-based RNA-seq datasets. DNA Res 25:61. https://doi.org/10.1093/dnares/dsx038

Liang Y, Zhang F, Wang J, Joshi T, Wang Y, Xu D (2011) Prediction of drought-resistant genes in *Arabidopsis thaliana* using SVM-RFE. PLoS One 6(7):e21750. https://doi.org/10.1371/journal.pone.0021750

Liang YH, Cai B, Chen F, Wang G, Wang M, Zhong Y, Cheng ZM (2014) Construction and validation of a gene co-expression network in grapevine (Vitis vinifera. L.). Hortic Res 1:14040. https://doi.org/10.1038/hortres.2014.40

Liao Y, Smyth GK, Shi W (2013) The Subread aligner: fast, accurate and scalable read mapping by seed-and-vote. Nucleic Acids Res 41(10):e108. https://doi.org/10.1093/nar/gkt214

Liao Y, Smyth GK, Shi W (2014) featureCounts: an efficient general purpose program for assigning sequence reads to genomic features. Bioinformatics 30(7):923–930. https://doi.org/10.1093/bioinformatics/btt656

Libbrecht MW, Noble WS (2015) Machine learning applications in genetics and genomics. Nat Rev Genet 16(6):321–332. https://doi.org/10.1038/nrg3920

Liesecke F, Daudu D, Duge de Bernonville R, Besseau S, Clastre M, Courdavault V, de Craene JO, Creche J, Giglioli-Guivarc'h N, Glevarec G, Pichon O, Duge de Bernonville T (2018) Ranking genome-wide correlation measurements improves microarray and

RNA-seq based global and targeted co-expression networks. Sci Rep 8(1):10885. https://doi.org/10.1038/s41598-018-29077-3

Lima-Mendez G, van Helden J (2009) The powerful law of the power law and other myths in network biology. Mol BioSyst 5(12):1482–1493. https://doi.org/10.1039/b908681a

Lister R, O'Malley RC, Tonti-Filippini J, Gregory BD, Berry CC, Millar AH, Ecker JR (2008) Highly integrated single-base resolution maps of the epigenome in Arabidopsis. Cell 133(3):523–536. https://doi.org/10.1016/j.cell.2008.03.029

Liu Y, Zhou J, White KP (2014) RNA-seq differential expression studies: more sequence or more replication? Bioinformatics 30(3):301–304. https://doi.org/10.1093/bioinformatics/btt688

Liu X, Mei W, Soltis PS, Soltis DE, Barbazuk WB (2017) Detecting alternatively spliced transcript isoforms from single-molecule long-read sequences without a reference genome. Mol Ecol Resour 17(6):1243–1256. https://doi.org/10.1111/1755-0998.12670

Love MI, Huber W, Anders S (2014) Moderated estimation of fold change and dispersion for RNA-seq data with DESeq2. Genome Biol 15(12):550. https://doi.org/10.1186/s13059-014-0550-8

Luo F, Yang Y, Zhong J, Gao H, Khan L, Thompson DK, Zhou J (2007) Constructing gene co-expression networks and predicting functions of unknown genes by random matrix theory. BMC Bioinformatics 8:299. https://doi.org/10.1186/1471-2105-8-299

Luo R, Liu B, Xie Y, Li Z, Huang W, Yuan J, He G, Chen Y, Pan Q, Liu Y, Tang J, Wu G, Zhang H, Shi Y, Yu C, Wang B, Lu Y, Han C, Cheung DW, Yiu SM, Peng S, Xiaoqian Z, Liu G, Liao X, Li Y, Yang H, Wang J, Lam TW (2012) SOAPdenovo2: an empirically improved memory-efficient short-read de novo assembler. GigaScience 1(1):18. https://doi.org/10.1186/2047-217X-1-18

Ma HM, Schulze S, Lee S, Yang M, Mirkov E, Irvine J, Moore P, Paterson A (2004) An EST survey of the sugarcane transcriptome. Theor Appl Genet 108(5):851–863. https://doi.org/10.1007/s00122-003-1510-y

Ma X, Zhao H, Xu W, You Q, Yan H, Gao Z, Su Z (2018) Co-expression gene network analysis and functional module identification in bamboo growth and development. Front Genet 9:574. https://doi.org/10.3389/fgene.2018.00574

Ma Y, Liu M, Stiller J, Liu C (2019) A pan-transcriptome analysis shows that disease resistance genes have undergone more selection pressure during barley domestication. BMC Genomics 20(1):12. https://doi.org/10.1186/s12864-018-5357-7

Magalhaes JV, Liu J, Guimaraes CT, Lana UG, Alves VM, Wang YH, Schaffert RE, Hoekenga OA, Pineros MA, Shaff JE, Klein PE, Carneiro NP, Coelho CM, Trick HN, Kochian LV (2007) A gene in the multidrug and toxic compound extrusion (MATE) family confers aluminum tolerance in sorghum. Nat Genet 39(9):1156–1161. https://doi.org/10.1038/ng2074

Mamoshina P, Vieira A, Putin E, Zhavoronkov A (2016) Applications of deep learning in biomedicine. Mol Pharm 13(5):1445–1454. https://doi.org/10.1021/acs.molpharmaceut.5b00982

Mao R, Raj Kumar PK, Guo C, Zhang Y, Liang C (2014) Comparative analyses between retained introns and constitutively spliced introns in *Arabidopsis thaliana* using random forest and support vector machine. PLoS One 9(8):e104049. https://doi.org/10.1371/journal.pone.0104049

Marioni JC, Mason CE, Mane SM, Stephens M, Gilad Y (2008) RNA-seq: an assessment of technical reproducibility and comparison with gene expression arrays. Genome Res 18(9):1509–1517. https://doi.org/10.1101/gr.079558.108

Marquez Y, Brown JW, Simpson C, Barta A, Kalyna M (2012) Transcriptome survey reveals increased complexity of the alternative splicing landscape in Arabidopsis. Genome Res 22(6):1184–1195. https://doi.org/10.1101/gr.134106.111

Masalia RR, Bewick AJ, Burke JM (2017) Connectivity in gene coexpression networks negatively correlates with rates of molecular evolution in flowering plants. PLoS One 12(7):e0182289. https://doi.org/10.1371/journal.pone.0182289

Matsumura H, Ito A, Saitoh H, Winter P, Kahl G, Reuter M, Kruger DH, Terauchi R (2005) SuperSAGE. Cell Microbiol 7(1):11–18. https://doi.org/10.1111/j.1462-5822.2004.00478.x

McCarthy DJ, Chen Y, Gordon KS (2012) Differential expression analysis of multifactor RNA-Seq experiments with respect to biological variation. Nucleic Acids Research 40(10):4288–4297

McKay JK, Leach JE (2011) Linkage illuminates a complex genome. Nat Biotechnol 29(8):717–718. https://doi.org/10.1038/nbt.1945

McKenna A, Hanna M, Banks E, Sivachenko A, Cibulskis K, Kernytsky A, Garimella K, Altshuler D, Gabriel S, Daly M, DePristo MA (2010) The Genome Analysis Toolkit: a MapReduce framework for analyzing next-generation DNA sequencing data. Genome Res 20(9):1297–1303. https://doi.org/10.1101/gr.107524.110

Meyer PE, Kontos K, Lafitte F, Bontempi G (2007) Information-theoretic inference of large transcriptional regulatory networks. EURASIP J Bioinform Syst Biol 2007:79879. https://doi.org/10.1155/2007/79879

Meyer PE, Lafitte F, Bontempi G (2008) minet: a R/Bioconductor package for inferring large transcriptional networks using mutual information. BMC Bioinformatics 9:461. https://doi.org/10.1186/1471-2105-9-461

Min S, Lee B, Yoon S (2017) Deep learning in bioinformatics. Brief Bioinform 18(5):851–869. https://doi.org/10.1093/bib/bbw068

Minio A, Massonnet M, Figueroa-Balderas R, Vondras AM, Blanco-Ulate B, Cantu D (2019) Iso-Seq allows genome-independent transcriptome profiling of grape berry development. G3 (Bethesda) 9(3):755–767. https://doi.org/10.1534/g3.118.201008

Mofatto LS, Carneiro Fde A, Vieira NG, Duarte KE, Vidal RO, Alekcevetch JC, Cotta MG, Verdeil JL, Lapeyre-Montes F, Lartaud M, Leroy T, De Bellis F, Pot D, Rodrigues GC, Carazzolle MF, Pereira GA, Andrade AC, Marraccini P (2016) Identification of candidate genes for drought tolerance in coffee by high-throughput sequencing in the shoot apex of different Coffea arabica cultivars. BMC Plant Biol 16:94. https://doi.org/10.1186/s12870-016-0777-5

Moreton J, Izquierdo A, Emes RD (2015) Assembly, assessment, and availability of de novo generated eukaryotic transcriptomes. Front Genet 6:361. https://doi.org/10.3389/fgene.2015.00361

Mortazavi A, Williams BA, McCue K, Schaeffer L, Wold B (2008) Mapping and quantifying mammalian transcriptomes by RNA-Seq. Nat Methods 5(7):621–628. https://doi.org/10.1038/nmeth.1226

Murphy KP (2012) Machine learning: a probabilistic perspective. Adaptive computation and machine learning series. MIT Press, Cambridge, MA

Mutwil M, Usadel B, Schutte M, Loraine A, Ebenhoh O, Persson S (2010) Assembly of an interactive correlation network for the Arabidopsis genome using a novel heuristic clustering algorithm. Plant Physiol 152(1):29–43. https://doi.org/10.1104/pp.109.145318

Nagalakshmi U, Wang Z, Waern K, Shou C, Raha D, Gerstein M, Snyder M (2008) The transcriptional landscape of the yeast genome defined by RNA sequencing. Science 320(5881):1344–1349. https://doi.org/10.1126/science.1158441

NuGen (n.d.) AnyDeplete

O'Neil D, Glowatz H, Schlumpberger M (2013) Ribosomal RNA depletion for efficient use of RNA-seq capacity. Curr Protoc Mol Biol Chapter 4:Unit 4.19. https://doi.org/10.1002/0471142727.mb0419s103

Oliver S (2000) Guilt-by-association goes global. Nature 403(6770):601–603. https://doi.org/10.1038/35001165

One Thousand Plant Transcriptomes Initiative (2019) One thousand plant transcriptomes and the phylogenomics of green plants. Nature 574(7780):679–685. https://doi.org/10.1038/s41586-019-1693-2

Oono Y, Yazawa T, Kanamori H, Sasaki H, Mori S, Matsumoto T (2017) Genome-wide analysis of rice cis-natural antisense transcription under cadmium exposure using strand-specific RNA-Seq. BMC Genomics 18(1):761. https://doi.org/10.1186/s12864-017-4108-5

Osato N, Yamada H, Satoh K, Ooka H, Yamamoto M, Suzuki K, Kawai J, Carninci P, Ohtomo Y, Murakami K, Matsubara K, Kikuchi S, Hayashizaki Y (2003) Antisense transcripts with rice full-length cDNAs. Genome Biol 5(1):R5. https://doi.org/10.1186/gb-2003-5-1-r5

Pan X, Xiong K (2015) PredcircRNA: computational classification of circular RNA from other long non-coding RNA using hybrid features. Mol BioSyst 11(8):2219–2226. https://doi.org/10.1039/c5mb00214a

Park JJ, Wang H, Gargouri M, Deshpande RR, Skepper JN, Holguin FO, Juergens MT, Shachar-Hill Y, Hicks LM, Gang DR (2015) The response of Chlamydomonas reinhardtii to nitrogen deprivation: a systems biology analysis. Plant J 81(4):611–624. https://doi.org/10.1111/tpj.12747

Parkinson J, Blaxter M (2009) Expressed sequence tags: an overview. Methods Mol Biol 533:1–12. https://doi.org/10.1007/978-1-60327-136-3_1

Patro R, Mount SM, Kingsford C (2014) Sailfish enables alignment-free isoform quantification from RNA-seq reads using lightweight algorithms. Nat Biotechnol 32(5):462–464. https://doi.org/10.1038/nbt.2862

Patro R, Duggal G, Love MI, Irizarry RA, Kingsford C (2017) Salmon provides fast and bias-aware quantification of transcript expression. Nat Methods 14(4):417–419. https://doi.org/10.1038/nmeth.4197

Pavy N, Paule C, Parsons L, Crow JA, Morency MJ, Cooke J, Johnson JE, Noumen E, Guillet-Claude C, Butterfield Y, Barber S, Yang G, Liu J, Stott J, Kirkpatrick R, Siddiqui A, Holt R, Marra M, Seguin A, Retzel E, Bousquet J, MacKay J (2005) Generation, annotation, analysis and database integration of 16,500 white spruce EST clusters. BMC Genomics 6:144. https://doi.org/10.1186/1471-2164-6-144

Pavy N, Boyle B, Nelson C, Paule C, Giguere I, Caron S, Parsons LS, Dallaire N, Bedon F, Berube H, Cooke J, Mackay J (2008) Identification of conserved core xylem gene sets: conifer cDNA microarray development, transcript profiling and computational analyses. New Phytol 180(4):766–786. https://doi.org/10.1111/j.1469-8137.2008.02615.x

Pearce S, Vazquez-Gross H, Herin SY, Hane D, Wang Y, Gu YQ, Dubcovsky J (2015) WheatExp: an RNA-seq expression database for polyploid wheat. BMC Plant Biol 15:299. https://doi.org/10.1186/s12870-015-0692-1

Peng Z, Gallo M, Tillman BL, Rowland D, Wang J (2016) Molecular marker development from transcript sequences and germplasm evaluation for cultivated peanut (Arachis hypogaea L.). Mol Gen Genomics 291(1):363–381. https://doi.org/10.1007/s00438-015-1115-6

Pereira-Santana A, Alvarado-Robledo EJ, Zamora-Briseno JA, Ayala-Sumuano JT, Gonzalez-Mendoza VM, Espadas-Gil F, Alcaraz LD, Castano E, Keb-Llanes MA, Sanchez-Teyer F, Rodriguez-Zapata LC (2017) Transcriptional profiling of sugarcane leaves and roots under progressive osmotic stress reveals a regulated coordination of gene expression in a spatiotemporal manner. PLoS One 12(12):e0189271. https://doi.org/10.1371/journal.pone.0189271

Pertea M, Pertea GM, Antonescu CM, Chang TC, Mendell JT, Salzberg SL (2015) StringTie enables improved reconstruction of a transcriptome from RNA-seq reads. Nat Biotechnol 33(3):290–295. https://doi.org/10.1038/nbt.3122

Pham GM, Newton L, Wiegert-Rininger K, Vaillancourt B, Douches DS, Buell CR (2017) Extensive genome heterogeneity leads to preferential allele expression and copy number-dependent expression in cultivated

potato. Plant J 92(4):624–637. https://doi.org/10.1111/tpj.13706
Pimentel H, Bray NL, Puente S, Melsted P, Pachter L (2017) Differential analysis of RNA-seq incorporating quantification uncertainty. Nat Methods 14(7):687–690. https://doi.org/10.1038/nmeth.4324
Piriyapongsa J, Kaewprommal P, Vaiwsri S, Anuntakarun S, Wirojsirasak W, Punpee P, Klomsa-Ard P, Shaw PJ, Pootakham W, Yoocha T, Sangsrakru D, Tangphatsornruang S, Tongsima S, Tragoonrung S (2018) Uncovering full-length transcript isoforms of sugarcane cultivar Khon Kaen 3 using single-molecule long-read sequencing. PeerJ 6:e5818. https://doi.org/10.7717/peerj.5818
Qiao Q, Xue L, Wang Q, Sun H, Zhong Y, Huang J, Lei J, Zhang T (2016) Comparative transcriptomics of strawberries (*Fragaria* spp.) provides insights into evolutionary patterns. Front Plant Sci 7:1839. https://doi.org/10.3389/fpls.2016.01839
Qiao L, Cao M, Zheng J, Zhao Y, Zheng ZL (2017) Gene coexpression network analysis of fruit transcriptomes uncovers a possible mechanistically distinct class of sugar/acid ratio-associated genes in sweet orange. BMC Plant Biol 17(1):186. https://doi.org/10.1186/s12870-017-1138-8
Quast C, Pruesse E, Yilmaz P, Gerken J, Schweer T, Yarza P, Peplies J, Glockner FO (2013) The SILVA ribosomal RNA gene database project: improved data processing and web-based tools. Nucleic Acids Res 41(Database issue):D590–D596. https://doi.org/10.1093/nar/gks1219
Quesada V, Ponce MR, Micol JL (1999) OTC and AUL1, two convergent and overlapping genes in the nuclear genome of *Arabidopsis thaliana*. FEBS Lett 461(1–2):101–106
Rabiee-Ghahfarrokhi B, Rafiei F, Niknafs AA, Zamani B (2015) Prediction of microRNA target genes using an efficient genetic algorithm-based decision tree. FEBS Open Bio 5:877–884. https://doi.org/10.1016/j.fob.2015.10.003
Ramirez-Gonzalez RH, Segovia V, Bird N, Fenwick P, Holdgate S, Berry S, Jack P, Caccamo M, Uauy C (2015) RNA-Seq bulked segregant analysis enables the identification of high-resolution genetic markers for breeding in hexaploid wheat. Plant Biotechnol J 13(5):613–624. https://doi.org/10.1111/pbi.12281
Ran JH, Shen TT, Wang MM, Wang XQ (2018) Phylogenomics resolves the deep phylogeny of seed plants and indicates partial convergent or homoplastic evolution between Gnetales and angiosperms. Proc Biol Sci 285(1881):20181012. https://doi.org/10.1098/rspb.2018.1012
Ranwez V, Holtz Y, Sarah G, Ardisson M, Santoni S, Glemin S, Tavaud-Pirra M, David J (2013) Disentangling homeologous contigs in allo-tetraploid assembly: application to durum wheat. BMC Bioinformatics 14(Suppl 15):S15. https://doi.org/10.1186/1471-2105-14-S15-S15
Ren Y, Wang D, Wang Y, Zhou J, Zhang H, Zhou Y, Liang Y (2010) Prediction of disease-resistant gene in rice based on SVM-RFE. In: 2010 3rd International Conference on Biomedical Engineering and Informatics, 16–18 October 2010, pp 2343–2346. https://doi.org/10.1109/bmei.2010.5640583
Rhoads A, Au KF (2015) PacBio sequencing and its applications. Genom Proteom Bioinformatics 13(5):278–289. https://doi.org/10.1016/j.gpb.2015.08.002
Riano-Pachon DM, Mattiello L, Cruz LP (2016) Surveying the complex polyploid sugarcane genome sequence using synthetic long reads. Laboratório Nacional de Ciência e Pesquisa do Bioetanol, Centro Nacional de Pesquisa em Energia e Materiais, Campinas. https://doi.org/10.13140/RG.2.1.3468.0565
Ritchie ME, Phipson B, Wu D, Hu Y, Law CW, Shi W, Smyth GK (2015) limma powers differential expression analyses for RNA-sequencing and microarray studies. Nucleic Acids Res 43(7):e47. https://doi.org/10.1093/nar/gkv007
Robertson G, Schein J, Chiu R, Corbett R, Field M, Jackman SD, Mungall K, Lee S, Okada HM, Qian JQ, Griffith M, Raymond A, Thiessen N, Cezard T, Butterfield YS, Newsome R, Chan SK, She R, Varhol R, Kamoh B, Prabhu AL, Tam A, Zhao Y, Moore RA, Hirst M, Marra MA, Jones SJ, Hoodless PA, Birol I (2010) De novo assembly and analysis of RNA-seq data. Nat Methods 7(11):909–912. https://doi.org/10.1038/nmeth.1517
Robinson MD, Oshlack A (2010) A scaling normalization method for differential expression analysis of RNA-seq data. Genome Biol 11(3):R25. https://doi.org/10.1186/gb-2010-11-3-r25
Robinson MD, Smyth GK (2007) Moderated statistical tests for assessing differences in tag abundance. Bioinformatics 23(21):2881–2887. https://doi.org/10.1093/bioinformatics/btm453
Robinson MD, Smyth GK (2008) Small-sample estimation of negative binomial dispersion, with applications to SAGE data. Biostatistics 9(2):321–332. https://doi.org/10.1093/biostatistics/kxm030
Robinson MD, McCarthy DJ, Smyth GK (2010) edgeR: a Bioconductor package for differential expression analysis of digital gene expression data. Bioinformatics 26(1):139–140. https://doi.org/10.1093/bioinformatics/btp616
Rosenow C, Saxena RM, Durst M, Gingeras TR (2001) Prokaryotic RNA preparation methods useful for high density array analysis: comparison of two approaches. Nucleic Acids Res 29(22):E112. https://doi.org/10.1093/nar/29.22.e112
Rothfels CJ, Larsson A, Li FW, Sigel EM, Huiet L, Burge DO, Ruhsam M, Graham SW, Stevenson DW, Wong GK, Korall P, Pryer KM (2013) Transcriptome-mining for single-copy nuclear markers in ferns. PLoS One 8(10):e76957. https://doi.org/10.1371/journal.pone.0076957
Sayols S, Scherzinger D, Klein H (2016) dupRadar: a Bioconductor package for the assessment of PCR artifacts in RNA-Seq data. BMC Bioinformatics 17(1):428. https://doi.org/10.1186/s12859-016-1276-2

Schulz MH, Zerbino DR, Vingron M, Birney E (2012) Oases: robust de novo RNA-seq assembly across the dynamic range of expression levels. Bioinformatics 28(8):1086–1092. https://doi.org/10.1093/bioinformatics/bts094

Schurch NJ, Schofield P, Gierlinski M, Cole C, Sherstnev A, Singh V, Wrobel N, Gharbi K, Simpson GG, Owen-Hughes T, Blaxter M, Barton GJ (2016) How many biological replicates are needed in an RNA-seq experiment and which differential expression tool should you use? RNA 22(6):839–851. https://doi.org/10.1261/rna.053959.115

Serin EA, Nijveen H, Hilhorst HW, Ligterink W (2016) Learning from co-expression networks: possibilities and challenges. Front Plant Sci 7:444. https://doi.org/10.3389/fpls.2016.00444

Serin EAR, Snoek LB, Nijveen H, Willems LAJ, Jimenez-Gomez JM, Hilhorst HWM, Ligterink W (2017) Construction of a high-density genetic map from RNA-Seq data for an Arabidopsis Bay-0 x Shahdara RIL population. Front Genet 8:201. https://doi.org/10.3389/fgene.2017.00201

Shang X, Cao Y, Ma L (2017) Alternative splicing in plant genes: a means of regulating the environmental fitness of plants. Int J Mol Sci 18(2):432. https://doi.org/10.3390/ijms18020432

Shao M, Ma J, Wang S (2017) DeepBound: accurate identification of transcript boundaries via deep convolutional neural fields. Bioinformatics 33(14):i267–i273. https://doi.org/10.1093/bioinformatics/btx267

Shao L, Xing F, Xu C, Zhang Q, Che J, Wang X, Song J, Li X, Xiao J, Chen LL, Ouyang Y (2019) Patterns of genome-wide allele-specific expression in hybrid rice and the implications on the genetic basis of heterosis. Proc Natl Acad Sci U S A 116(12):5653–5658. https://doi.org/10.1073/pnas.1820513116

Shi L, Guo Y, Dong C, Huddleston J, Yang H, Han X, Fu A, Li Q, Li N, Gong S, Lintner KE, Ding Q, Wang Z, Hu J, Wang D, Wang F, Wang L, Lyon GJ, Guan Y, Shen Y, Evgrafov OV, Knowles JA, Thibaud-Nissen F, Schneider V, Yu CY, Zhou L, Eichler EE, So KF, Wang K (2016) Long-read sequencing and de novo assembly of a Chinese genome. Nat Commun 7:12065. https://doi.org/10.1038/ncomms12065

Shi R, Wang JP, Lin YC, Li Q, Sun YH, Chen H, Sederoff RR, Chiang VL (2017) Tissue and cell-type co-expression networks of transcription factors and wood component genes in Populus trichocarpa. Planta 245(5):927–938. https://doi.org/10.1007/s00425-016-2640-1

Simpson JT, Wong K, Jackman SD, Schein JE, Jones SJ, Birol I (2009) ABySS: a parallel assembler for short read sequence data. Genome Res 19(6):1117–1123. https://doi.org/10.1101/gr.089532.108

siTOOLsBiotech (2018) riboPOOL: affordable ribosomal/custom RNA depletion for any species

Smith-Unna R, Boursnell C, Patro R, Hibberd JM, Kelly S (2016) TransRate: reference-free quality assessment of de novo transcriptome assemblies. Genome Res 26(8):1134–1144. https://doi.org/10.1101/gr.196469.115

Song L, Langfelder P, Horvath S (2012) Comparison of co-expression measures: mutual information, correlation, and model based indices. BMC Bioinformatics 13:328. https://doi.org/10.1186/1471-2105-13-328

Srivastava A, Malik L, Sarkar H, Zakeri M, Soneson C, Love MI, Kingsford C, Patro R (2019) Alignment and mapping methodology influence transcript abundance estimation. bioRxiv:657874. https://doi.org/10.1101/657874

Stumpf MPH, Ingram PJ (2005) Probability models for degree distributions of protein interaction networks. Europhys Lett 71(1):152–158

Sun X, Yang Q, Deng Z, Ye X (2014) Digital inventory of Arabidopsis transcripts revealed by 61 RNA sequencing samples. Plant Physiol 166(2):869–878. https://doi.org/10.1104/pp.114.241604

Sveinsson S, McDill J, Wong GK, Li J, Li X, Deyholos MK, Cronk QC (2014) Phylogenetic pinpointing of a paleopolyploidy event within the flax genus (Linum) using transcriptomics. Ann Bot 113(5):753–761. https://doi.org/10.1093/aob/mct306

Tai Y, Liu C, Yu S, Yang H, Sun J, Guo C, Huang B, Liu Z, Yuan Y, Xia E, Wei C, Wan X (2018) Gene co-expression network analysis reveals coordinated regulation of three characteristic secondary biosynthetic pathways in tea plant (Camellia sinensis). BMC Genomics 19(1):616. https://doi.org/10.1186/s12864-018-4999-9

Tian J, Feng S, Liu Y, Zhao L, Tian L, Hu Y, Yang T, Wei A (2018) Single-molecule long-read sequencing of zanthoxylum bungeanum maxim. transcriptome: identification of aroma-related genes. Forests 9(12):765

Trapnell C, Salzberg SL (2009) How to map billions of short reads onto genomes. Nat Biotechnol 27(5):455–457. https://doi.org/10.1038/nbt0509-455

Trapnell C, Pachter L, Salzberg SL (2009) TopHat: discovering splice junctions with RNA-Seq. Bioinformatics 25(9):1105–1111. https://doi.org/10.1093/bioinformatics/btp120

Trapnell C, Williams BA, Pertea G, Mortazavi A, Kwan G, van Baren MJ, Salzberg SL, Wold BJ, Pachter L (2010) Transcript assembly and quantification by RNA-Seq reveals unannotated transcripts and isoform switching during cell differentiation. Nat Biotechnol 28(5):511–515. https://doi.org/10.1038/nbt.1621

Trapnell C, Hendrickson DG, Sauvageau M, Goff L, Rinn JL, Pachter L (2013) Differential analysis of gene regulation at transcript resolution with RNA-seq. Nat Biotechnol 31(1):46–53. https://doi.org/10.1038/nbt.2450

Trick M, Long Y, Meng J, Bancroft I (2009) Single nucleotide polymorphism (SNP) discovery in the polyploid Brassica napus using Solexa transcriptome sequencing. Plant Biotechnol J 7(4):334–346. https://doi.org/10.1111/j.1467-7652.2008.00396.x

Tsaparas P, Marino-Ramirez L, Bodenreider O, Koonin EV, Jordan IK (2006) Global similarity and local divergence in human and mouse gene co-expression networks. BMC Evol Biol 6:70. https://doi.org/10.1186/1471-2148-6-70

Ungaro A, Pech N, Martin JF, McCairns RJS, Mevy JP, Chappaz R, Gilles A (2017) Challenges and advances for transcriptome assembly in non-model species. PLoS One 12(9):e0185020. https://doi.org/10.1371/journal.pone.0185020

Velculescu VE, Zhang L, Vogelstein B, Kinzler KW (1995) Serial analysis of gene expression. Science 270(5235):484–487

Vettore AL, da Silva FR, Kemper EL, Souza GM, da Silva AM, Ferro MI, Henrique-Silva F, Giglioti EA, Lemos MV, Coutinho LL, Nobrega MP, Carrer H, Franca SC, Bacci Junior M, Goldman MH, Gomes SL, Nunes LR, Camargo LE, Siqueira WJ, Van Sluys MA, Thiemann OH, Kuramae EE, Santelli RV, Marino CL, Targon ML, Ferro JA, Silveira HC, Marini DC, Lemos EG, Monteiro-Vitorello CB, Tambor JH, Carraro DM, Roberto PG, Martins VG, Goldman GH, de Oliveira RC, Truffi D, Colombo CA, Rossi M, de Araujo PG, Sculaccio SA, Angella A, Lima MM, de Rosa Junior VE, Siviero F, Coscrato VE, Machado MA, Grivet L, Di Mauro SM, Nobrega FG, Menck CF, Braga MD, Telles GP, Cara FA, Pedrosa G, Meidanis J, Arruda P (2003) Analysis and functional annotation of an expressed sequence tag collection for tropical crop sugarcane. Genome Res 13(12):2725–2735. https://doi.org/10.1101/gr.1532103

Visser EA, Wegrzyn JL, Steenkmap ET, Myburg AA, Naidoo S (2015) Combined de novo and genome guided assembly and annotation of the Pinus patula juvenile shoot transcriptome. BMC Genomics 16:1057. https://doi.org/10.1186/s12864-015-2277-7

Vital-Lopez FG, Memišević V, Dutta B (2012) Tutorial on biological networks. Wiley Interdiscipl Rev Data Mini Knowl Discov 2(4):298–325. https://doi.org/10.1002/widm.1061

Wagner GP, Kin K, Lynch VJ (2012) Measurement of mRNA abundance using RNA-seq data: RPKM measure is inconsistent among samples. Theory Biosci 131(4):281–285. https://doi.org/10.1007/s12064-012-0162-3

Walters RW, Matheny T, Mizoue LS, Rao BS, Muhlrad D, Parker R (2017) Identification of NAD+ capped mRNAs in Saccharomyces cerevisiae. Proc Natl Acad Sci U S A 114(3):480–485. https://doi.org/10.1073/pnas.1619369114

Wang XJ, Gaasterland T, Chua NH (2005) Genome-wide prediction and identification of cis-natural antisense transcripts in *Arabidopsis thaliana*. Genome Biol 6(4):R30. https://doi.org/10.1186/gb-2005-6-4-r30

Wang Z, Gerstein M, Snyder M (2009) RNA-Seq: a revolutionary tool for transcriptomics. Nat Rev Genet 10(1):57–63. https://doi.org/10.1038/nrg2484

Wang L, Feng Z, Wang X, Zhang X (2010) DEGseq: an R package for identifying differentially expressed genes from RNA-seq data. Bioinformatics 26(1):136–138. https://doi.org/10.1093/bioinformatics/btp612

Wang L, Wang S, Li W (2012) RSeQC: quality control of RNA-seq experiments. Bioinformatics 28(16):2184–2185. https://doi.org/10.1093/bioinformatics/bts356

Wang B, Tseng E, Regulski M, Clark TA, Hon T, Jiao Y, Lu Z, Olson A, Stein JC, Ware D (2016) Unveiling the complexity of the maize transcriptome by single-molecule long-read sequencing. Nat Commun 7:11708. https://doi.org/10.1038/ncomms11708

Wang Y, Xie J, Zhang H, Guo B, Ning S, Chen Y, Lu P, Wu Q, Li M, Zhang D, Guo G, Zhang Y, Liu D, Zou S, Tang J, Zhao H, Wang X, Li J, Yang W, Cao T, Yin G, Liu Z (2017) Mapping stripe rust resistance gene YrZH22 in Chinese wheat cultivar Zhoumai 22 by bulked segregant RNA-Seq (BSR-Seq) and comparative genomics analyses. Theor Appl Genet 130(10):2191–2201. https://doi.org/10.1007/s00122-017-2950-0

Wang H, Gu L, Zhang X, Liu M, Jiang H, Cai R, Zhao Y, Cheng B (2018a) Global transcriptome and weighted gene co-expression network analyses reveal hybrid-specific modules and candidate genes related to plant height development in maize. Plant Mol Biol 98(3):187–203. https://doi.org/10.1007/s11103-018-0763-4

Wang M, Wang P, Liang F, Ye Z, Li J, Shen C, Pei L, Wang F, Hu J, Tu L, Lindsey K, He D, Zhang X (2018b) A global survey of alternative splicing in allopolyploid cotton: landscape, complexity and regulation. New Phytol 217(1):163–178. https://doi.org/10.1111/nph.14762

Washburn JD, Schnable JC, Conant GC, Brutnell TP, Shao Y, Zhang Y, Ludwig M, Davidse G, Pires JC (2017) Genome-guided phylo-transcriptomic methods and the nuclear phylogentic tree of the paniceae grasses. Sci Rep 7(1):13528. https://doi.org/10.1038/s41598-017-13236-z

Waterhouse RM, Seppey M, Simao FA, Manni M, Ioannidis P, Klioutchnikov G, Kriventseva EV, Zdobnov EM (2017) BUSCO applications from quality assessments to gene prediction and phylogenomics. Mol Biol Evol 35:543. https://doi.org/10.1093/molbev/msx319

Wen J, Egan AN, Dikow RB, Zimmer EA (2015) Utility of transcriptome sequencing for phylogenetic inference and character evolution. In: Hörandl E, Appelhans MS (eds) Next-generation sequencing in plant systematics. International Association for Plant Taxonomy (IAPT), Bratislava, pp 51–91

Weng JK, Tanurdzic M, Chapple C (2005) Functional analysis and comparative genomics of expressed sequence tags from the lycophyte *Selaginella moellendorffii*. BMC Genomics 6:85. https://doi.org/10.1186/1471-2164-6-85

Wickett NJ, Mirarab S, Nguyen N, Warnow T, Carpenter E, Matasci N, Ayyampalayam S, Barker MS, Burleigh JG, Gitzendanner MA, Ruhfel BR, Wafula E, Der JP, Graham SW, Mathews S, Melkonian M, Soltis DE, Soltis PS, Miles NW, Rothfels CJ, Pokorny L, Shaw AJ, DeGironimo L, Stevenson DW, Surek B, Villarreal JC, Roure B, Philippe H, dePamphilis CW, Chen T, Deyholos MK, Baucom RS, Kutchan TM, Augustin MM, Wang J, Zhang Y, Tian Z, Yan

Z, Wu X, Sun X, Wong GK, Leebens-Mack J (2014) Phylotranscriptomic analysis of the origin and early diversification of land plants. Proc Natl Acad Sci U S A 111(45):E4859–E4868. https://doi.org/10.1073/pnas.1323926111

Wilhelm BT, Marguerat S, Watt S, Schubert F, Wood V, Goodhead I, Penkett CJ, Rogers J, Bahler J (2008) Dynamic repertoire of a eukaryotic transcriptome surveyed at single-nucleotide resolution. Nature 453(7199):1239–1243. https://doi.org/10.1038/nature07002

Williams PH, Eyles R, Weiller G (2012) Plant microRNA prediction by supervised machine learning using C5.0 decision trees. J Nucleic Acids 2012:652979. https://doi.org/10.1155/2012/652979

Wu TD, Nacu S (2010) Fast and SNP-tolerant detection of complex variants and splicing in short reads. Bioinformatics 26(7):873–881. https://doi.org/10.1093/bioinformatics/btq057

Wu P, Xie J, Hu J, Qiu D, Liu Z, Li J, Li M, Zhang H, Yang L, Liu H, Zhou Y, Zhang Z, Li H (2018) Development of molecular markers linked to powdery mildew resistance gene Pm4b by combining SNP discovery from transcriptome sequencing data with bulked segregant analysis (BSR-Seq) in wheat. Front Plant Sci 9:95. https://doi.org/10.3389/fpls.2018.00095

Xiao YL, Smith SR, Ishmael N, Redman JC, Kumar N, Monaghan EL, Ayele M, Haas BJ, Wu HC, Town CD (2005) Analysis of the cDNAs of hypothetical genes on Arabidopsis chromosome 2 reveals numerous transcript variants. Plant Physiol 139(3):1323–1337. https://doi.org/10.1104/pp.105.063479

Xie Y, Wu G, Tang J, Luo R, Patterson J, Liu S, Huang W, He G, Gu S, Li S, Zhou X, Lam TW, Li Y, Xu X, Wong GK, Wang J (2014) SOAPdenovo-Trans: de novo transcriptome assembly with short RNA-Seq reads. Bioinformatics 30(12):1660–1666. https://doi.org/10.1093/bioinformatics/btu077

Xu Z, Peters RJ, Weirather J, Luo H, Liao B, Zhang X, Zhu Y, Ji A, Zhang B, Hu S, Au KF, Song J, Chen S (2015) Full-length transcriptome sequences and splice variants obtained by a combination of sequencing platforms applied to different root tissues of Salvia miltiorrhiza and tanshinone biosynthesis. Plant J 82(6):951–961. https://doi.org/10.1111/tpj.12865

Yamamoto K, Sasaki T (1997) Large-scale EST sequencing in rice. Plant Mol Biol 35(1–2):135–144

Yang Y, Dong C, Yang S, Li X, Sun X (2015) Physiological and proteomic adaptation of the alpine grass Stipa purpurea to a drought gradient. PLoS One 10(2):e0117475. https://doi.org/10.1371/journal.pone.0117475

Yang G, Liu Z, Gao L, Yu K, Feng M, Yao Y, Peng H, Hu Z, Sun Q, Ni Z, Xin M (2018a) Genomic imprinting was evolutionarily conserved during wheat polyploidization. Plant Cell 30(1):37–47. https://doi.org/10.1105/tpc.17.00837

Yang L, Jin Y, Huang W, Sun Q, Liu F, Huang X (2018b) Full-length transcriptome sequences of ephemeral plant Arabidopsis pumila provides insight into gene expression dynamics during continuous salt stress. BMC Genomics 19(1):717. https://doi.org/10.1186/s12864-018-5106-y

Young MD, Wakefield MJ, Smyth GK, Oshlack A (2010) Gene ontology analysis for RNA-seq: accounting for selection bias. Genome Biol 11(2):R14. https://doi.org/10.1186/gb-2010-11-2-r14

Yu X, Yang D, Guo C, Gao L (2018) Plant phylogenomics based on genome-partitioning strategies: progress and prospects. Plant Divers 40(4):158–164. https://doi.org/10.1016/j.pld.2018.06.005

Zerbino DR, Birney E (2008) Velvet: algorithms for de novo short read assembly using de Bruijn graphs. Genome Res 18(5):821–829. https://doi.org/10.1101/gr.074492.107

Zhang B, Horvath S (2005) A general framework for weighted gene co-expression network analysis. Stat Appl Genet Mol Biol 4:17. https://doi.org/10.2202/1544-6115.1128

Zhang L, Yu S, Zuo K, Luo L, Tang K (2012) Identification of gene modules associated with drought response in rice by network-based analysis. PLoS One 7(5):e33748. https://doi.org/10.1371/journal.pone.0033748

Zhang N, Liu B, Ma C, Zhang G, Chang J, Si H, Wang D (2014) Transcriptome characterization and sequencing-based identification of drought-responsive genes in potato. Mol Biol Rep 41(1):505–517. https://doi.org/10.1007/s11033-013-2886-7

Zhang F, Zhu G, Du L, Shang X, Cheng C, Yang B, Hu Y, Cai C, Guo W (2016a) Genetic regulation of salt stress tolerance revealed by RNA-Seq in cotton diploid wild species, Gossypium davidsonii. Sci Rep 6:20582. https://doi.org/10.1038/srep20582

Zhang ZF, Li YY, Xiao BZ (2016b) Comparative transcriptome analysis highlights the crucial roles of photosynthetic system in drought stress adaptation in upland rice. Sci Rep 6:19349. https://doi.org/10.1038/srep19349

Zhang C, Zhang B, Lin LL, Zhao S (2017) Evaluation and comparison of computational tools for RNA-seq isoform quantification. BMC Genomics 18(1):583. https://doi.org/10.1186/s12864-017-4002-1

Zhang H, Wang H, Zhu Q, Gao Y, Zhao L, Wang Y, Xi F, Wang W, Yang Y, Lin C, Gu L (2018) Transcriptome characterization of moso bamboo (Phyllostachys edulis) seedlings in response to exogenous gibberellin applications. BMC Plant Biol 18(1):125. https://doi.org/10.1186/s12870-018-1336-z

Zhang H, Zhong H, Zhang S, Shao X, Ni M, Cai Z, Chen X, Xia Y (2019a) NAD tagSeq reveals that NAD+-capped RNAs are mostly produced from a large number of protein-coding genes in *Arabidopsis*. Proc Natl Acad Sci 116(24):12072–12077. https://doi.org/10.1073/pnas.1903683116

Zhang T, Liu C, Huang X, Zhang H, Yuan Z (2019b) Land-plant phylogenomic and pomegranate transcriptomic analyses reveal an evolutionary scenario of CYP75 genes subsequent to whole genome duplications. J Plant Biol 62(1):48–60. https://doi.org/10.1007/s12374-018-0319-9

Zhao QY, Wang Y, Kong YM, Luo D, Li X, Hao P (2011) Optimizing de novo transcriptome assembly from short-read RNA-Seq data: a comparative study. BMC Bioinformatics 12(Suppl 14):S2. https://doi.org/10.1186/1471-2105-12-S14-S2

Zhao S, Fung-Leung WP, Bittner A, Ngo K, Liu X (2014) Comparison of RNA-Seq and microarray in transcriptome profiling of activated T cells. PLoS One 9(1):e78644. https://doi.org/10.1371/journal.pone.0078644

Zhao X, Li J, Lian B, Gu H, Li Y, Qi Y (2018) Global identification of *Arabidopsis* lncRNAs reveals the regulation of MAF4 by a natural antisense RNA. Nat Commun 9(1):5056. https://doi.org/10.1038/s41467-018-07500-7

Zhao L, Zhang H, Kohnen MV, Prasad K, Gu L, Reddy ASN (2019) Analysis of transcriptome and epitranscriptome in plants using PacBio Iso-Seq and nanopore-based direct RNA sequencing. Front Genet 10:253. https://doi.org/10.3389/fgene.2019.00253

Zhou X, Wang G, Sutoh K, Zhu JK, Zhang W (2008) Identification of cold-inducible microRNAs in plants by transcriptome analysis. Biochim Biophys Acta 1779(11):780–788. https://doi.org/10.1016/j.bbagrm.2008.04.005

Zhou Q, Su X, Jing G, Chen S, Ning K (2018) RNA-QC-chain: comprehensive and fast quality control for RNA-Seq data. BMC Genomics 19(1):144. https://doi.org/10.1186/s12864-018-4503-6

Plant Proteomics and Systems Biology

3

Flavia Vischi Winck, André Luis Wendt dos Santos, and Maria Juliana Calderan-Rodrigues

Abstract

Proteome analysis of model and non-model plants is a genuine scientific field in expansion. Several technological advances have contributed to the implementation of different proteomics approaches for qualitative and quantitative analysis of the dynamics of cellular responses at the protein level. The design of time-resolved experiments and the emergent use of multiplexed proteome analysis using chemical or isotopic and isobaric labeling strategies as well as label-free approaches are generating a vast amount of proteomics data that is going to be essential for analysis of protein posttranslational modifications and implementation of systems biology approaches. Through the target proteomics analysis, especially the ones that combine the untargeted methods, we should expect an improvement in the completeness of the identification of proteome and reveal nuances of regulatory cellular mechanisms related to plant development and responses to environmental stresses. Both genomic sequencing and proteomic advancements in the last decades coupled to integrative data analysis are enriching biological information that was once confined to model plants. Therewith, predictions of a changing environment places proteomics as an especially useful tool for crops performance.

Keywords

Mass spectrometry · Protein · 2-DE · Quantitative · Expression

F. V. Winck (✉)
Institute of Chemistry, University of São Paulo, São Paulo, Brazil

Center for Nuclear Energy in Agriculture, University of São Paulo,
Piracicaba, Brazil
e-mail: winck@cena.usp.br

A. L. W. dos Santos
Laboratory of Plant Cellular Biology, Instituto de Biociências, Universidade de São Paulo, São Paulo, Brazil

M. J. Calderan-Rodrigues
Max Planck Institute of Molecular Plant Physiology, Potsdam, Germany
e-mail: calderan@mpimp-golm.mpg.de

3.1 Introduction

Proteins are fundamental macromolecules that execute a vast number of biological functions in the organisms, ranging from structural activities to exceptionally precise regulatory roles inside the cells. These characteristics intrigued scientists for centuries and the protein sequences started to be detailed in the twentieth century, with the sequencing of the first protein, phenylalanyl chain of insulin 1, in 1951 by Sanger and

F. V. Winck (ed.), *Advances in Plant Omics and Systems Biology Approaches*, Advances in Experimental Medicine and Biology 1346, https://doi.org/10.1007/978-3-030-80352-0_3

Tuppy (1951) using partial hydrolysis techniques. For the first time, they presented a structured amino acid sequence composition encoded by the DNA of the living organisms. Many other proteins were sequenced thereof using biochemical sequencing analysis established by Edman (1949). At the same pace, with the development of ionization methods used for mass spectrometry (MS), specially the electrospray ionization (Fenn et al. 1989), the analysis of intact or digested proteins increased significantly. Several biochemical and molecular properties of thousands of proteins were identified and are currently well known, with the precise description of their amino acid sequence, tridimensional structure, activities, and chemical modifications. The unique functional and molecular characteristic of each protein species is outstandingly provocative and has revealed the enormous complexity of the possible molecular interactions these molecules can perform in a biological system. This also revealed the need for systems analysis in plant species, focusing in quantifying the changes in abundance and modifications of a large number of proteins simultaneously, and if possible, with a cell or tissue spatial and time resolution.

Today, we understand that a single-protein sequence can execute single or multiple biological roles, depending on the chemical modifications these proteins present in their structure and their sub-cellular compartmentalization.

With the vast information about how cellular systems function, it has become imperative since the beginning of the 1990s the study of comprehensive amount of proteins simultaneously in a cell or tissue. The need for the identification of all proteins was urgent, and the term Proteome was, therefore, first mentioned in 1995 by Wilkins et al. (1996), indicating an age of dramatic change on biology scale which has begun with the availability of complete genomic sequences of many organisms. As stated by Wilkins, the proteome is the entire protein complement expressed by a genome. By nature, the proteome is dynamic, being representative of the whole-protein repertoire of a cell or tissue in a certain time and condition. For instance, the proteome is different every period of the lifetime of a plant, or in response to any challenging environmental stress. The proteome also includes, a priori, all different chemical modifications proteins may present, which are usually present as posttranslational modifications (PTMs) of the synthesized proteins or may be introduced by dynamic processes of chemical modifications of amino acids during cellular signaling events. The alterations on the protein abundance and PTMs are especially important for defining the cell fate in response to changes in environmental conditions, and current proteome research addresses many of these alterations using highly-sensitive analytical tools and sophisticated computational data analysis. The proteome of different species has been explored using different techniques, ranging from the use of two-dimensional polyacrylamide gel electrophoresis (2-DE) and its modified form 2D DIGE (two-dimensional difference gel electrophoresis) (Jorrin-Novo et al. 2019; Martins de Souza et al. 2008), passing through isotope and isobaric labeling (Pappireddi et al. 2019), leading to the massive proteome analysis using shotgun approaches and stable isotope labeling or label-free relative quantification (de Godoy et al. 2008), which further was followed by target approaches (Rodiger and Baginsky 2018), MS imaging (Boughton et al. 2016; Kaspar et al. 2011) and, more recently, single-cell proteomics (Marx 2019).

In the current days, there is an understanding that the identification of all intricated connections of the proteins inside the cells must be revealed. However, to detect the proteome of a cell or to represent it in a systems analysis is not a trivial task, but many efforts have been done to address several technical challenges. The availability of large datasets of plant proteomes in different experimental and field conditions generated by the unprecedent analysis of the proteome of several plant species is opening an exceptional path to the discovery of previous unknown cellular mechanisms and molecular emergent patterns.

In the last 21 years, the analysis of plant or plant-related proteome has increased from four scientific papers a year in 1998 to 1483 scientific papers a year in 2019 according to the PubMed repository (https://pubmed.ncbi.nlm.nih.gov/) queried through the keywords "plant proteomics." During this period of time, several technical advances have been established for the analysis of plant proteomics, including the optimization of techniques and instrumentation. The proteome analysis in general had begun basically through gel-based approaches, usually with Two-Dimensional Polyacrylamide gel Electrophoresis (2D-PAGE) coupled to peptide sequencing or Mass Spectrometry (MS) (Shevchenko et al. 1996a, b). Further approaches expanded to applications using liquid chromatography (LC) coupled to tandem MS, including shotgun analysis (Neubauer et al. 1998; Ong and Mann 2005), selected/parallel reaction monitoring (SRM/PRM) (Picotti and Aebersold 2012) and target data acquisition (TDA) (Schmidt et al. 2009). Recent applications suggest the combination of the TDA with tandem MS/MS analysis using Data-Dependent Acquisition (DDA) or Data-Independent Acquisition (DIA) as a valuable way to get hypothesis-driven and non-hypothesis-driven quantitative data collection from the same sample (Hart-Smith et al. 2017), even for the analysis of protein post-translational modifications (Pappireddi et al. 2019).

The analysis of the proteome for many different plant species had applied many of the approaches mentioned and increased our knowledge of plant molecular physiology and evolution in an enormous way. The level of details identified by proteomics analysis revealed a large proportion of the repertoire of cellular mechanisms and proteins that define the underlying principles of plant development and metabolism, including seed germination (He and Yang 2013), root growth (Li et al. 2019), stress responses (Kosova et al. 2014), senescence (Kim et al. 2016), light responses (Mettler et al. 2014), among others.

Plant proteome analysis can be performed using many different techniques and computational resources through diverse approaches. Each of these approaches will complement our knowledge of plant phenotypes and influence in the future directions of plant proteomics and systems biology. The information available today and further development of novel techniques for proteome analysis will definitely pave the way to a more comprehensive understanding of the complex phenomena that take place in plants and how the numerous set of molecular interactions and mechanisms are built and retained in the cells in response to varying environmental and intracellular conditions.

3.2 Research and Technical Approaches

3.2.1 The Gel-Electrophoresis-Based Plant Proteome Analysis

At the end of the 1990s, beginning of 2000s, there was a modest racing into defining the so-called reference gel or reference proteome of an organism or organismal phenotype. Two-dimensional polyacrylamide gel electrophoresis (2-DE) has been used as one of the most powerful techniques of protein separation in a couple of technical steps (O'Farrell 1975). In 2-DE, the proteins contained in a biological sample are usually separated in a polyacrylamide gel by their isoelectric point (pI) and molecular weight (mW). Usually proteins are solubilized by chaotropic chemicals, such as urea and thiourea and solvated by non-ionic detergents. The first dimension of separation occurs with the migration of proteins in their intact form through an electric field, permeating an inert gel matrix soaked with amphoteric substances that form a transient gradient, allowing the proteins to move in these conditions toward the electrodes until they reach to their isoelectric point (pI), where they achieve their minimal solubility, thus high precipitation. In the second dimension of the 2-DE, the proteins separated and immobilized in the first gel dimension are separated by their molecular weight through SDS-PAGE. This will result in a two-dimensional plan with *y* and *x* axis, with proteins separated, respectively, by molecular mass (MW) and isoelectric point (pI) visualized as separated gel protein "spots" that can be visualized through dying techniques using, for instance, Coomassie Blue or Silver nitrate chemicals (Shevchenko et al.

1996a, b). In an optimal situation, the results of the gel staining will reveal the whole set of proteins from the biological sample analyzed, represented by protein spots. Each spot may contain one or more proteins with the approximate same biochemical characteristics. In proteome approaches, each spot is cut from the gel and digested with specific enzymes (e.g., Trypsin, Lys-C) that are going to generate peptides which can be identified through sequencing or MS analysis. Using gel-based approaches, the identification of the repertoire of proteins from different parts of the plants and those involved in complex developmental processes was possible and contributed to elucidate how proteins interconnect to each other to define a systemic function. The initial efforts into the analysis of the proteome of several important plant species had been done using gel-based approaches, and revealed important aspects of the plant structure and molecular physiology, including the analysis of green and etiolated shoots of rice (Komatsu et al. 1999), *Arabidopsis* seed germination and priming (Gallardo et al. 2001), maize leaves (Porubleva et al. 2001), Medicago mycorrhizal roots (Bestel-Corre et al. 2002), among others.

However, the gel-based proteomics analysis has several limitations and are difficult to be implemented in a large-scale proteomics analysis. Two-dimensional gels are quite laborious and time-consuming to be performed, which renders extra difficulty levels to other limitations in 2-DE gels that include problems with protein spot resolution, gel reproducibility, and low throughput. Usually, in a 2-DE gel, it is possible to separate approximately one thousand proteins simultaneously, but, due to the possible overlap between protein spots with similar mW and pI, the protein separation of some protein spots is not complete.

Nowadays, still, 2-DE is considered a complementary approach for some applications such as subcellular proteomics, analysis of low-complexity samples, and study of protein isoforms and their modifications (Jorrin-Novo et al. 2019). The approaches in subcellular proteomics are not going to be addressed here since this topic is presented in a specific chapter of this book.

The 2-DE-based proteomics coupled to protein sequencing or MS techniques has contributed greatly to the development of plant proteomic analysis since the 1990s and has continuously adding important knowledge.

The diversity of 2-DE-based proteomic applications is vast, including analysis of plant structure (Giavalisco et al. 2005), plant–pathogen interactions and molecular signaling (Delaunois et al. 2014), and plant metabolism (Chang et al. 2017), to mention a few examples. Many of the proteomics results (including 2-DE-gel image) were also made available through public repositories, such as GABI Primary Database of Arabidopsis (URL: https://www.gabipd.org/).

One important aspect of the use 2-DE-based comparative proteomics resides in the fact that this approach can be implemented for any plant species, no matter if a genome sequence is available to this species. With that in mind, it is of great importance that comparisons between proteome profiles of non-model plants are investigated, elucidating the differences in the cellular responses of the many plant species that are of regionally importance to local agriculture and may not have their genome sequenced in the near future. The 2-DE-based proteomics may be a way to sustain this endeavor.

3.2.2 Mass Spectrometry-Based Proteomics

Among the methods that succeed in answering the increased demand for high-throughput proteomics approaches, the MS is currently the most applied and developed technique. Mass spectrometry is an analytical technique that is applied in many different fields of applications, including elemental analysis, organic and bio-organic analysis, structure elucidation, characterization of ionic species and reactions, and spectral imaging. In all these applications, the use of MS is aiming to identify a compound from the molecular or atomic mass(es) of its constituents (Gross 2017). So, an important aspect of proteome analysis resides in the fact that the chemical species, in this case proteins or peptides, must be ionized to be analyzed through a mass spectrometer.

From the first attempts to understand electric discharges in gases and charged ions made by Joseph John Thomson with his first instrument that separated ions by mass-to-charge ratio to the most recent mass spectrometers that use high-resolution mass analyzers, there was a huge improvement and expansion of the applications of MS in Biochemistry and Biological Sciences (Gross 2017).

The mass spectrometry is the analytical technique that significantly generated our current substantial amount of plant proteomics data available. The development and evolution of ionization methods for proteins and peptides had contributed drastically to the change in proteomics scale, allowing the easy ionization of peptides and proteins into gas phase. These technical developments were complemented by the co-evolution of the sample preparation for proteomics, types of ionization techniques, and mass analyzers integrated into mass spectrometers, which significantly enhanced proteome coverage, the mass resolution, and throughput obtained for the mass-to-charge measurements. A recent review presented an overview about the protein extraction and preparation for proteomics analysis, discussing many approaches applicable for plant proteomics (Patole and Bindschedler 2019).

For qualitative and quantitative proteome analysis, two types of MS applications have been used more frequently to the identification of proteins: the matrix-assisted laser desorption/ionization mass spectrometry (MALDI-MS) (Hillenkamp and Karas 1990), and liquid-chromatography (LC) coupled to electrospray ionization mass spectrometry (ESI-MS) (Smith et al. 1990). In the LC-MS analysis, the peptides obtained from the digestion of the total proteome are separated by liquid chromatography prior to the injection into the ion source of the MS instruments, where the peptides are ionized. After that the peptides are analyzed by the instruments resulting in the identification of their mass, and usually thousands of proteins can be identified in a shotgun MS analysis.

The implementation of high-throughput shotgun MS analysis in plant proteomics is usually dependent on the existence of a genome sequence or customized protein sequence databases derived from transcriptomic analyses (RNA-Seq data) for the plant species under investigation. Protein identification has evolved through the development of different computational tools and strategies that allowed the MS data to be analyzed by a series of steps that permit the identification of the best protein hits (or protein groups) in a statistical-based computational analysis. Protein identification is basically performed computationally by comparing the mass of the peptides analyzed (precursor ions) experimentally from the mass spectra data with the mass of the precursor ions generated in silico based on the translated genome of the plant species under study. Therefore, the protein identification in this case results from a probability analysis of the best protein hit or group of protein hits. The abundance of the peptide ions analyzed is determined, usually, by the arbitrary signal intensity measures for each peptide ion analyzed in a MS scan or by the counts of MS spectra for a given precursor ion, which gives the information useful for quantitative proteomics, using label-free or labeled approaches.

Recent analysis using sensitive and efficient mass spectrometers that have high-resolution mass analyzers has revealed the identity of thousands of proteins, increasing the capacity to observe broad mechanisms that take place intra and intercellularly. However, it is still highly challenging to analyze the whole set of proteins from a cell or tissue. Thus, the plant proteome analysis performed nowadays is still representing a partial view of the total cellular proteome.

Nevertheless, more than 10 years ago, the impact of proteomics analysis to the understanding of plant development and responses to biotic and abiotic stresses and to the identification of the protein repertoire of organelles, tissues and sub-cellular compartments was already clear (Wienkoop et al. 2010). Several initiatives including the creation of LC-MS/MS spectral library repository ProMEX with 116,364 tryptic peptide product ion spectra entries of 13 plant species

(Wienkoop et al. 2012) and the creation of the Multinational Arabidopsis Steering Subcommittee to coordinate Arabidopsis proteome international research were efforts toward the consolidation of databases and data resources that contributed to address some of the main challenges on plant proteomics through the creation of integrated proteome repositories and data analysis platforms, including MASCP Gator (Joshi et al. 2011). Other efforts such as International Plant Proteomics Organization (INPPO) (http://www.inppo.com/) were also on place. Currently, some of the most complete information about Arabidopsis proteome are organized in the ProteomicsDB (https://www.proteomicsdb.org), which contains MS-based proteomics meta-data and proteomics expression profiles for 30 different tissues, that can be visualized through body maps (Samaras et al. 2020). Another source of plant proteomics data is the PlantPReS (URL www.proteome.ir), an online database of plant proteome related to stresses, containing more than 20,413 protein entries and their expression patterns, extracted from 456 manually curated articles (Mousavi et al. 2016). Alternatively, there is the ATHENA (Arabidopsis THaliana ExpressioN Atlas; http://athena.proteomics.wzw.tum.de:5002/master_arabidopsisshiny/) database, which has a collection of more than 18,000 proteins and their expression profiles for a set of 30 matching tissues from *Arabidopsis thaliana* (Col-0). It allows the user to explore the comparative expression analysis of proteins in different tissues, to visualize enriched pathways, phosphorylation sites, and similar global gene expression profiles in different tissues of *Arabidopsis thaliana*.

Very recently, a detailed MS-based draft of the *Arabidopsis* proteome was described (Mergner et al. 2020) based on the information retrieved from the proteomics databases ProteomicsDB and ATHENA. In this draft, the molecular data retrieved from more than 18,000 proteins identified revealed that most transcripts and proteins are actually expressed in a non-tissue-specific manner, with only a few transcripts or proteins being expressed in a tissue-specific manner, as it was evidenced for the proteins exclusively identified in pollen. The authors have found that different tissue types may have distinct quantitative abundance patterns of proteins, showing a positive correlation (Pearson's correlation $r = 0.28$–0.7) between the transcript and protein levels in most tissues (Mergner et al. 2020). However, in recent single-cell proteomics approaches, this behavior was not observed for most of the genes, and there was low correlation of mRNA and protein levels in root hair analysis (Wang et al. 2016), with higher positive correlation of transcript and protein levels found mostly for highly expressed genes.

Some other multi-omics databases and frameworks are operating to integrate functional genomics data providing annotation and visualization options for diverse plant species, including platforms such as ePlant (http://bar.utoronto.ca/eplant/), Virtual Plant (http://virtualplant.bio.nyu.edu/cgi-bin/vpweb/), MapMan (https://mapman.gabipd.org/mapmanstore) (Schwacke et al. 2019), PLAZA (https://bioinformatics.psb.ugent.be/plaza/), Plant Metabolic Network (PMN) (https://plantcyc.org/), Plant Reactome (https://plantreactome.gramene.org/), among others. Right now, some of these platforms contain mostly genomics and transcriptomics datasets, with few proteomics datasets. However, the integration of several layers of plant omics information seems to be in the horizon of these platforms, which is essential for integrative analysis and modeling approaches and the continuous development of systems biology studies in plants (Falter-Braun et al. 2019).

The permanent development and improvement of such databases is essential to guarantee the availability of plant proteomics data for future integrative omics analysis. As the proteomics data is everyday more abundant, it would be important to include omics data of all plant species in these repositories, not only model plants, since the proteomics data is currently the linkage point between the transcriptome information and the cell metabolism.

Quantitative MS-based proteomics seems to be currently the most prominent proteomics approach for investigating plant molecular phenotypes. Some of the most recent approaches for discovery quantitative proteomics (non-targeted proteomics) include chemical labeling of the proteins or peptides. The most common strategies for relative and quantitative analysis of plant proteomics include one of the following techniques: isobaric tags for relative and absolute quantitation (iTRAQ) (Unwin et al. 2005), tandem mass tags (TMT) (Thompson et al. 2003), metabolic labeling such as stable isotope labeling by amino acids (SILAC) (Ong et al. 2002), and stable isotope ^{15}N (Oda et al. 1999) or label-free quantification (Washburn et al. 2001).

In many different studies performed using shotgun proteomics with LC and tandem MS (MS/MS), the use of data-dependent acquisition (DDA) was the method of choice, combined or not with peptide labeling with tandem mass tag (TMT) for quantitative or absolute quantification in proteomic analysis.

Multiplexing TMT or the use of other peptide labeling techniques, such as isobaric labeling, has been recently applied for the comparative proteome analysis, revealing cellular mechanisms and elucidating the role of candidate genes in diverse cellular responses. The time-resolved analysis of plant processes combined with the quantitative comparative proteomics is revealing novel aspects of plant molecular physiology and will certainly serve as basis for modeling and recognition of regulatory principles of cell response and metabolism.

For instance, in a time-resolved redox proteome analysis using thiol-specific iodo TMTs and LC-MS/MS analysis, the evidences for a fundamental strategy of rapid control of the cell metabolism during seed germination through the modifications of cysteines were confirmed. This biological process seems to be controlled by the function of hundreds of Cys-based redox switches which are operational even when hormonal and genetic programs are not yet functional. With 741 Cys peptides identified and quantified, it was demonstrated that tricarboxylic acid cycle is regulated by thioredoxins (TRX), through the TRX-mediated redox modulation of the activity of succinate dehydrogenase and fumarase. The proteome analysis also indicated that all shifts observed were reductive, indicating the influx of electrons into the complex cellular thiol redox systems during the early stages of seed germination (Nietzel et al. 2020).

A recent proteome analysis of autophagy-deficient *Arabidopsis* seedlings indicated that this process is a response rapidly activated under diverse stimuli, including microbial elicitors, danger signals, and hormones. Quantitative comparative proteome analysis using TMTs identified more than 11 000 proteins and showed that autophagy is associated with cellular phenotypic plasticity. In autophagy-deficient cells, the plasticity was reduced, and the cell dedifferentiation impaired, indicating that autophagy is an essential mechanism for cells to reprogramming their development in response to several different stimulus, being necessary for wound-induced dedifferentiation and tissue repair (Rodriguez et al. 2020).

In another approach, the information about the differences between plant tissues (roots, aboveground parts, cauline leaves, 13 stages of flowers, and organs/stages) was investigated through proteome and phosphoproteome analysis performed with LC-MS/MS and iTRAQ labeling of six developmental stages of 16-day-old *Arabidopsis* plants. This work revealed the identity of 2187 proteins and the pattern of protein expression phosphorylated peptides (Lu et al. 2020). The comparative proteomics analysis indicated that reproductive organs expressed, in a more prominent manner, proteins related to translation and metabolic processes, while plant seedlings present a protein repertoire enriched in proteins related to oxidation-reduction and responses to different stresses. In the same work, the transcription factor and transcriptional regulator proteins were identified and a network of putative kinase substrates was generated, reinforcing evidences that bZIP16 transcription factor may be the substrate of Map kinase 6 (MAPK6) during floral development, integrating light and hormone signaling pathways during early seedling development (Lu et al. 2020).

Some plant proteomic approaches also included the implementation of Data-Independent Acquisition (DIA) (Venable et al. 2004), which acquires data from MS1 and MS2 spectra in MS/

MS applications, without pre-selection of precursor ions. The acquired data is usually analyzed by computational creation of spectral libraries and/or by using these libraries or consulting preexisting libraries for searching characteristic mass spectra (Zhang et al. 2020).

A complementation between discovery and targeted proteomics approaches can be initially addressed by sequential window acquisition of all theoretical mass spectra (SWATH)/DIA (data-independent acquisition) and by comparison of these spectra with constructed spectral libraries. For instance, by developing a mass spectra library, the effects of abscisic acid in *Arabidopsis* were investigated (Zhang et al. 2019). An extensive analysis of *Arabidopsis* proteome performed the quantification of 8793 proteins using a combination of untargeted LC-MS/MS using DDA and DIA and generation of a spectral library. The effects of the hormone abscisic acid (ABA) in *Arabidopsis* proteome was investigated, rendering detailed evidences of the previously described role of oxidative-reduction processes induced by ABA in the cells, and the transitory or gradual response of plant metabolism exposed to ABA with an initial reduction of the metabolic cellular activity and increase in ribosome biogenesis after 2 h of ABA posttreatment followed by an increased metabolism of carbohydrate and sucrose, and reduced metabolism of *N*-acetylglucosamine and macromolecules at 72 h posttreatment (Zhang et al. 2019).

Within an integrative omics analysis, a quantitative label-free proteomics approach investigated the role of autophagy in maize (*Zea mays*). Wild-type maize background plants (W22) where compared with mutant plants for ATG12 protein, one of the proteins identified as AUTOPHAGY-RELATED (ATG) (McLoughlin et al. 2018). The study was conducted in selected two leaves (two and four), which are rapidly expanding sink tissues, in response to short nitrogen stress. The quantitative proteome analysis revealed main biological processes affected in nitrogen starved and non-starved plants, indicating that autophagy is involved in many nutrient recycling mechanisms, but it is also an important mechanism in the maintenance of normal protein abundance in plants. Other effects of the mutation of *atg12* were observed in the alteration of the abundance of proteins related to biosynthesis of phenylpropanoids, fatty acids, aromatic amino acids, and the positive correlation of the proteome and transcriptome data for genes associated with phenylpropanoid metabolism and glutathione transferase activities. These results suggested that alterations in the autophagic turnover of molecules such as pigments, antioxidants, lipids, among others, are essential for the resulting plant leaf phenotype, even under non-stress, nitrogen-rich conditions (McLoughlin et al. 2018). Further studies of the maize proteome under carbon-stress conditions reinforced the aspects of autophagy as a critical process for proteostasis in plants, likely by the recycling of proteins and organelles. To increase the number of proteins identified, the authors performed two MS runs in a data-dependent mode and another two runs of the same extracts using an exclusion list of the 5000 most prominent peptides from the first analysis, which increased the depth of the proteome analysis of the further runs (McLoughlin et al. 2020). The proteome results of this approach indicated that leaves of maize *atg12* mutant have increased levels of ribosome-associate proteins, and proteins related to redox homeostasis and catabolism of fatty acids, amino acids, small molecules, nucleotides, and glutathione (GSH), suggesting that even under conditions of impaired autophagy, the cells respond partially in a similar way as with full autophagy systems. These effects may be part of a complex compensatory mechanism that may take place in plants serving as alternative for the deficient autophagy (McLoughlin et al. 2020).

The quantitative proteome analysis is significantly contributing to increase our understanding of several complex phenomena in plants. Large studies using time-resolved in-depth quantitative approaches will certainly disclose regulatory aspects of the cellular mechanisms related to the control of cell growth, carbon usage, and stress responses in plants.

3.2.3 Proteomics Analyses of Posttranslational Modifications (PTMs)

With the increased throughput and mass resolution of the current MS instruments, the identification of chemical modifications of peptides or intact proteins occurs through the analysis of posttranslational modifications (PTMs), which do refer to the covalent and generally enzymatic modification of proteins following protein biosynthesis. These modifications generally include potential changes in protein sub-cellular localization, protein stabilization/degradation, enzyme activity, and interactions with protein partners or other biomolecules (Friso and van Wijk 2015; Spoel 2018). The occurrence of PTMs greatly expands proteome complexity and diversity, affecting numerous cellular signaling events and responses to the environment (Friso and van Wijk 2015; Larsen et al. 2006). Based on the type of modification, more than 300 potential PTMs can occur in vivo including: (a) reversible/irreversible addition of chemical groups (phosphorylation, acetylation, methylation, and redox-based modifications), (b) reversible addition of polypeptides (ubiquitination, SUMOylation, and other modifications by ubiquitin-like (Ubl) polypeptide), (c) reversible addition of complex molecules (glycosylation, attachment of lipids, ADP-ribosylation, and AMPylation), and (d) irreversible direct modification of amino acids (deamidation, eliminylation) or protein cleavage by proteolysis (Larsen et al. 2006; Spoel 2018; Vu et al. 2018). In addition, apart from a single regulatory PTM role, there is also potential cross-talk with other PTMs, making the mechanisms and dynamics of protein modification still more complex (Arsova et al. 2018; Du et al. 2019; Vu et al. 2018).

The study of PTMs is considered technically demanding due to the labile nature, low stoichiometry, and abundance of protein modification when analyzing whole-cell lysates (Larsen et al. 2006; Swaney and Villen 2016). Therefore, a PTM enrichment step for modified peptides is normally necessary before MS analysis (Murray et al. 2012; Swaney and Villen 2016). Dealing with complex plant proteomes of model and non-model species, it is important to select adequate proteomic approaches aiming the identification of proteins and analysis of their dynamic PTMs (Hu et al. 2015). Traditionally, PTMs have been identified by Edman degradation, amino acid analysis, isotopic labeling, or immunochemistry (Larsen et al. 2006). Nowadays, shotgun proteomics is one of the most widely used approaches to analyze PTMs (Yu et al. 2020). The adoption of large-scale quantitative proteomic approaches including isotope-coded affinity tags (ICATs), tandem mass tags (TMTs), and isobaric tags for relative, and absolute quantitation (iTRAQ) is enabling a more confident identification and multiplexed quantitation of PTMs (Hu et al. 2015; Liu et al. 2019; Murray et al. 2012). Once a PTM site is identified, biological characterization of protein modifications can be addressed with targeted MS-based quantitative approaches such as multiplexed selective reaction monitoring (SRM) or Sequential Windowed Acquisition of All Theoretical Fragment Ion Mass Spectra (SWATH MS) (Arsova et al. 2018; Sidoli et al. 2015).

Several aspects of plant metabolism and development involve signaling events. Different inorganic or organic compounds can function as signaling molecules that are going to regulate the plant cell behavior. Among the many possible molecular events that may take place in plant signaling mechanisms, the posttranslational modification of proteins is of upmost interest as it controls a vast set of cellular responses, including growth, membrane trafficking, gene expression, degradation, to mention a few.

The use of proteomics to uncover signaling mechanisms is step by step taking advantage of the high resolution and throughput of the MS instruments to uncover the signaling networks of the plant cells. In the current investigations, the analysis of protein targets of the reactive oxygen species (ROS) and reactive nitrogen species (RNS) is gaining more attention due to their promising role in plant growth through the redox homeostasis. The investigation of the cellular role of these chemical species can be done through the analysis of the chemical modifications in the target molecules generated by the reactions with reactive oxygen and/or nitrogen species.

Among the several possible protein oxidoreduction PTMs that can be identified in plants, the protein nitrosation (or nitrosylation) which represents target protein alterations in response to the function of nitric oxide has gained more attention as central process in defining carbon and nitrogen metabolism in land plants (Nabi et al. 2020) and microalgae under stress conditions (De Mia et al. 2019; Morisse et al. 2014).

Cysteine *S*-nitrosation is a redox-based PTM that mediates major physiological and biochemical effects of nitric oxide (NO). Emerging evidence indicates that protein *S*-nitrosation is ubiquitously involved in the regulation of plant development and stress responses (Feng et al. 2019; Gong et al. 2019). Despite its importance in plants, studies exploring protein signaling pathways that are regulated by *S*-nitrosation during plant development and embryogenesis are still scarce, particularly in non-model plant species. For instance, using redox proteome analysis of *S*-nitrosation, a considerable array of proteins associated with a large variety of molecular functions were identified in Brazilian pine proteome, generating novel insights into the roles of *S*-nitrosation during somatic and zygotic embryo development. Using a method adapted to the Biotin-Switch (Forrester et al. 2009), replacing biotin with the iodo-TMT126 (iodo-Tandem Mass Tag), the occurrence of in vivo and in vitro *S*-nitrosation was investigated during somatic and zygotic embryo formation of Brazilian pine (*Araucaria angustifolia* (Bertol.) Kuntze), an endangered native conifer of South America.

Previous analyses using physiological, transcriptomic, and quantitative proteomics approaches (dos Santos et al. 2016; Elbl et al. 2015; Silveira et al. 2006) suggested a potential influence of the redox environment and nitric oxide production during Brazilian pine embryo formation. The *S*-nitrosoproteome analyses identified 158 *S*-nitrosylated proteins in vitro (i.e., via the incorporation of a NO donor), with 36 proteins detected during seed development (globular embryo until late cotyledonal stage) and 122 proteins detected during somatic embryo formation (transition from proembryogenic masses to early somatic embryos). This study indicated that most *S*-nitrosylated proteins were involved in metabolism of primary compounds (carbohydrates and nucleic acids) and cellular processes and signaling (turnover of proteins and chaperones). For late-stage embryogenesis, functions associated with *S*-nitrosylated proteins were stress resistance (abiotic stress response), cellular processes (signal translation, chaperones, and protein turnover), and metabolism (energy production, transport, and carbohydrate metabolism). Interestingly, 47 proteins were identified as endogenous *S*-nitrosylated during early and late-stage embryogenesis, suggesting a role of this PTM during Brazilian pine embryo formation. The possibility of generating a stable bond between Cys and iodo-TMT126 enabled the identification of labeled peptides using mass spectrometry and the determination of the position of nitrosylated Cys residues. The identification of the possible cellular biological processes affected in non-model plants by NO reinforce the observations about the myriad of functions regulated by the protein PTMs generated by ROS and RNS, which are likely coordinated by a complex regulatory network (Leon and Costa-Broseta 2020). The cellular effects coordinated by these networks are possibly accomplished by the interplay between redox homeostasis systems which involve cysteine oxidative modifications and the systems of thioredoxins and disulfide reductases, which are expressed by several genes in plants, composing a highly complex regulatory mechanism of cellular response (Navrot et al. 2011).

3.3 Proteomics of Non-model Species

The many challenges already present in the plant proteomics analysis are more pronounced in the investigation of non-model plant species. This is especially true after the past decade, when the number of species with a fully sequenced genome grew substantially, with non-model plant species distinguished by the lack of experimentally validated functional evidence for a great variety of annotated genes. This scarcity on information may result from long life cycle, large genome size (often polyploid) and recalcitrance to laboratorial cultivation, whose characteristics are opposed to model plants, such as *Arabidopsis thaliana*. The study on non-model plants greatly relied on establishing gene and protein sequence homology analyses with model organisms, but this approach is questionable when there is only far phylogenetic relation between the species analyzed. As the more distant organisms are phylogenetically lesser there can be functional extrapolation (Heck and Neely 2020). Plants are physiologically diverse, implying that sometimes homology-based annotation does not accurately indicate the function of groups of genes or their relation to complex cellular mechanisms. Additional co-expression or regression analyses have been shown as powerful tools to be used for the search of functional categories and novel pathways inherent of non-model plant proteomes, which may guide functional investigations in non-model plant species.

Currently, the progress on genome sequencing and annotation, followed by advances in bioinformatics and data integration offer analytical tools similar to the ones used for model plants (Bolger et al. 2017). We are now living the post-model organism era, whose resources have been being well utilized and essential for non-model plants research, such as crops (Heck and Neely 2020). One example is sugarcane, a C4 grass able to accumulate large amounts of sucrose and considered the world's leading biomass crop (Souza et al. 2019). For a long time, rice (also a monocot but with a C3 metabolism) and maize were the "model" organisms phylogenetically closer to sugarcane. The first sugarcane proteomic studies tested different extraction methods (Amalraj et al. 2010), identified specific classes (Cesarino et al. 2012), and provided the characterization of abiotic stress-response (Zhou et al. 2012). Until 2013, proteomes surveys of this crop were scarce (Boaretto and Mazzafera 2013). However, after several perseverant efforts on *Saccharum* spp. genome sequencing and assembly (Boaretto and Mazzafera 2013; Garsmeur et al. 2018; Grativol et al. 2014; Miller et al. 2017; Okura et al. 2016; Riano-Pachon and Mattiello 2017; Souza et al. 2019; Vettore et al. 2003; Vilela et al. 2017), the catalog of proteins identified by mass spectrometry largely increased for both descriptive and several sorts of biological conditions (for a review see Miller et al. 2017). In this highly polyploid species, the identification of glycoside hydrolases (GH) at the protein level paved the way for target experimentation. Domain prediction and homology investigation indicated that members of the GH families are numerous in sugarcane, but only dedicated extraction and analyses of proteomic pipelines could point to which exact proteins and at which amount were present in different organs and developmental stages (Calderan-Rodrigues et al. 2014, 2016; Fonseca et al. 2018). Combined omics and protein activity essays from these retrieved GHs showed that the cell wall degradation occurring on a particular tissue was timely orchestrated by different proteins that tackled pectin, callose, hemicelluloses, and cellulose at last (Grandis et al. 2019). The GHs identified in this cell disassembly mechanism could be manipulated in a timely controlled fashion to produce plants more amenable to saccharification resulting in augmented ethanol yield. For dedicated functional studies, an alternative that has been used for non-model plants is the transference of the target gene to a model host, allowing a better comprehension of the metabolic changes. This approach was successfully employed for the sugarcane cell wall-related transcription factor SHINE (Martins et al. 2018) and a Dirigent-Jacalin (Andrade et al. 2019).

So far, the selection of target genes for non-model plants has been performed mostly based on genomic, transcriptomic, or in silico data. Furthermore, the possibility to promptly generate RNAseq data can provide draft proteome databases for MS-identification, which makes the use of proteogenomics a powerful tool to study non-model plant species (Armengaud et al. 2014). One of the frontiers to be crossed relies on a more straightforward use of plant proteomics as an instrument to select the expressing alleles and thus markers more successfully related to phenotype. However, proteomic pipelines did not reach the same advancement levels as transcriptomic ones, and we urge to have sensitive, high-throughput, and low-cost technologies in order to make this possible (Peace et al. 2019).

The study of non-model plants is extensively laborious and sometimes does not attract as many attention as the possibility to deepen the investigation on known biological pathways by using model plants. However, proteomic investigations of non-model species have brought light to exciting discoveries related to plant metabolic and developmental specificities (Grandis et al. 2019; Sergeant et al. 2019), biomass accumulation (Calderan-Rodrigues et al. 2019), disease responses (Diaz-Vivancos et al. 2006; He et al. 2012; Tahara et al. 2003; Zhang et al. 2015), and several sorts of abiotic stresses (Aghaei and Komatsu 2013; Cia et al. 2018; Huang and Sethna 1991; Wang et al. 2017). Research in this area not only contributed to biological insights from these specific conditions and species but also provided a more complete picture of the whole autotrophic nature, allowing to deepen the evolutionary discussion as well. Once confined to conditional conclusions, non-model plant proteomics can now take a step further, and instead of conflictive, data integration between model and non-model ones and target experimentation will allow these two groups to be considered of the same level. This will attract interest for non-model species research and will fuel a virtuous cycle to provide more and more biological information.

3.4 Future Directions

The recent advances in proteomics analysis of model and non-model plant species have demonstrated that the evolution of techniques and instrumentation had revolutionized how fast we can identify the plant molecular phenotypes. Emergent characteristics have been uncovered by the proteomics, and the proper discovery of many linkages between transcriptomics and other omics through the quantification of proteins is revealing several processes and cellular mechanisms that may play fundamental role in regulating cell growth and metabolism.

It is clear that a big challenge in the proteomics field is the limitation of analyzing, at the same time, all aspects of the plant proteome, specially the regulation of cellular responses, such as the function of transcription factors and regulators, the roles of PTMs, and redox homeostasis in cellular phenotypes. The analysis of how individual cells respond and contribute to an organismic phenotype are revealing how complex the interaction between cells are in defining a phenotype. For instance, there are already indications of molecular similarities between different tissues or parts of a plant but vast differences at the level of cell types or individual cells. The combination of the total proteome analysis of organisms with targeted cell type proteome may drive comprehensive systems analysis of plant phenotypes, mixing the power of holistic analysis with the observation of cell type specificities, specially the regulatory ones.

The analysis of plant phenotypes has been performed using different proteomics approaches, which in many cases integrate cell biology and analytical techniques. This knowledge, mainly based on the application of shotgun mass spectrometry-based proteomics, has brought to light specific quantitative and qualitative information of the molecular physiology of plants.

Nevertheless, plant proteomics at the single-cell level or single-cell type can and will likely contribute to the understanding of the dynamic changes that occur in cells in response to environmental alterations and to regulatory aspects of the plant differentiation or development. It may also disclose the importance, rules, and the degree of heterogeneous responses cell population exert in defining a plant phenotype, elucidating how different mechanisms are coordinated to generate a phenotype from a series of individual cellular responses (Libault et al. 2017).

The use of gel-based proteomics approach is probably taking a more consistent path toward the comparative analysis of subcellular proteomes and targeted analysis of protein complexes, while been extensively used for comparative proteomics analysis of non-model plant species coupled to protein identification by MS.

Time-resolved plant shotgun MS-based proteomics analysis is still the most prominent approach currently applied in the plant proteomic science and will remain like that for the near future, unless a large-scale protein sequencing strategy is developed that could compete or substitute the MS-based proteomics, for some analysis at least.

This technical advance would contribute to drastically expand the number of proteome studies of non-model plant species, bringing more information for a future large comparative systems biology analysis, which may reveal emergent properties of the plant development and responses to environmental changes.

References

Aghaei K, Komatsu S (2013) Crop and medicinal plants proteomics in response to salt stress. Front Plant Sci 4:8

Amalraj RS et al (2010) Sugarcane proteomics: establishment of a protein extraction method for 2-DE in stalk tissues and initiation of sugarcane proteome reference map. Electrophoresis 31(12):1959–1974

Andrade LM et al (2019) Biomass accumulation and cell wall structure of rice plants overexpressing a Dirigent-Jacalin of sugarcane (ShDJ) under varying conditions of water availability. Front Plant Sci 10:65

Armengaud J et al (2014) Non-model organisms, a species endangered by proteogenomics. J Proteome 105:5–18

Arsova B, Watt M, Usadel B (2018) Monitoring of plant protein post-translational modifications using targeted proteomics. Front Plant Sci 9:1168

Bestel-Corre G et al (2002) Proteome analysis and identification of symbiosis-related proteins from Medicago truncatula Gaertn. by two-dimensional electrophoresis and mass spectrometry. Electrophoresis 23(1):122–137

Boaretto LF, Mazzafera P (2013) The proteomes of feedstocks used for the production of second-generation ethanol: a lacuna in the biofuel era. Ann Appl Biol 163(1):12–22

Bolger M et al (2017) From plant genomes to phenotypes. J Biotechnol 261:46–52

Boughton BA et al (2016) Mass spectrometry imaging for plant biology: a review. Phytochem Rev 15:445–488

Calderan-Rodrigues MJ et al (2014) Cell wall proteomics of sugarcane cell suspension cultures. Proteomics 14(6):738–749

Calderan-Rodrigues MJ et al (2016) Cell wall proteome of sugarcane stems: comparison of a destructive and a non-destructive extraction method showed differences in glycoside hydrolases and peroxidases. BMC Plant Biol 16:14

Calderan-Rodrigues MJ et al (2019) Plant cell wall proteomics: a focus on monocot species, brachypodium distachyon, Saccharum spp. and Oryza sativa. Int J Mol Sci 20(8):1975

Cesarino I et al (2012) Enzymatic activity and proteomic profile of class III peroxidases during sugarcane stem development. Plant Physiol Biochem 55:66–76

Chang TS et al (2017) Mapping and comparative proteomic analysis of the starch biosynthetic pathway in rice by 2D PAGE/MS. Plant Mol Biol 95(4–5):333–343

Cia MC et al (2018) Novel insights into the early stages of ratoon stunting disease of sugarcane inferred from transcript and protein analysis. Phytopathology 108(12):1455–1466

De Mia M et al (2019) Nitric oxide remodels the photosynthetic apparatus upon S-starvation in Chlamydomonas reinhardtii. Plant Physiol 179(2):718–731

Delaunois B et al (2014) Uncovering plant-pathogen crosstalk through apoplastic proteomic studies. Front Plant Sci 5:249

Diaz-Vivancos P et al (2006) The apoplastic antioxidant system in Prunus: response to long-term plum pox virus infection. J Exp Bot 57(14):3813–3824

dos Santos AL et al (2016) Quantitative proteomic analysis of Araucaria angustifolia (Bertol.) Kuntze cell lines with contrasting embryogenic potential. J Proteome 130:180–189

Du H et al (2019) A new insight to explore the regulation between S-nitrosylation and N-glycosylation. Plant Direct 3(2):e00110

Edman P (1949) A method for the determination of amino acid sequence in peptides. Arch Biochem 22(3):475

Elbl P et al (2015) Comparative transcriptome analysis of early somatic embryo formation and seed development in Brazilian pine, Araucaria angustifolia (Bertol.) Kuntze. Plant Cell Tissue Organ Cult 120(3):903–915

Falter-Braun P et al (2019) iPlant Systems Biology (iPSB): an international network hub in the plant community. Mol Plant 12(6):727–730

Feng J, Chen L, Zuo J (2019) Protein S-nitrosylation in plants: current progresses and challenges. J Integr Plant Biol 61(12):1206–1223

Fenn JB et al (1989) Electrospray ionization for mass spectrometry of large biomolecules. Science 246(4926):64–71

Fonseca JG et al (2018) Cell wall proteome of sugarcane young and mature leaves and stems. Proteomics:18(2)

Forrester MT et al (2009) Detection of protein S-nitrosylation with the biotin-switch technique. Free Radic Biol Med 46(2):119–126

Friso G, van Wijk KJ (2015) Posttranslational protein modifications in plant metabolism. Plant Physiol 169(3):1469–1487

Gallardo K et al (2001) Proteomic analysis of arabidopsis seed germination and priming. Plant Physiol 126(2):835–848

Garsmeur O et al (2018) A mosaic monoploid reference sequence for the highly complex genome of sugarcane. Nat Commun 9(1):2638
Giavalisco P et al (2005) Proteome analysis of Arabidopsis thaliana by two-dimensional gel electrophoresis and matrix-assisted laser desorption/ionisation-time of flight mass spectrometry. Proteomics 5(7):1902–1913
de Godoy LM et al (2008) Comprehensive mass-spectrometry-based proteome quantification of haploid versus diploid yeast. Nature 455(7217):1251–1254
Gong B et al (2019) Unravelling GSNOR-mediated S-nitrosylation and multiple developmental programs in tomato plants. Plant Cell Physiol 60(11):2523–2537
Grandis A et al (2019) Cell wall hydrolases act in concert during aerenchyma development in sugarcane roots. Ann Bot 124(6):1067–1089
Grativol C et al (2014) Sugarcane genome sequencing by methylation filtration provides tools for genomic research in the genus Saccharum. Plant J 79(1):162–172
Gross JH (2017) Mass spectrometry: a textbook. Springer International Publishing, Cham
Hart-Smith G et al (2017) Improved quantitative plant proteomics via the combination of targeted and untargeted data acquisition. Front Plant Sci 8:1669
He D, Yang P (2013) Proteomics of rice seed germination. Front Plant Sci 4:246
He L et al (2012) Proteomic analysis of the effects of exogenous calcium on hypoxic-responsive proteins in cucumber roots. Proteome Sci 10(1):42
Heck M, Neely BA (2020) Proteomics in non-model organisms: a new analytical frontier. J Proteome Res 19(9):3595–3606
Hillenkamp F, Karas M (1990) Mass spectrometry of peptides and proteins by matrix-assisted ultraviolet laser desorption/ionization. Methods Enzymol 193:280–295
Hu J, Rampitsch C, Bykova NV (2015) Advances in plant proteomics toward improvement of crop productivity and stress resistancex. Front Plant Sci 6:209
Huang M, Sethna JP (1991) History dependence of a two-level system. Phys Rev B Condens Matter 43(4):3245–3254
Jorrin-Novo JV et al (2019) Gel electrophoresis-based plant proteomics: past, present, and future. Happy 10th anniversary Journal of Proteomics! J Proteome 198:1–10
Joshi HJ et al (2011) MASCP Gator: an aggregation portal for the visualization of Arabidopsis proteomics data. Plant Physiol 155(1):259–270
Kaspar S et al (2011) MALDI-imaging mass spectrometry - an emerging technique in plant biology. Proteomics 11(9):1840–1850
Kim J, Woo HR, Nam HG (2016) Toward systems understanding of leaf senescence: an integrated multi-omics perspective on leaf senescence research. Mol Plant 9(6):813–825
Komatsu S, Muhammad A, Rakwal R (1999) Separation and characterization of proteins from green and etiolated shoots of rice (Oryza sativa L.): towards a rice proteome. Electrophoresis 20(3):630–636
Kosova K, Vitamvas P, Prasil IT (2014) Proteomics of stress responses in wheat and barley-search for potential protein markers of stress tolerance. Front Plant Sci 5:711
Larsen MR et al (2006) Analysis of posttranslational modifications of proteins by tandem mass spectrometry. BioTechniques 40(6):790–798
Leon J, Costa-Broseta A (2020) Present knowledge and controversies, deficiencies, and misconceptions on nitric oxide synthesis, sensing, and signaling in plants. Plant Cell Environ 43:1
Li L et al (2019) Comparative proteomic analysis provides insights into the regulatory mechanisms of wheat primary root growth. Sci Rep 9(1):11741
Libault M et al (2017) Plant systems biology at the single-cell level. Trends Plant Sci 22(11):949–960
Liu Y et al (2019) Proteomics: a powerful tool to study plant responses to biotic stress. Plant Methods 15:135
Lu J et al (2020) Global quantitative proteomics studies revealed tissue-preferential expression and phosphorylation of regulatory proteins in Arabidopsis. Int J Mol Sci 21(17):6116
Martins de Souza D et al (2008) The untiring search for the most complete proteome representation: reviewing the methods. Brief Funct Genomic Proteomic 7(4):312–321
Martins APB et al (2018) Ectopic expression of sugarcane SHINE changes cell wall and improves biomass in rice. Biomass Bioenergy 119:322–334
Marx V (2019) A dream of single-cell proteomics. Nat Methods 16(9):809–812
McLoughlin F et al (2018) Maize multi-omics reveal roles for autophagic recycling in proteome remodelling and lipid turnover. Nat Plants 4(12):1056–1070
McLoughlin F et al (2020) Autophagy plays prominent roles in amino acid, nucleotide, and carbohydrate metabolism during fixed-carbon starvation in maize. Plant Cell 32(9):2699–2724
Mergner J et al (2020) Mass-spectrometry-based draft of the Arabidopsis proteome. Nature 579(7799):409–414
Mettler T et al (2014) Systems analysis of the response of photosynthesis, metabolism, and growth to an increase in irradiance in the photosynthetic model organism Chlamydomonas reinhardtii. Plant Cell 26(6):2310–2350
Miller JR et al (2017) Initial genome sequencing of the sugarcane CP 96-1252 complex hybrid. F1000Res 6:688
Morisse S et al (2014) Insight into protein S-nitrosylation in Chlamydomonas reinhardtii. Antioxid Redox Signal 21(9):1271–1284
Mousavi SA et al (2016) PlantPReS: a database for plant proteome response to stress. J Proteome 143:69–72
Murray CI et al (2012) Identification and quantification of S-nitrosylation by cysteine reactive tandem mass tag switch assay. Mol Cell Proteomics 11(2):M111.013441
Nabi RBS et al (2020) Functional insight of nitric-oxide induced DUF genes in Arabidopsis thaliana. Front Plant Sci 11:1041
Navrot N et al (2011) Plant redox proteomics. J Proteome 74(8):1450–1462

Neubauer G et al (1998) Mass spectrometry and EST-database searching allows characterization of the multi-protein spliceosome complex. Nat Genet 20(1):46–50

Nietzel T et al (2020) Redox-mediated kick-start of mitochondrial energy metabolism drives resource-efficient seed germination. Proc Natl Acad Sci U S A 117(1):741–751

O'Farrell PH (1975) High resolution two-dimensional electrophoresis of proteins. J Biol Chem 250(10):4007–4021

Oda Y et al (1999) Accurate quantitation of protein expression and site-specific phosphorylation. Proc Natl Acad Sci U S A 96(12):6591–6596

Okura VK et al (2016) BAC-pool sequencing and assembly of 19 Mb of the complex sugarcane genome. Front Plant Sci 7:342

Ong SE, Mann M (2005) Mass spectrometry-based proteomics turns quantitative. Nat Chem Biol 1(5):252–262

Ong SE et al (2002) Stable isotope labeling by amino acids in cell culture, SILAC, as a simple and accurate approach to expression proteomics. Mol Cell Proteomics 1(5):376–386

Pappireddi N, Martin L, Wuhr M (2019) A review on quantitative multiplexed proteomics. ChemBioChem 20(10):1210–1224

Patole C, Bindschedler LV (2019) Chapter 4 - Plant proteomics: a guide to improve the proteome coverage. In: Meena SN, Naik MM (eds) Advances in biological science research. Academic Press, New York, NY, pp 45–67

Peace CP et al (2019) Apple whole genome sequences: recent advances and new prospects. Hortic Res 6:59

Picotti P, Aebersold R (2012) Selected reaction monitoring-based proteomics: workflows, potential, pitfalls and future directions. Nat Methods 9(6):555–566

Porubleva L et al (2001) The proteome of maize leaves: use of gene sequences and expressed sequence tag data for identification of proteins with peptide mass fingerprints. Electrophoresis 22(9):1724–1738

Riano-Pachon DM, Mattiello L (2017) Draft genome sequencing of the sugarcane hybrid SP80-3280. F1000Res 6:861

Rodiger A, Baginsky S (2018) Tailored use of targeted proteomics in plant-specific applications. Front Plant Sci 9:1204

Rodriguez E et al (2020) Autophagy mediates temporary reprogramming and dedifferentiation in plant somatic cells. EMBO J 39(4):e103315

Samaras P et al (2020) ProteomicsDB: a multi-omics and multi-organism resource for life science research. Nucleic Acids Res 48(D1):D1153–D1163

Sanger F, Tuppy H (1951) The amino-acid sequence in the phenylalanyl chain of insulin. I. The identification of lower peptides from partial hydrolysates. Biochem J 49(4):463–481

Schmidt A, Claassen M, Aebersold R (2009) Directed mass spectrometry: towards hypothesis-driven proteomics. Curr Opin Chem Biol 13(5–6):510–517

Schwacke R et al (2019) MapMan4: a refined protein classification and annotation framework applicable to multi-omics data analysis. Mol Plant 12(6):879–892

Sergeant K et al (2019) The dynamics of the cell wall proteome of developing alfalfa stems. Biology (Basel) 8(3):60

Shevchenko A et al (1996a) Linking genome and proteome by mass spectrometry: large-scale identification of yeast proteins from two dimensional gels. Proc Natl Acad Sci U S A 93(25):14440–14445

Shevchenko A et al (1996b) Mass spectrometric sequencing of proteins silver-stained polyacrylamide gels. Anal Chem 68(5):850–858

Sidoli S et al (2015) Sequential window acquisition of all theoretical mass spectra (SWATH) analysis for characterization and quantification of histone post-translational modifications. Mol Cell Proteomics 14(9):2420–2428

Silveira V et al (2006) Polyamine effects on the endogenous polyamine contents, nitric oxide release, growth and differentiation of embryogenic suspension cultures of Araucaria angustifolia (Bert.) O. Ktze. Plant Sci 171(1):91–98

Smith RD et al (1990) New developments in biochemical mass spectrometry: electrospray ionization. Anal Chem 62(9):882–899

Souza GM et al (2019) Assembly of the 373k gene space of the polyploid sugarcane genome reveals reservoirs of functional diversity in the world's leading biomass crop. GigaScience 8(12):giz129

Spoel SH (2018) Orchestrating the proteome with post-translational modifications. J Exp Bot 69(19):4499–4503

Swaney DL, Villen J (2016) Proteomic analysis of protein posttranslational modifications by mass spectrometry. Cold Spring Harb Protoc 2016(3):pdb.top077743

Tahara ST, Mehta A, Rosato YB (2003) Proteins induced by Xanthomonas axonopodis pv. passiflorae with leaf extract of the host plant (Passiflorae edulis). Proteomics 3(1):95–102

Thompson A et al (2003) Tandem mass tags: a novel quantification strategy for comparative analysis of complex protein mixtures by MS/MS. Anal Chem 75(8):1895–1904

Unwin RD et al (2005) Quantitative proteomic analysis using isobaric protein tags enables rapid comparison of changes in transcript and protein levels in transformed cells. Mol Cell Proteomics 4(7):924–935

Venable JD et al (2004) Automated approach for quantitative analysis of complex peptide mixtures from tandem mass spectra. Nat Methods 1(1):39–45

Vettore AL et al (2003) Analysis and functional annotation of an expressed sequence tag collection for tropical crop sugarcane. Genome Res 13(12):2725–2735

Vilela MM et al (2017) Analysis of three sugarcane homo/homeologous regions suggests independent polyploidization events of Saccharum officinarum and Saccharum spontaneum. Genome Biol Evol 9(2):266–278

Vu LD, Gevaert K, De Smet I (2018) Protein language: post-translational modifications talking to each other. Trends Plant Sci 23(12):1068–1080

Wang H, Lan P, Shen RF (2016) Integration of transcriptomic and proteomic analysis towards understanding the systems biology of root hairs. Proteomics 16(5):877–893

Wang Y et al (2017) Proteomic analysis of Camellia sinensis (L.) reveals a synergistic network in the response to drought stress and recovery. J Plant Physiol 219:91–99

Washburn MP, Wolters D, Yates JR III. (2001) Large-scale analysis of the yeast proteome by multidimensional protein identification technology. Nat Biotechnol 19(3):242–247

Wienkoop S, Baginsky S, Weckwerth W (2010) Arabidopsis thaliana as a model organism for plant proteome research. J Proteome 73(11):2239–2248

Wienkoop S et al (2012) ProMEX - a mass spectral reference database for plant proteomics. Front Plant Sci 3:125

Wilkins MR et al (1996) Progress with proteome projects: why all proteins expressed by a genome should be identified and how to do it. Biotechnol Genet Eng Rev 13:19–50

Yu F et al (2020) Identification of modified peptides using localization-aware open search. Nat Commun 11(1):4065

Zhang CX, Tian Y, Cong PH (2015) Proteome analysis of pathogen-responsive proteins from apple leaves induced by the alternaria blotch alternaria alternata. PLoS One 10(6):e0122233

Zhang H et al (2019) Arabidopsis proteome and the mass spectral assay library. Sci Data 6(1):278

Zhang F et al (2020) Data-independent acquisition mass spectrometry-based proteomics and software tools: a glimpse in 2020. Proteomics 20(17–18):e1900276

Zhou G et al (2012) Proteomic analysis of osmotic stress-responsive proteins in sugarcane leaves. Plant Mol Biol Report 30(2):349–359

Subcellular Proteomics as a Unified Approach of Experimental Localizations and Computed Prediction Data for Arabidopsis and Crop Plants

4

Cornelia M. Hooper, Ian R. Castleden, Sandra K. Tanz, Sally V. Grasso, and A. Harvey Millar

Abstract

In eukaryotic organisms, subcellular protein location is critical in defining protein function and understanding sub-functionalization of gene families. Some proteins have defined locations, whereas others have low specificity targeting and complex accumulation patterns. There is no single approach that can be considered entirely adequate for defining the in vivo location of all proteins. By combining evidence from different approaches, the strengths and weaknesses of different technologies can be estimated, and a location consensus can be built. The Subcellular Location of Proteins in Arabidopsis database (http://suba.live/) combines experimental data sets that have been reported in the literature and is analyzing these data to provide useful tools for biologists to interpret their own data. Foremost among these tools is a consensus classifier (SUBAcon) that computes a proposed location for all proteins based on balancing the experimental evidence and predictions. Further tools analyze sets of proteins to define the abundance of cellular structures. Extending these types of resources to plant crop species has been complex due to polyploidy, gene family expansion and contraction, and the movement of pathways and processes within cells across the plant kingdom. The Crop Proteins of Annotated Location database (http://crop-pal.org/) has developed a range of subcellular location resources including a species-specific voting consensus for 12 plant crop species that offers collated evidence and filters for current crop proteomes akin to SUBA. Comprehensive cross-species comparison of these data shows that the subcellular proteomes (subcellulomes) depend only to some degree on phylogenetic relationship and are more conserved in major biosynthesis than in metabolic pathways. Together SUBA and cropPAL created reference subcellulomes for plants as well as species-specific subcellulomes for cross-species data mining. These data collections are increasingly used

C. M. Hooper · I. R. Castleden · S. K. Tanz · S. V. Grasso · A. H. Millar (✉)
The Centre of Excellence in Plant Energy Biology, The University of Western Australia, Crawley, WA, Australia
e-mail: cornelia.hooper@uwa.edu.au; ian.castleden@uwa.edu.au; sandra.tanz@uwa.edu.au; sally.grasso@uwa.edu.au; harvey.millar@uwa.edu.au

F. V. Winck (ed.), *Advances in Plant Omics and Systems Biology Approaches*, Advances in Experimental Medicine and Biology 1346, https://doi.org/10.1007/978-3-030-80352-0_4

by the research community to provide a subcellular protein location layer, inform models of compartmented cell function and protein–protein interaction network, guide future molecular crop breeding strategies, or simply answer a specific question—where is my protein of interest inside the cell?

Keywords

Arabidopsis · SUBA · CropPAL · SUBAcon · Crops · Protein localization · Subcellular location · System biology · Protein location · Subcellular compartment

Abbreviation

AMPDB	Arabidopsis Mitochondrial Protein Database
ASV	Alternative splice variant
CAT	Co-expression Adjacency Tool
cropPAL	Database for crop proteins with annotated locations
ER	Endoplasmic reticulum
FLAG	Epitope polypeptide DYKDDDDK
FP	Fluorescent protein
GO	Gene Ontology http://www.geneontology.org
GUI	Graphical user interface
LOPIT	Localization of Organelle Protein by Isotope Tagging
MMAP	Multiple Marker Abundance Profiling
MS	Mass spectrometry
MS/MS	Tandem mass spectrometry
NPAS	Normalized protein abundance scores
PAT	PPI adjacency tool
PPI	Protein–protein interaction
SRM	Selected reaction monitoring
SUBA	Subcellular localization database for Arabidopsis proteins
TAP	Tandem affinity purification

4.1 Introduction

4.1.1 The Historical Context of Subcellular Location in Proteomics

Subcellular proteomics are an integral part of plant proteomics due to the tight connection between protein location within cells and their function (Weckwerth et al. 2008; Millar et al. 2009; Joshi et al. 2011). The importance of subcellular location of different isozymes of proteins in plants has been long recognized (Gottlieb 1982). However, until recently, most conventional methodologies for determining protein locations in the cell have been labor intensive. Methodologies such as gold particle marking and immunological tagging followed by microscopy were used over decades, but both approaches are very labor and cost intensive. More recently, PCR and cloning techniques, overexpression vectors, fluorescent protein (FP) chimerics and expression in alternative hosts allowed the localization of lowly expressed proteins in difficult to study organisms (Chiu et al. 1996). These studies, although not high throughput, have been a big step forward for rare proteins.

Using mass spectrometry (MS) of tandem mass spectrometry (MS/MS) to build comprehensive subcellular proteome sets came into the picture with improving organellar extractions and MS detection for peptides (Heazlewood et al. 2005). The organelles and compartments that attracted most interest were the major energy organelles such as the mitochondrion (Kruft et al. 2001; Millar et al. 2001) and the chloroplast (Peltier et al. 2001), which had to be physically extracted and purified away from the cytosol, nucleus, and secretory parts of the cells. Following on, research focusing on the secretory system needed to distinguish between endoplasmic reticulum (ER), Golgi apparatus, and tonoplast (vascular membrane) within the cell and the plasma membrane surrounding the cell. Finally, extracellular proteins secreted from cells into the apoplast were studied. With higher resolution of microscopes and better separation techniques, finer points of differentiation began to distinguish suborganellar compartments of these organelles, including protein localizations in the inner and outer membranes, the intracellular space and the matrix of mitochondria (Millar et al. 2001; Werhahn et al. 2001), thylakoids in chloroplasts (Schubert et al. 2002), cytoskeletal structures (Hamada et al. 2013), and specific sections of the secretory system (Drakakaki et al. 2012; Heard et al. 2015). With the improvement of MS/MS sensitivity, over time more proteins could be identified and traced in these purified organellar separations.

The Localization of Organelle Protein by Isotope Tagging (LOPIT) approach offered an alternative for defining the localization of membrane proteins without physical subcellular separation to purity (Dunkley et al. 2004). This method, based on profiling protein abundances in size separated fractions, has been used to map a significant number of proteins from mitochondria, plastids, ER, and secretory compartments (Nikolovski et al. 2012). LOPIT has recently been optimized for subcellular protein mapping and is used in a variety of global subcellular mapping projects spanning plant and disease biology (Mulvey et al. 2017).

When considering all subcellular localization methods as a whole, the majority of data today are derived from proteomic MS methodology due to the high-throughput nature of the approach. A subcellular proteomics mapping study typically produces more than 1000 subcellular localizations (Heard et al. 2015; Nguyen-Kim et al. 2016; Senkler et al. 2017). In contrast, high-throughput fluorescent protein (FP) studies report between 10 and 100 (Cutler et al. 2000; Boruc et al. 2010; Inze et al. 2012) with the largest study to date offering 148 protein localizations (Koroleva et al. 2005). The emerging importance of large data mining requires that subcellular location data are available as a global data set using all available information, making data aggregations for subcellular proteomics increasingly popular resources. A growing number of databases available have generated subcellular proteomics data sets containing over 40,000 experimental localizations spanning different methodologies. These resources often fill gaps in experimental data and compare the advantages as well as disadvantages of each method using computational strategies.

4.1.2 Collation of Arabidopsis Subcellular Data Established Subcellular Proteomics

Subcellular proteomics has been refined and improved through the aggregation of localization data. In plant biology, this first started to take shape for the model plant Arabidopsis. Today, the SUBcellular location database for Arabidopsis proteins (SUBA, http://suba.live) in its fourth generation is a substantial collection of manually curated published data sets of large-scale subcellular proteomics (MS/MS), FP visualization, protein–protein interaction (PPI), and subcellular targeting calls from 22 prediction programs as well as a consensus algorithm (SUBAcon).

The collection started with studies on the mitochondrial proteome more than 10 years ago (Heazlewood et al. 2004) when an MS study of the mitochondrial organelle revealed a large number of low-abundance proteins that had been predicted to localize elsewhere in the cell. This led to the generation of the Arabidopsis Mitochondrial Protein Database (AMPDB) that offered an overview of the detailed MS data sets from 17 published mitochondrial studies as well as predictions from six subcellular location algorithms (Heazlewood and Millar 2005). While similar efforts for the plastid were realized as the Plastid Proteome DataBase (PPDB) at the same time (Friso et al. 2004), it became clear that rapid expansion of data for many organelles required the establishment of a one-stop data collection hub for subcellular compartmentalization data. An initial data collation and categorization showed surprisingly little overlap between data sets from different researchers, and their combination seemed crucial for large-scale data mining (Heazlewood et al. 2005). It was then when the SUBA acronym was introduced and the data aggregation quickly revealed that protein families, subsets, and isoforms with distinct subcellular location patterns existed. The analysis of the data also meta-assessed the reliability of experimental data and indicated that most experimental methods are more variable and error-prone than the wider research community presumed.

Since then, the Arabidopsis subcellular proteome data set and applications associated with the collection have been increasing in terms of gene annotations, experimental, and computational data types (Heazlewood et al. 2007; Tanz et al. 2013a), as well as high-confidence subsets (Arabidopsis Subcellular Reference—ASURE) and location consensus classifications (SUBAcon) (Hooper et al. 2014) to arrive as the current version of SUBA4 that includes a separate section with an interactive analysis toolbox (Hooper et al. 2017a). The core experimental subcellular location data are now more than ten times the volume of the original SUBA1 (Heazlewood et al. 2007). SUBA started out categorizing ten distinct subcellular locations cytosol, cytoskeleton, endoplasmic reticulum (ER), Golgi, mitochondrion, nucleus, plastid, plasma membrane, peroxisome, and vacuole. With improving sensitivity of methods used to pinpoint protein locations, SUBA4 expanded this into sub-organellar compartments including differentiations into distinct membranes and aqueous compartments (Fig. 4.1). Experimental data pinpointing sub-organellar locations are now searchable within SUBA4.

The subcellular proteome data for Arabidopsis has increased from representing information on only 7% of predicted proteins in 2007 to over 32% of predicted proteins in 2017 arriving at the current >40% coverage in 2021 (Fig. 4.2a). Since the foundation of SUBA, Dr JL Heazlewood, Dr SK Tanz, Dr CM Hooper, Dr N Ayamanesh, and Ms Sally Grasso have been the key curators, while Dr J Tonti-Filippini, Dr CM Hooper, and Dr IR Castleden have developed most of the GUI and database services to enable the user experience. A small fraction of manual subcellular

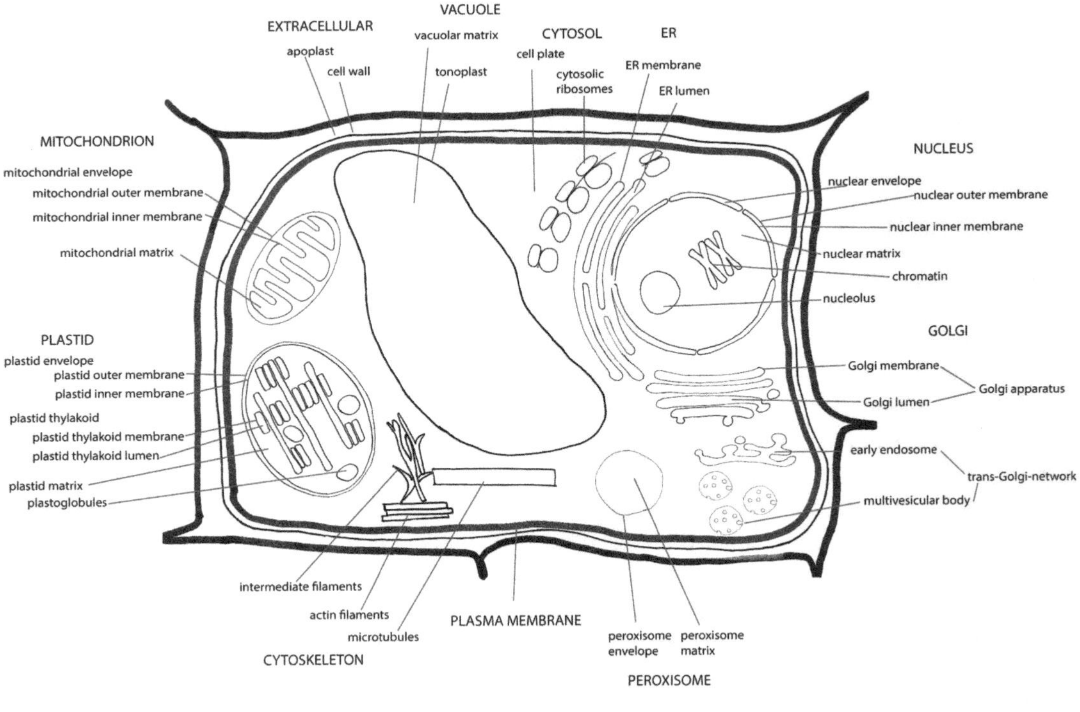

Fig. 4.1 Subcellular and suborganellar structures searchable within SUBA and cropPAL. The subcellular categories previously assessed by SUBA1-3 and cropPAL included 11 major subcellular locations (UPPER CASE). SUBA4 has increased location definition into suborganellar locations (lower case). *ER* endoplasmic reticulum, *SUBA* subcellular location database for Arabidopsis proteins (http://suba.live/)

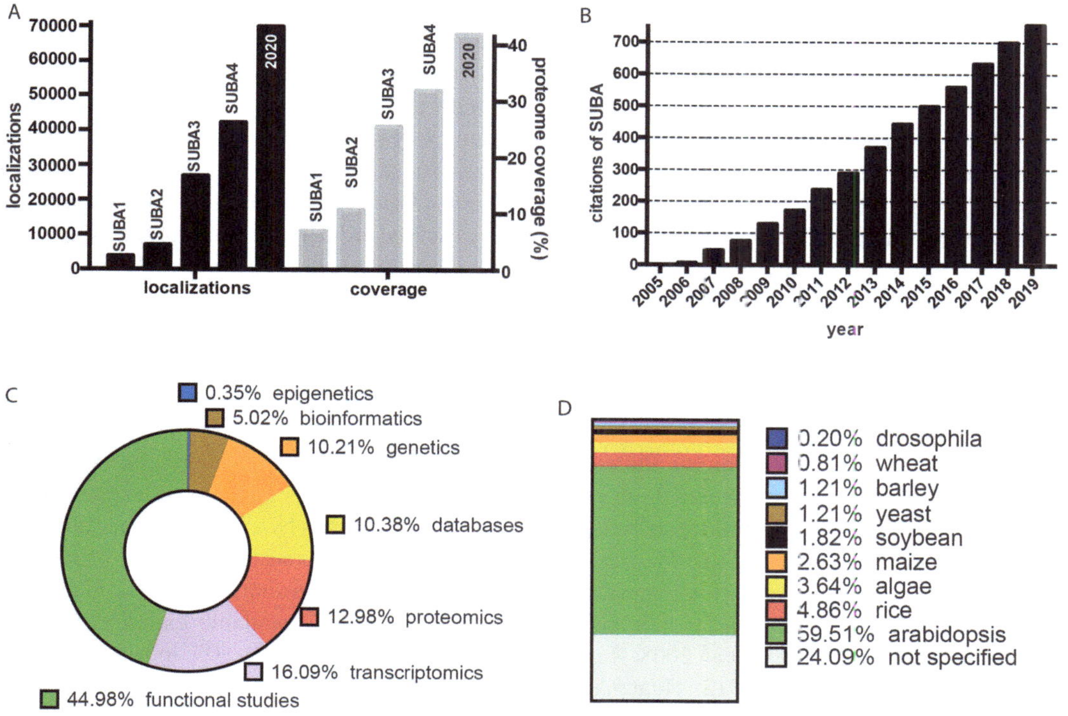

Fig. 4.2 SUBA database expansion and data use in plant biology. (**a**) Continuous curation and integration of subcellular localization data has increased the number of localizations and the coverage of the Arabidopsis proteome throughout SUBA releases from 2005 to 2017. The size of SUBA4 at the time of writing in 2020 is indicated. (**b**) The accumulative citation record of published SUBA and SUBA tools indicates increasing importance of subcellular proteomics resources (source: Scopus). (**c**) The area of research and (**d**) species studied was determined using keyword-based text mining to show the distribution of fields of research that SUBA was used for in the past decade. A total of 465 studies that cited SUBA were examined. (Figure modified from Hooper et al. 2017a)

curations were independently derived from TAIR, GO, and Swissprot (Lamesch et al. 2012; Croft et al. 2014).

4.1.3 The Collation of Plant Subcellular Data Progressed into Crop Plants by Establishing cropPAL

High-throughput genome sequencing technologies, computing, and database management have made the protein sequences available through http://www.gramene.org/ for a range of non-model plant species of economic importance (Gupta et al. 2016; Tello-Ruiz et al. 2018). This has led to the exponential growth in the number of available reference plant genomes in recent years (Monaco et al. 2014). Notably, this includes the improved coverage of the bread wheat genome where researchers tackled a number of significant problems that occur when annotating highly polyploid genome sets (Bolser et al. 2015). Proteins across species share important similarities in their functional motifs, and this has driven linking information on orthologous proteins from model plants to less studied crop plants (Otto et al. 2008). Cross-species comparison highlights the amount of gene and genome duplication and gene loss throughout angiosperm evolution that has led to a huge variation in genome size and proteome composition between even close relatives (Tang et al. 2008). Researching protein specialization and sub-functionalization across and within species provides new insights into why plants differ so extensively in their growth, yield, and response to the environment.

While duplicated proteins may perform similar functions, small differences in versions that reside in distinct subcellular compartments can allow distinct optima, better suited to individual subcellular compartments. For example, small differences in protein sequence can improve function in differing pH environments (Scheibe et al. 2005). Therefore, the cost of protein or pathway duplication can more than compensate the energy investment in transporting them across membranes (Wu et al. 2006; Cheung et al. 2013). Knowing key turning points between energy budget and protein location during plant evolution has become a crucial consideration for studying plant product yields and determining the energy production of cells. With crop breeding in mind, data from the model plant Arabidopsis has been useful to bridge knowledge gaps for rice and maize through combining information with the independent subcellular proteomics data sets that exist for these species (Natera et al. 2008; Reiland et al. 2011; Majeran et al. 2012; Huang et al. 2013). This led to the generation of two further species-specific subcellular proteomics databases; riceDB for rice (Narsai et al. 2013) and PPDB for maize (Friso et al. 2004; Huang et al. 2013). Nevertheless, many more crops species exists for which Arabidopsis data presented the only resource for subcellular location information (Hooper et al. 2017a).

Subcellular proteomics and other localization data for most crop species exist scattered across published scientific reports and is often linked to obsolete protein accession annotations that are not concurrent with recent genome annotations. In this format a significant body of experimental subcellular proteomics data for barley (Endler et al. 2006; Ploscher et al. 2011) and wheat (Kamal et al. 2012; Suliman et al. 2013) are difficult to access for most researchers. The need for a cross species protein localization database emerged and was formulated in the formation of a new resource, crop Proteins with Annotated Location (cropPAL, https://crop-pal.org/). The cropPAL1 database contained just under 18,000 of the scattered experimental localizations for four mono-cotyledon crops including rice (*Oryza sativa*), maize (*Zea mays*), wheat (*Triticum aestivum*), and barley (*Hordeum vulgare*) connected to each other and to Arabidopsis (Hooper et al. 2016). CropPAL underwent major upgrades to include the two additional monocotyledon sorghum (*Sorghum bicolor*) and banana (*Musa acuminata*) and six additional dicotyledon species granola (*Brassica napus*), field mustard (*Brassica rapa*), soybean (*Glycine max*), tomato (*Solanum lycopersicum*), potato (*Solanum tuberosum*), and grape vine (*Vitis vinifera*). In cropPAL2020, subcellular proteomics data from MS/MS and FP localization data as well as pre-computed subcellular localizations from 11 predictors were collated (Hooper et al. 2020). Aligning to the Ensembl Plants/Gramene identifiers, experimental data in cropPAL was linked to the current genome annotations by a custom semi-automated pipeline. This offers sustainable links of research data that had increasingly obsolete identifiers. Using this system, available experimental data more than tripled to 61,505 localizations and generated large enough data sets for statistical comparisons between mono- and dicotyledon species or cross-species data mining opportunities between legumes and fruiting crops. Altogether, cropPAL2020 collates more than 800 scientific peer-reviewed studies. These data represent the collective work of >700 scientists from 600 organizations in 45 countries showcasing a global effort in elucidating protein subcellular location divergence and conservation across crop species.

The SUBA and cropPAL resources have been used for cultivar discrimination, engineering salt-resistant crops, increasing protein content, as well as improving yield and market value of grains, legumes, palm, mango, and tomato (Bajpai et al. 2018; Lau et al. 2018; Matamoros et al. 2018; Jiang et al. 2019; Schneider et al. 2019; McKenzie et al. 2020). If subcellular protein distributions (subcellulomes) are not cataloged for a species, scientists often fall back on data in Arabidopsis of the nearest species. This raises the questions around the validity of these discoveries if they are based on the assumption that we can borrow cross-species information. On the one hand, homology-linking protein subcellular location data is widely accepted on the basis that the metabolic and biosynthetic pathways in plants are highly conserved. On the other hand, reports exist that highlight the divergence in protein subcellular location between species by mechanisms of dual targeted proteins or protein family expansion (Carrie and Whelan 2013; Xu et al. 2013; One Thousand Plant Transcriptomes 2019). In context of the diversity of species physiology, metabolism capacity, and their ability to adapt to different environments, subcellular location diversification offers a potential starting point for plant performance improvement through biotechnological applications. The combination of data as well as the linking of 12 economically important crop species with Arabidopsis has placed plant subcellular data at the forefront of subcellular proteomics combining the skills of laboratory methodologies, data management and bioinformatics. These comprehensive data resources are now ready to aid current research questions around crop cell compartmentalization and crop biology.

4.2 Research and Technical Approach

4.2.1 Visualization and Separation of Proteins for Subcellular Localization Are Improving

Subcellular compartments and structures in plants were first defined by microscopy; what could be seen inside cells. Their separation and characterization have focused on attempts to recover these observed structures, free of contaminants. Initial separations are often based on the use of empirically derived speeds and times of differential centrifugation to enrich components of specific size ranges. Second, the use of density gradients separates structures based on their isopycnic point (buoyant density) which enables further purification of subcellular structures (for review see Taylor and Millar 2017).

Other physical and chemical properties have been developed as supplemental or even primary methods of isolation of specific structures. Electrical processes to separate organelles started with laminar-flow electrophoresis and lead to development of free-flow electrophoresis to purify subcellular particles like endosomes, lysosomes, peroxisomes, and ER-vesicles based on differences in surface charge. In plants, free-flow electrophoresis has been used to purify Arabidopsis plasma membrane and the tonoplast (Bardy et al. 1998), mitochondria (Eubel et al. 2007), and the Golgi apparatus (Parsons et al. 2012). Solid-phase separations through chromatography can also be used but has been typically limited to smaller sub-cellular structures such as mega Dalton protein complexes, for example, the pyruvate dehydrogenase complex (4–10 MDa), ribosomes (3–4 MDa), and the proteasome (2.5 MDa). Addition of affinity tags to target proteins by chemistry or genetic engineering allows isolation of many structures of interest from cells using the same affinity system. The DYKDDDDK epitope (FLAG) and tandem affinity purification (TAP) tagging are typical approaches performed in plants, for example, for the Arabidopsis proteasome (Book et al. 2010) and cytosolic ribosome (Reynoso et al. 2015). While this is typically expensive compared to other approaches, it can provide access to structures that either cannot be separated or are labile during the sequential physical processes of traditional isolations.

A number of different techniques can then be used for the assessment of organelle and structure contamination including microscopy, the use of marker enzyme activity assays, antibodies raised to marker proteins, selected reaction monitoring (SRM) MS and quantitative MS, or comparisons to literature claims of subcellular protein locations (Taylor et al. 2014; Millar et al. 2009; Taylor and Millar 2017). The use of stable isotope labeling or quantitation tags during MS can help screen out unknown contaminants by ensuring the target proteins are quantitatively enriched during organellar purification or are enriched more than other co-enriched cellular structures (Eubel et al. 2008; Mueller et al. 2014).

MS of compartment-enriched samples remains the most popular large-scale approach for defining subcellular localization of proteins, despite ongoing questions around the purity of these lists (Joshi et al. 2011). The FP-tagging approach is generally more accurate, but labor- and time-intensive, resulting in small study sizes with only a few high-throughput studies (Dunkley et al. 2006; Boruc et al. 2010; Lee et al. 2011). While the low coverage makes FP insufficient as a stand-alone large-scale data set, the collation of such studies over the last 15 years has generated a sizable subcellular proteome data set that remains one of the most widely accepted by biologists. Notably FP studies show both, where targeted proteins *are* and *are not* present, providing an internal control to evaluate competing claims of location that is sorely missing from most MS/MS data sets.

4.2.2 Subcellular Proteomics Can Be Supplemented with Homology Gap-Filling and Subcellular Protein Location Predictions

The varying number and type of subcellular proteomic studies performed among each plant species has led to uneven coverage between species and subcellular compartments. Of the 12 crop species, the largest number of subcellular localization experimentation to date has been performed in rice, maize, and soybean with tomato and wheat catching up steadily (Hooper et al. 2020). The proteome coverage of the most comprehensive experimental data sets collated for rice and tomato reached ~18% followed by soybean and maize with 5–10%. High-throughput MS/MS cataloging commonly focused on nucleus, plasma membrane, and extracellular extractions, for assessing proteins induced and secreted during host defense (Shah et al. 2012; Shinano et al. 2013). In contrast, plastidial and mitochondrial purifications for mass spectrometry analyses are often studying biogenesis and metabolic functions (Huang et al. 2009; Barsan et al. 2012; Salvato et al. 2014; Xing et al. 2016). Compartment catalogs existing for Arabidopsis (Reumann et al. 2009; Ito et al. 2011; Parsons et al. 2012; Heard et al. 2015) but not yet in crops include the cytosol, Golgi (Chateigner-Boutin et al. 2015), endoplasmic reticulum (ER) (Komatsu et al. 2017), and peroxisome (Arai et al. 2008). No crop species has been cataloged across all compartments, which means experimental data for any systems biology study is too sparse for downstream applications.

A recent effort pooled all experimental data across 12 crop species and Arabidopsis into biological MapMan categories to reveal that the percentage coverage of biological functions with experimental localizations in crops was similar to that observed in Arabidopsis (Schwacke et al. 2019; Hooper et al. 2020). Thereafter, the integrated data from the 12 crops increases coverage to >70% in most biological categories, showing that this can aid coverage of functional categories for less researched crops. However, the majority of data leading to this coverage derived from the well-researched species rice or maize and thus remains as valid as the two assumptions that (1) experimental error rates are small and that (2) subcellular locations are typically conserved. Since researchers often study proteins from the same compartments and functional categories in crops as well as Arabidopsis, gap-filling across species reaches a limit. To fill gaps beyond experimental data, predictors are necessary to achieve complete subcellulome coverage. A variety of proteome-wide subcellular location predictors have been developed based on sequence properties (Shen et al. 2007; Chou and Shen 2010; Yu et al. 2010). This includes various machine-learning and pattern recognition approaches (Chou and Shen 2007), such as support vector machines (Hua and Sun 2001), *k*-nearest neighbor (Horton et al. 2007), neural networks (Small et al. 2004), and hidden Markov models (Lin et al. 2011). Similar to different experimental techniques, these individual approaches have their own advantages and shortcomings in terms of the number of required features, the danger of over-fitting, and the ability to handle multiple optima. In order to improve accuracy, single machine-learning approaches have been stacked into multi-layer algorithms (Petsalaki et al. 2006; Pierleoni et al. 2006; Blum et al. 2009). Predictors typically derive their subcellular location calculations using protein sequence features, associated properties and/or gene ontology (GO) (Shen et al. 2007), and curator annotations (Briesemeister et al. 2010). Thereafter, predictors based on protein sequence identify sequence patterns in the primary protein sequence that target to individual organelles (Zybailov et al. 2008; Blum et al. 2009). Using similar or identical inputs, distinct machine-learning algorithms often yield different results that have shown a surprisingly poor overlap (Tanz and Small 2011). This variability is the main reason why experimental data are still seen as the gold standard by most biologists despite the unresolved difficulties associated with the experimental approaches themselves (Millar et al. 2009).

For Arabidopsis, SUBA contains the subcellular location outputs of 22 computational predictors including: AdaBoost (Niu et al. 2008), ATP (Mitschke et al. 2009), BaCelLo (Pierleoni et al. 2006), ChloroP 1.1 (Emanuelsson et al. 1999), EpiLoc (Brady and Shatkay 2008), iPSORT (Bannai et al. 2002), MitoPred (Guda et al. 2004), MitoProt (Claros and Vincens 1996), MultiLoc2 (Blum et al. 2009), Nucleo (Hawkins et al. 2007), PCLR 0.9 (Schein et al. 2001), Plant-mPLoc (Chou and Shen 2010), PProwler 1.2 (Hawkins and Boden 2006), Predotar v1.03 (Small et al. 2004), PredSL (Petsalaki et al. 2006), PTS1 (Neuberger et al. 2003), SLPFA (Tamura and Akutsu 2007), SLP-Local (Matsuda et al. 2005), SubLoc (Hua and Sun 2001), TargetP 1.1 (Emanuelsson et al. 2000), WoLF PSORT (Horton et al. 2007), and YLoc (Briesemeister et al. 2010). For Arabidopsis, the targeting predictions were carried out on the full-length protein sequences obtained from TAIR10 (Lamesch et al. 2012) or Ensembl plants (Kersey et al. 2018). The performance of the 22 predictors was tested on a high-confidence subcellular location reference data set (ASURE) that is accessible through SUBA4. The assessment of the predictors indicated that for Arabidopsis some predictors perform better for particular compartments and sometimes even better than experimental data. This is most visual for nuclear proteins where MultiLoc and Yloc outperformed unified contradicting MS and FP data (Fig. 4.3). In the years since their development, some of the above predictors have become unavailable. Hence, for cropPAL, the integration of only 11 out of 22 predictors was suitable or possible (Hooper et al. 2020). In total, predictive data sets in cropPAL span >6 million predictions, but for complete proteomes of all 12 species, only the six predictors MultiLoc2, TargetP, Predotar, YLoc, iPSORT, and WolfPSORT were available. The performance of predictors and experimental data in crop plants remains to be tested when data collections and high-confidence reference proteomes become available. However, a comparative analysis of crop experimental and predictive data to Arabidopsis suggests that the error rates of these methods in crops are similar to those seen in SUBA (Hooper et al. 2020).

4.2.3 An Objective Collation and Unification Strategy Can Resolve Varied and Conflicting Subcellular Location Information

Despite significant technological progress, errors in experimental data contribute to overlapping or contradicting data sets (Ito et al. 2011; Elmore et al. 2012; Nikolovski et al. 2012). As experimental data resources grow and the number of predictors increases so do the contradictions, and strategies are needed to integrate this multifaceted information. SUBAcon was developed to integrate the multi- and single-compartment predictor outputs with all available experimental data resources to generate an overall estimation of subcellular location for Arabidopsis proteins (Hooper et al. 2014).

SUBAcon uses FP and MS/MS data as a feature and determines the classification strength on their location calls when trained on a reference data set compared against other available localization features. When searching for other biological data that can be used as features, association data has become a popular choice due to the "guilt by association" principle. Associations like protein–protein interaction (PPI) and transcript co-expression aid the identification of functional clusters within the proteome. Considering that proteins in functional

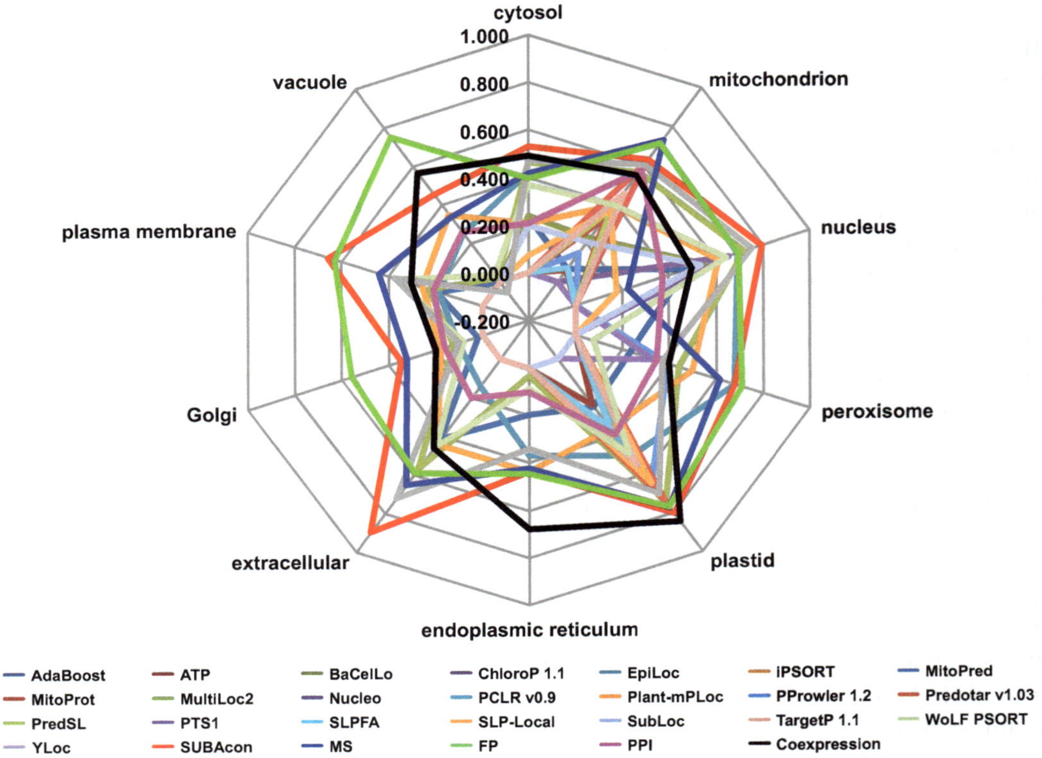

Fig. 4.3 Performance of individual predictors, experimental data, and SUBAcon. The classification performance of six top scoring separate component (lines with symbols) in SUBAcon was compared to the overall SUBAcon consensus classification using all components (gray fill). The comparison of performance indicator MCC indicates superior subcellular location classification of Arabidopsis proteins when all information was used. Individual components generally perform well for some compartments allowing choice of best predictor for target compartments for Arabidopsis proteins. *FP* fluorescent proteins, *MCC* Matthews correlation coefficient, *MS* mass spectrometry, *PPI* protein-protein interaction. (Figure modified from Hooper et al. 2014)

pathways more often co-locate, the location of a PPI partner can be used as indirect experimental evidence for protein location. Co-expression and PPI-associated protein sets are known for containing higher numbers of same-location protein groups than expected by random (Huh et al. 2003; Geisler-Lee et al. 2007). PPI data in particular have been suggested previously to be resources for predicting sub-cellular location of proteins in multiple eukaryotic species (Shin et al. 2009; Jiang and Wu 2012). Less is known about the true value of co-expression data for predicting co-location. These voluminous expression data sets have been used widely for predicting function (Stuart et al. 2003; Heyndrickx and Vandepoele 2012) and are typically the largest data sets available for most species. We showed that co-expression was useful to infer subcellular location for proteins with little experimental evidence, suggesting that such data alone can be highly informative for some compartments, rivaling sequence-based prediction (Hooper et al. 2014).

The lack of a single best method for inferring subcellular location has prompted using all available knowledge about proteins and is an attractive approach for forming a consensus view. Integrating a number of varied data sources has been used in yeast mitochondrial studies for some time, where this approach revealed promising new insights into genes involved in mitochondrial functions (Prokisch et al. 2004). The strategy of SUBAcon was to unify FP, MS, PPI, co-expression, and prediction data objectively to have one output with the highest probability of being correct (Fig. 4.4). This generated a data set where one protein was assigned one location or a set of locations that can be used for downstream omics applications. SUBAcon integrated 22 selected computational predictors into a two-phase naive Bayes classifier, which equaled or surpassed the classification accuracy for most compartments in comparison to single predictors even before integrating subcellular proteomics data (Figs. 4.3 and 4.4). The assessment of single and stepwise integration confirmed that SUBAcon objectively weighs individual predictors and experimental data to assign proteins to a location (or locations) more accurately than any of the input predictors or data did separately (Hooper et al. 2014). The analysis also confirmed a strong influence of experimental data on classification outcome; high proportions of FP protein localizations (~78%) and MS data (~65%) agreed with the ASURE locations. In both methodologies, the proportion of location mismatches was compartment-specific (Fig. 4.3). Consequently, the ongoing collation of experimental localization data will remain a key aspect of up-to-date classification by tools like SUBAcon.

In order to produce a classifier taking crop species-specific differences as well as error rates

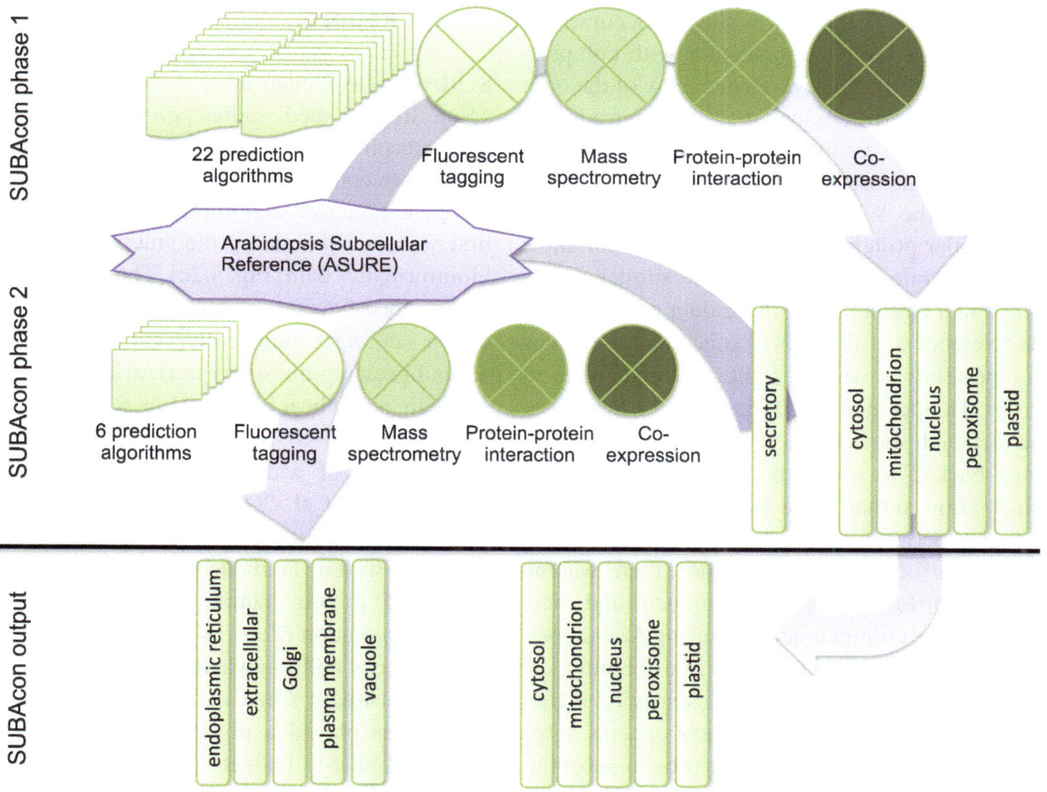

Fig. 4.4 The SUBAcon prediction and unification strategy. The subcellular location information from 22 predictors, fluorescent tagging, mass spectrometry, protein–protein interaction and co-expression (experimental data) and the ASURE standard was used to train a naive Bayes algorithm in two phases. Phase 1 (top) distinguished cytosol, mitochondrion, nucleus, peroxisome, and plastid from secretory proteins. Phase 2 (middle) used six secretory predictors and the experimental data to classify endoplasmic reticulum, extracellular, Golgi, plasma membrane and vacuole proteins. The secretory classifications were combined with the phase 1 locations and present the final SUBAcon output (bottom). *ASURE* Arabidopsis Subcellular Reference standard. (Figure modified from Hooper et al. 2014)

of methodologies into account, a cross-species gold standard not biased by experimental and homology inference like ASURE is essential. Such a gold standard does not yet exist for these crops. Therefore, within cropPAL, a winner-takes-all (WTA) uniform location call was derived for each crop protein by vote-counting experimental locations and adding all predictors as one vote with each predictor a fraction of that vote (Hooper et al. 2020). This weights the final location call toward experimental locations when available and allowing predictors to gap-fill when not. When comparing the error rates, the accuracy of both, MS/MS and FP methodologies in the cropPAL collection, was comparable to that previously reported in Arabidopsis (Hooper et al. 2014). As expected, FP localizations were overall more accurate (61–87%), while the accuracy of MS/MS data varied significantly with compartment (22–85%). The overall accuracy of the crop WTA calls when compared to the inferred crop reference was estimated as greater than 67%, supporting the use of the voting system.

Using the WTA, individual species-specific subcellular protein distributions (subcellulomes) were generated that proved to be similar across crops and Arabidopsis. These data resources act as reference distributions of subcellular locations for proteins as well as biological categories in crops (Hooper et al. 2020) marking the beginning of detailed species-specific subcellulome catalogs backed by experimental data. Indeed, the study showed that while close evolutionary relationship between species is evident in the higher agreement of their subcellulomes, there was no obvious difference between monocot and dicot species. The current practice of using Arabidopsis data for dicot research versus rice for monocot research and arguing for species-specific differences is challenging. While most species agree substantially (60–80%) in their protein subcellular locations, the agreement is not equally distributed across biological functions. Underexplored subcellular divergence information was found in particular in metabolic categories. Metabolic diversification has been reported in a number pathways including enzymes of the amino acid metabolism (Schenck and Last 2020) as well as subcellular partitioning of effector or signaling proteins specific to tissues or metabolic changes upon stimuli (Powers et al. 2019). On the proteomics level, the differences between species are often subtle shifts in the distribution of a number of proteins with similar functions between two or more subcellular locations (Hooper et al. 2020). Such shifts are mainly due to alternative splice variants (ASV) or duplications of genes that are more likely to be retained if subcellularly diverse (Avelange-Macherel et al. 2018).

4.2.4 Subcellular Proteomics Data Resources in SUBA Have Contributed to Over 900 Downstream Scientific Reports

SUBA has been cited more than 700 times averaging 30 published studies per year (Fig. 4.2b). The subcellular proteomics data of SUBA has been more commonly used for exploring protein and gene functions and improving the interpretation of transcriptomics, proteomics, genetics, and bioinformatics data (Fig. 4.2c). The use of SUBA has reached beyond Arabidopsis showing application in agricultural hypothesis formation around pressing questions in rice, barley, maize, soybean, and wheat biology (Fig. 4.2d).

In research, SUBA has contributed to the development of widely used organelle marker sets (Nelson et al. 2007), protein family clone collections for functional genomics (Lao et al. 2014), as well as facilitated the functional elucidation of protein families involved in plant growth regulation (Zentella et al. 2007). The latter resources and knowledge were used in over 900 downstream studies. Over the last decade, SUBA has played a pivotal role in estimating plant cell energy budgets (Cheung et al. 2013) and the costs of maintaining the plant proteome in different compartments (Li et al. 2017). In the context of systems biology approaches, knowledge of proteome-wide subcellular locations is an important component for defining functional neighborhoods and deducing metabolic and signaling networks within complex eukaryotic cells

(Waese et al. 2017). It has also been used for exploring sugar metabolism networks in barley (Lunn et al. 2014) and demonstrating subfunctionalization of gene family expansions (Tanz et al. 2013b). This shaped our understanding of the subcellular plant metabolism in order to resolve diurnal relationships of plant metabolism (Furtauer et al. 2019) as well as contributed to increasing the resolution and accuracy of mathematical representations of plant cell and tissue metabolism during the last 5 years (Shi and Schwender 2016). The in silico estimation of organellar protein abundance (Hooper et al. 2017b) influenced concepts of subcellular phenotyping that helped achieve cultivar-specific discrimination through rapid estimation of organellar differences (Schneider et al. 2019). The breath of work benefitting from SUBA highlights the importance of ongoing efforts in developing this central subcellular resource.

The smaller fraction of crop research using SUBA highlights the importance of improving the linkage of SUBA across species-specific borders as well as the need to improve linkage of comprehensive subcellular data collections for more crop species. The compendium of cropPAL begins to address this challenge across 12 crop species, and it has provided protein localization data since 2015. Right from the start, cropPAL the subcellular proteomics data contributed to the characterization of protein families across species (Chen et al. 2016a) as well as to a high confidence training set used for a novel plant and effector protein localization prediction algorithm (Sperschneider et al. 2017). Increasing awareness put cropPAL forward as a valuable resource for developing accurate proteomics pathways and network maps in economically important crops (Larrainzar and Wienkoop 2017).

SUBA and cropPAL also hold considerable contributions to molecular breeding concepts for increased crop quality and global food security. Highlights include the recent report listing the use of SUBA4 for unraveling crucial adaptation mechanisms for salinity tolerance in plants that provide promising genetic targets for engineering salt-resistant crops (Jiang et al. 2019) as well as the reviewed importance of SUBA resources for molecular biomarker identification for addressing a variety of diseases, yield quality and sustainability challenges within the palm oil industry (Lau et al. 2018). It is particularly exciting to see the rising influence of cropPAL on agricultural breeding strategies including the identification of genetic breeding targets for improved mango peel features increasing the mango market value (Bajpai et al. 2018) and the identification of molecular targets that expand flowering duration for increasing pollination opportunities and yield in rice (Chen et al. 2016b).

Tackling a serious global concern about nitrogen integration and protein increase in crops, SUBA4 was used in guiding the discovery of proteins regulating the nitrogen metabolism in root nodules (Matamoros et al. 2018) while cropPAL was named one of the "key aspects that need to be strengthened in the future" considering the large number of proteins involved in nitrogen-fixation efficiency (Larrainzar and Wienkoop 2017). These resources will be crucial in the near future for unraveling the complexity of nitrate metabolism in plants with the aim to guide molecular breeding strategies toward securing nutrition of global food crops under changing environments.

4.2.5 The Collation and Integration of Arabidopsis Subcellular Proteomics Data Presents Opportunities for New Approaches for In Silico Analysis

Both SUBA and cropPAL subcellular data collections are data warehouses publicly available through http://suba.live and http://crop-pal.org, respectively, that provide easy GUI-based data search and filtering functions. The web portals enable biologically meaningful subcellular location annotations and integrations by APIs or focused list creation through the web query builder. Users do not need computational expertise to mine the data sets for subcellular locations, methodology, protein properties, gene associations, authorship, or country of data origin. In addition, the interface offers a BLAST function for scientists researching alternative crops that enables to link their sequence of interest to the closest match in the SUBA or cropPAL data set. Within SUBA4, a separate toolbox exists that contains tools that allow immediate access and analysis of the core Arabidopsis subcellular location data in linkage to external user data sources. The toolbox currently offers the Multiple Marker Abundance Profiling tool (MMAP), the Co-expression Adjacency Tool (CAT), and the PPI Adjacency Tool (PAT). Each tool provides a unique link to the subcellular location consensus (SUBAcon).

For Arabidopsis, the co-expression data and PPI data sets were linked to the unified SUBAcon calls. The SUBAcon calls of each protein partner were joined and categorized into proximity relationships according to their biological interpretation, such as location within the same organelle, neighboring organelle, or distant organelles. The subcellular locations, proximity relationships, mutual rank and average correlation coefficient data allow user-lead prioritizing of strong associations. With the CAT tool, the user can assess a list of proteins for their subcellular location in context with their gene expression association, allowing to discover potential relationships between proteins based on the vast amount of expression data available.

The PAT tool uses experimental evidence from 26,327 unique PPI and assigns subcellular location derived from SUBAcon to each protein. This allows the interpretation of protein associations in context of proteins proximity in the cell. Thereafter, proteins located within the same subcellular compartment can interact and proteins on outer membranes or interfaces of one organelle can also interact with proteins from neighboring organelles. The PAT tool allows filtering for location pairings to target specific organellar or interorganellar interactions for hypotheses formation around PPI networks that influence biological processes.

The newest showcase using the full range of proteomics integrated into subcellular proteomics is the MMAP tool (Hooper et al. 2017b). This tool can estimate the proportion of different subcellular protein structures in a user-provided list of Arabidopsis Gene Identifiers (AGIs). It is based on combining localization information from SUBA and quantitative MS observations of proteins collated in the MASCP gator database (Joshi et al. 2011; Mann et al. 2013). While relative protein quantitation is possible using quantitative MS such approaches are expensive or moderately accurate (Thompson et al. 2003; Cox and Mann 2008; Arike and Peil 2014; Christoforou et al. 2016). Using available quantitative tissue proteome data indicates that such data can be standardized to achieve a more true representation of an Arabidopsis protein observation (Wang et al. 2012, 2015). In order to achieve a subcellular proteome quantitation including low-abundance proteins in specific organelles, data from over 100 publications describing enriched subcellular proteomes, organelles and protein complexes derived from public databases were added and normalized to an in silico protein abundance score (Sun et al. 2009; Ferro et al. 2010; Hooper et al. 2017a). The novel way of using normalized protein abundance scores (NPAS) for 23,191 proteins contained 2602 proteins that had not previously been scored, and it covers a total of 85% of the predicted Arabidopsis proteome (Wang et al. 2012).

The user can submit custom AGI lists to the MMAP tool and receive the number of distinct proteins per each organelle as well as an estimate of relative protein abundance composition compared to expected subcellular abundances. The tool was developed because conventional methods for determining organellar abundance rely on a few marker proteins, which can be hit and miss considering the variety of biological conditions the data are derived from. A high-confidence subcellular marker lists for Arabidopsis was generated using SUBAcon, which can gather a high probability list of proteins for each organelle. Extensive manual curation and cross-examination against experimental data verified these data for use in the in silico tool enabling ad hoc estimates of relative organelle abundance. Thereafter, the tool allows assessment of experimental data before committing to further experimentation. While the latter was the original aim of the developers, the tool offers additional opportunities for subcellular proteomics to drive big data questions in the near future.

While the MMAP tool is relatively new and not yet widely applied, its function has been demonstrated on data derived from Golgi (Parsons et al. 2012), chloroplast (Zybailov et al. 2008), and plasma membrane proteomes (de Michele et al. 2016; Elmore et al. 2012), where the tool output directly corresponds to the changes in organellar protein abundance measured by spectral counting (Hooper et al. 2017b). The MMAP tool was able to retrospectively demonstrate progressive Golgi enrichment in silico equivalent to immunoblotting but was also able to show how other compartments were enriched or depleted at the same time without any further experimentation (Cox and Mann 2008). In better-known organelles, such as chloroplasts (Zybailov et al. 2008), the MMAP tool revealed that current plastidial isolation procedures deplete all other organelle fractions and only plastidial proteins enrich (Uberegui et al. 2015; Yin et al. 2015). Using the MMAP tool to compare two plasma membrane isolations demonstrated that the free-flow electrophoresis was better able to decrease the contamination of plastid, Golgi, and mitochondria (Elmore et al. 2012; de Michele et al. 2016).

Using this tool, a cross analysis of all proteomes in SUBA can be done very rapidly, showing that mitochondrial, plastidial, and peroxisomal protein isolations are among the purest, whereas current plasma membrane, cytosol, and vacuole separation techniques cannot achieve the same purities (Fig. 4.5). The data also highlight the problem that attempts to detect low abundance proteins within organellar extractions lead to a near exponential increase in captured impurities.

One of the unexplored functions of subcellular proteomics and quantitative proteomics is its use in analyzing tissue proteomes and potentially proteomes from different biological conditions in terms of subcellular proteome shifts or relative subcellular structure abundances. Using the MMAP tool, the plasmodesmata proteome (Fernandez-Calvino et al. 2011) showed predominance in the fraction of Golgi, vacuole, peroxisome, mitochondria, cytosolic, and plasma membrane compared to standard values allowing

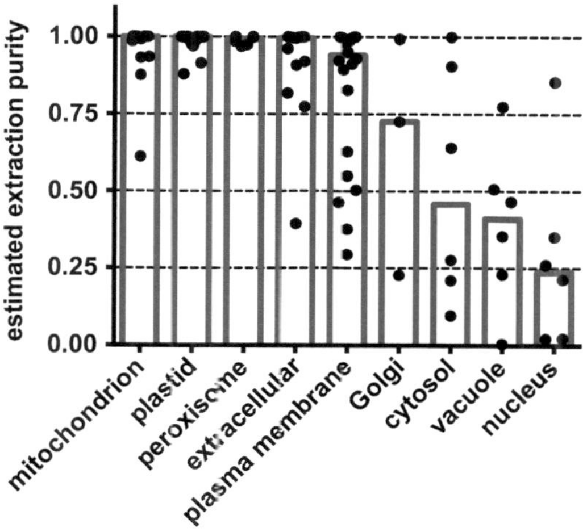

Fig. 4.5 Subcellular proteomics purity for Arabidopsis preparations. Published organellar separations during the last 15 years were retrieved from SUBA4 (http://suba.live). The lists of organellar proteins were loaded into the MMAP tool (http://suba.live/toolbox-app.html) and the obtained enrichment data was extrapolated to cover the whole protein list. Obtained fractions were graphed for each organelle as median bar showing purity of individual studies in dots. *MMAP* multiple marker abundance profiling

a superior analytical interpretation to the methods used by the authors in the study.

The MMAP tool allows the rapid generation of a holistic overview of relative organelle abundance for different tissues (Fig. 4.6), the same tissue following a treatment or environmental stimuli, or in mutant proteome phenotyping. Such analyses give an insight into how organelle proportions relate to tissue function. Analysis of MS/MS data from different tissues confirmed observed and biological relevant differences in organelles in cotyledons, leaf, root, and pollen tissues that would have taken considerable experimental efforts to otherwise confirm (Dunand et al. 2007; Grobei et al. 2009; Piques et al. 2009; Baerenfaller et al. 2011). The MMAP tool is open access and only requires a list of protein identifiers, thus a broad range of conditions can be queried beyond the ones listed above. It is yet to be seen how this tool can help interpret a variety of biological data including available proteomics, gene expression (Birnbaum et al. 2003) as well as protein turnover rate data (Li et al. 2012, 2017).

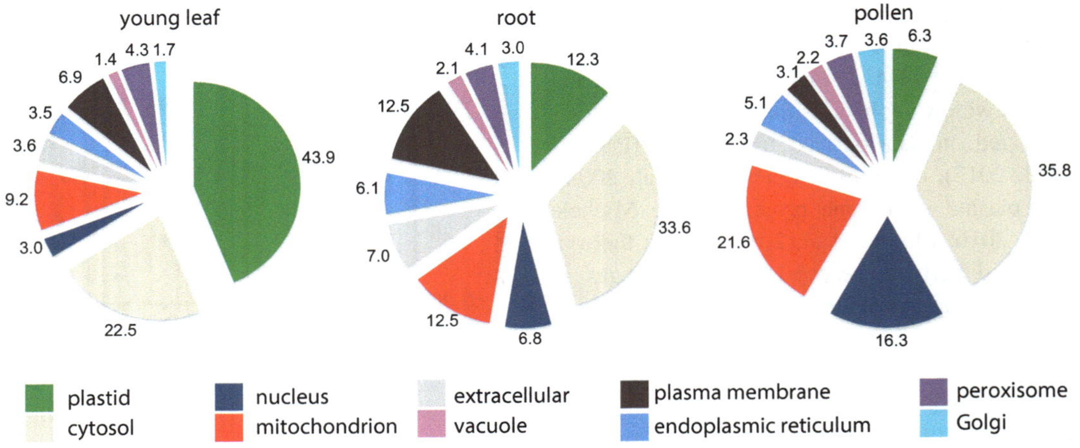

Fig. 4.6 In silico estimation of subcellular proteome distribution in tissues. Arabidopsis protein identifiers were obtained from mass spectrometry experimentation by Baerenfaller and colleagues (Baerenfaller et al. 2011) and submitted to the MMAP tool in the SUBA4 toolbox. The output was extrapolated across the unknown protein fraction and estimated subcellular abundance were graphed as percentage of total protein abundance. *MMAP* multiple marker abundance profiling, *SUBA* subcellular localization database for Arabidopsis proteins

4.3 Future Directions in the Field

Despite the reasonable coverage of subcellular proteomics in Arabidopsis, recent tools have pinpointed that there are large gaps in the analysis of subcellular proteomes that require specialized attention. This includes the nucleus where only 22% of the proteome have experimental data attached to it, whereas more than 50% of the plastid and peroxisome proteins have been experimentally observed. Similarly, when using the MMAP tool, the unknown proteins are generally low-expressed proteins that are not easy to measure by MS/MS or FP. The organelle-specific coverage was reflected in the struggle to find enough markers for the MMAP lists (Hooper et al. 2017b). Only two data sets describing ER (Dunkley et al. 2006; Nikolovski et al. 2012) and five data sets describing Golgi or Trans-Golgi network enrichments exist (Dunkley et al. 2006; Drakakaki et al. 2012; Parsons et al. 2012; Nikolovski et al. 2014; Heard et al. 2015), compared to over 30 plastidial and mitochondrial separations that are available in SUBA. Using the MMAP tool to assess the subproteomes has shown that we need to pay particular attention to the nuclear and ER proteomes, as they have the poorest coverage and a poor purity (Figs. 4.5 and 4.6). Targeted subcellular proteome mapping may be a much-needed focus for generating more complete subcellular proteomes for these less covered organelles or low abundance protein families.

While Arabidopsis subcellular proteomics has developed a solid omics presence, for crops this field is only now emerging. This is reflected in the much lower experimental coverage of crop proteomes for subcellular location. Considering that subcellular location has been regarded as crucial for determining protein function and belonging to biological processes and pathways (Cook and Cristea 2019), this is surprising and unsatisfying. The recent development of SUBAcon for Arabidopsis and WTA for crops enabled to fill these gaps and generated large enough data sets to assess the conservation of subcellular locations across 12 crop species and Arabidopsis. Subcellular location divergence of proteins is species-specific and harbors unexplored potential for data-driven agricultural breeding strategies. An increased understanding of how the subcellular location differences influence plant metabolism would be beneficial for designing breeding strategies toward more sustainable varieties. Protein subcellular location shifts have shown to increase plant growth, biosynthesis of secondary metabolites relevant to industrial production and therapeutic application (Shen et al. 2019) and comparable strategies in protein biosynthesis has the potential to achieve crop varieties with higher protein content in crops. Computational modeling approaches are emerging as a promising way to test current hypotheses around crop metabolic traits based on protein subcellular location shifts (Terasawa et al. 2016; Tabbita et al. 2017). However, such metabolic models currently rely on available subcellular data (Vinga et al. 2010) mainly derived from Arabidopsis through SUBA (Mintz-Oron et al. 2012). Other species data has been too sparse causing errors and redundancies in crop metabolic models resulting in the removal of potentially species-specific reactions (Seaver et al. 2012). The growing subcellular location resources SUBA and cropPAL will be an exciting contribution to achieving a better species-specific representation of such models in the near future.

References

Arai Y, Hayashi M, Nishimura M (2008) Proteomic identification and characterization of a novel peroxisomal adenine nucleotide transporter supplying ATP for fatty acid beta-oxidation in soybean and Arabidopsis. Plant Cell 20:3227–3240

Arike L, Peil L (2014) Spectral counting label-free proteomics. Methods Mol Biol 1156:213–222

Avelange-Macherel MH, Candat A, Neveu M, Tolleter D, Macherel D (2018) Decoding the divergent subcellular location of two highly similar paralogous LEA proteins. Int J Mol Sci 19:1620

Baerenfaller K, Hirsch-Hoffmann M, Svozil J, Hull R, Russenberger D, Bischof S, Lu Q, Gruissem W, Baginsky S (2011) pep2pro: a new tool for comprehensive proteome data analysis to reveal information about organ-specific proteomes in Arabidopsis thaliana. Integr Biol 3:225–237

Bajpai A, Khan K, Muthukumar M, Rajan S, Singh NK (2018) Molecular analysis of anthocyanin biosynthesis pathway genes and their differential expression in mango peel. Genome 61:157–166

Bannai H, Tamada Y, Maruyama O, Nakai K, Miyano S (2002) Extensive feature detection of N-terminal protein sorting signals. Bioinformatics 18:298–305

Bardy N, Carrasco A, Galaud JP, Pont-Lezica R, Canut H (1998) Free-flow electrophoresis for fractionation of Arabidopsis thaliana membranes. Electrophoresis 19:1145–1153

Barsan C, Zouine M, Maza E, Bian W, Egea I, Rossignol M, Bouyssie D, Pichereaux C, Purgatto E, Bouzayen M, Latche A, Pech JC (2012) Proteomic analysis of chloroplast-to-chromoplast transition in tomato reveals metabolic shifts coupled with disrupted thylakoid biogenesis machinery and elevated energy-production components. Plant Physiol 160:708–725

Birnbaum K, Shasha DE, Wang JY, Jung JW, Lambert GM, Galbraith DW, Benfey PN (2003) A gene expression map of the Arabidopsis root. Science 302:1956–1960

Blum T, Briesemeister S, Kohlbacher O (2009) MultiLoc2: integrating phylogeny and Gene Ontology terms improves subcellular protein localization prediction. BMC Bioinformatics 10:274

Bolser DM, Kerhornou A, Walts B, Kersey P (2015) Triticeae resources in ensembl plants. Plant Cell Physiol 56:e3

Book AJ, Gladman NP, Lee SS, Scalf M, Smith LM, Vierstra RD (2010) Affinity purification of the Arabidopsis 26 S proteasome reveals a diverse array of plant proteolytic complexes. J Biol Chem 285:25554–25569

Boruc J, Mylle E, Duda M, De Clercq R, Rombauts S, Geelen D, Hilson P, Inze D, Van Damme D, Russinova E (2010) Systematic localization of the Arabidopsis core cell cycle proteins reveals novel cell division complexes. Plant Physiol 152:553–565

Brady S, Shatkay H (2008) EpiLoc: a (working) text-based system for predicting protein subcellular location. Pac Symp Biocomput 13:604–615

Briesemeister S, Rahnenfuhrer J, Kohlbacher O (2010) YLoc--an interpretable web server for predicting subcellular localization. Nucleic Acids Res 38:W497–W502

Carrie C, Whelan J (2013) Widespread dual targeting of proteins in land plants: when, where, how and why. Plant Signal Behav 8:e25034

Chateigner-Boutin AL, Suliman M, Bouchet B, Alvarado C, Lollier V, Rogniaux H, Guillon F, Larre C (2015) Endomembrane proteomics reveals putative enzymes involved in cell wall metabolism in wheat grain outer layers. J Exp Biol 66:2649–2658

Chen BX, Li WY, Gao YT, Chen ZJ, Zhang WN, Liu QJ, Chen Z, Liu J (2016a) Involvement of polyamine oxidase-produced hydrogen peroxide during coleorhiza-limited germination of rice seeds. Front Plant Sci 7:1219

Chen Y, Ma J, Miller AJ, Luo B, Wang M, Zhu Z, Ouwerkerk PB (2016b) OsCHX14 is involved in the K+ homeostasis in rice (Oryza sativa) flowers. Plant Cell Physiol 57:1530–1543

Cheung CY, Williams TC, Poolman MG, Fell DA, Ratcliffe RG, Sweetlove LJ (2013) A method for accounting for maintenance costs in flux balance analysis improves the prediction of plant cell metabolic phenotypes under stress conditions. Plant J 75:1050–1061

Chiu W, Niwa Y, Zeng W, Hirano T, Kobayashi H, Sheen J (1996) Engineered GFP as a vital reporter in plants. Curr Biol 6:325–330

Chou KC, Shen HB (2007) Recent progress in protein subcellular location prediction. Anal Biochem 370:1–16

Chou KC, Shen HB (2010) Plant-mPLoc: a top-down strategy to augment the power for predicting plant protein subcellular localization. PLoS One 5:e11335

Christoforou A, Mulvey CM, Breckels LM, Geladaki A, Hurrell T, Hayward PC, Naake T, Gatto L, Viner R, Martinez Arias A, Lilley KS (2016) A draft map of the mouse pluripotent stem cell spatial proteome. Nat Commun 7:8992

Claros MG, Vincens P (1996) Computational method to predict mitochondrially imported proteins and their targeting sequences. Eur J Biochem 241:779–786

Cook KC, Cristea IM (2019) Location is everything: protein translocations as a viral infection strategy. Curr Opin Chem Biol 48:34–43

Cox J, Mann M (2008) MaxQuant enables high peptide identification rates, individualized p.p.b.-range mass accuracies and proteome-wide protein quantification. Nat Biotechnol 26:1367–1372

Croft D, Mundo AF, Haw R, Milacic M, Weiser J, Wu G, Caudy M, Garapati P, Gillespie M, Kamdar MR, Jassal B, Jupe S, Matthews L, May B, Palatnik S, Rothfels K, Shamovsky V, Song H, Williams M, Birney E, Hermjakob H, Stein L, D'Eustachio P (2014) The Reactome pathway knowledgebase. Nucleic Acids Res 42:D472–D477

Cutler SR, Ehrhardt DW, Griffitts JS, Somerville CR (2000) Random GFP::cDNA fusions enable visualization of subcellular structures in cells of Arabidopsis at a high frequency. Proc Natl Acad Sci U S A 97:3718–3723

Drakakaki G, van de Ven W, Pan S, Miao Y, Wang J, Keinath NF, Weatherly B, Jiang L, Schumacher K, Hicks G, Raikhel N (2012) Isolation and proteomic analysis of the SYP61 compartment reveal its role in exocytic trafficking in Arabidopsis. Cell Res 22:413–424

Dunand C, Crevecoeur M, Penel C (2007) Distribution of superoxide and hydrogen peroxide in Arabidopsis root and their influence on root development: possible interaction with peroxidases. New Phytol 174:332–341

Dunkley TP, Watson R, Griffin JL, Dupree P, Lilley KS (2004) Localization of organelle proteins by isotope tagging (LOPIT). Mol Cell Proteomics 3:1128–1134

Dunkley TP, Hester S, Shadforth IP, Runions J, Weimar T, Hanton SL, Griffin JL, Bessant C, Brandizzi F, Hawes C, Watson RB, Dupree P, Lilley KS (2006) Mapping the Arabidopsis organelle proteome. Proc Natl Acad Sci U S A 103:6518–6523

Elmore JM, Liu J, Smith B, Phinney B, Coaker G (2012) Quantitative proteomics reveals dynamic changes in the plasma membrane during Arabidopsis immune signaling. Mol Cell Proteomics 11(M111):014555

Emanuelsson O, Nielsen H, von Heijne G (1999) ChloroP, a neural network-based method for predicting chloroplast transit peptides and their cleavage sites. Protein Sci 8:978–984

Emanuelsson O, Nielsen H, Brunak S, von Heijne G (2000) Predicting subcellular localization of proteins based on their N-terminal amino acid sequence. J Mol Biol 300:1005–1016

Endler A, Meyer S, Schelbert S, Schneider T, Weschke W, Peters SW, Keller F, Baginsky S, Martinoia E, Schmidt UG (2006) Identification of a vacuolar sucrose transporter in barley and Arabidopsis mesophyll cells by a tonoplast proteomic approach. Plant Physiol 141:196–207

Eubel H, Lee CP, Kuo J, Meyer EH, Taylor NL, Millar AH (2007) Free-flow electrophoresis for purification of plant mitochondria by surface charge. Plant J 52:583–594

Eubel H, Meyer EH, Taylor NL, Bussell JD, O'Toole N, Heazlewood JL, Castleden I, Small ID, Smith SM, Millar AH (2008) Novel proteins, putative membrane transporters, and an integrated metabolic network are revealed by quantitative proteomic analysis of Arabidopsis cell culture peroxisomes. Plant Physiol 148:1809–1829

Fernandez-Calvino L, Faulkner C, Walshaw J, Saalbach G, Bayer E, Benitez-Alfonso Y, Maule A (2011) Arabidopsis plasmodesmal proteome. PLoS One 6:e18880

Ferro M, Brugiere S, Salvi D, Seigneurin-Berny D, Court M, Moyet L, Ramus C, Miras S, Mellal M, Le Gall S, Kieffer-Jaquinod S, Bruley C, Garin J, Joyard J, Masselon C, Rolland N (2010) AT_CHLORO, a comprehensive chloroplast proteome database with subplastidial localization and curated information on envelope proteins. Mol Cell Proteomics 9:1063–1084

Friso G, Giacomelli L, Ytterberg AJ, Peltier JB, Rudella A, Sun Q, Wijk KJ (2004) In-depth analysis of the thylakoid membrane proteome of Arabidopsis thaliana chloroplasts: new proteins, new functions, and a plastid proteome database. Plant Cell 16:478–499

Furtauer L, Kustner L, Weckwerth W, Heyer AG, Nagele T (2019) Resolving subcellular plant metabolism. Plant J 100:438–455

Geisler-Lee J, O'Toole N, Ammar R, Provart NJ, Millar AH, Geisler M (2007) A predicted interactome for Arabidopsis. Plant Physiol 145:317–329

Gottlieb LD (1982) Conservation and duplication of isozymes in plants. Science 216:373–380

Grobei MA, Qeli E, Brunner E, Rehrauer H, Zhang R, Roschitzki B, Basler K, Ahrens CH, Grossniklaus U (2009) Deterministic protein inference for shotgun proteomics data provides new insights into Arabidopsis pollen development and function. Genome Res 19:1786–1800

Guda C, Guda P, Fahy E, Subramaniam S (2004) MITOPRED: a web server for the prediction of mitochondrial proteins. Nucleic Acids Res 32:W372–W374

Gupta P, Naithani S, Tello-Ruiz MK, Chougule K, D'Eustachio P, Fabregat A, Jiao Y, Keays M, Lee YK, Kumari S, Mulvaney J, Olson A, Preece J, Stein J, Wei S, Weiser J, Huerta L, Petryszak R, Kersey P, Stein LD, Ware D, Jaiswal P (2016) Gramene database: navigating plant comparative genomics resources. Curr Plant Biol 7-8:10–15

Hamada T, Nagasaki-Takeuchi N, Kato T, Fujiwara M, Sonobe S, Fukao Y, Hashimoto T (2013) Purification and characterization of novel microtubule-associated proteins from Arabidopsis cell suspension cultures. Plant Physiol 163:1804–1816

Hawkins J, Boden M (2006) Detecting and sorting targeting peptides with neural networks and support vector machines. J Bioinforma Comput Biol 4:1–18

Hawkins J, Davis L, Boden M (2007) Predicting nuclear localization. J Proteome Res 6:1402–1409

Heard W, Sklenar J, Tome DF, Robatzek S, Jones AM (2015) Identification of regulatory and cargo proteins of endosomal and secretory pathways in Arabidopsis thaliana by proteomic dissection. Mol Cell Proteomics 14:1796–1813

Heazlewood JL, Millar AH (2005) AMPDB: the Arabidopsis mitochondrial protein database. Nucleic Acids Res 33:D605–D610

Heazlewood JL, Tonti-Filippini JS, Gout AM, Day DA, Whelan J, Millar AH (2004) Experimental analysis of the Arabidopsis mitochondrial proteome highlights signaling and regulatory components, provides assessment of targeting prediction programs, and indicates plant-specific mitochondrial proteins. Plant Cell 16:241–256

Heazlewood JL, Tonti-Filippini J, Verboom RE, Millar AH (2005) Combining experimental and predicted datasets for determination of the subcellular location of proteins in Arabidopsis. Plant Physiol 139:598–609

Heazlewood JL, Verboom RE, Tonti-Filippini J, Small I, Millar AH (2007) SUBA: the Arabidopsis subcellular database. Nucleic Acids Res 35:D213–D218

Heyndrickx KS, Vandepoele K (2012) Systematic identification of functional plant modules through the integration of complementary data sources. Plant Physiol 159:884–901

Hooper CM, Tanz SK, Castleden IR, Vacher MA, Small ID, Millar AH (2014) SUBAcon: a consensus algorithm for unifying the subcellular localization data of the Arabidopsis proteome. Bioinformatics 30:3356–3364

Hooper CM, Castleden IR, Aryamanesh N, Jacoby RP, Millar AH (2016) Finding the subcellular location of barley, wheat, rice and maize proteins: the compendium of crop proteins with annotated locations (cropPAL). Plant Cell Physiol 57:e9

Hooper CM, Castleden IR, Tanz SK, Aryamanesh N, Millar AH (2017a) SUBA4: the interactive data analysis centre for Arabidopsis subcellular protein locations. Nucleic Acids Res 45:D1064–D1074

Hooper CM, Stevens TJ, Saukkonen A, Castleden IR, Singh P, Mann GW, Fabre B, Ito J, Deery MJ, Lilley KS, Petzold CJ, Millar AH, Heazlewood JL, Parsons HT (2017b) Multiple marker abundance profiling: combining selected reaction monitoring and data-dependent acquisition for rapid estimation of organelle abundance in subcellular samples. Plant J 92:1202–1217

Hooper CM, Castleden IR, Aryamanesh N, Black K, Grasso SV, Millar AH (2020) CropPAL for discovering divergence in protein subcellular location in crops to support strategies for molecular crop breeding. Plant J 104:812

Horton P, Park KJ, Obayashi T, Fujita N, Harada H, Adams-Collier CJ, Nakai K (2007) WoLF PSORT: protein localization predictor. Nucleic Acids Res 35:W585–W587

Hua S, Sun Z (2001) Support vector machine approach for protein subcellular localization prediction. Bioinformatics 17:721–728

Huang S, Taylor NL, Narsai R, Eubel H, Whelan J, Millar AH (2009) Experimental analysis of the rice mitochondrial proteome, its biogenesis, and heterogeneity. Plant Physiol 149:719–734

Huang M, Friso G, Nishimura K, Qu X, Olinares PD, Majeran W, Sun Q, van Wijk KJ (2013) Construction of plastid reference proteomes for maize and Arabidopsis and evaluation of their orthologous relationships; the concept of orthoproteomics. J Proteome Res 12:491–504

Huh WK, Falvo JV, Gerke LC, Carroll AS, Howson RW, Weissman JS, O'Shea EK (2003) Global analysis of protein localization in budding yeast. Nature 425:686–691

Inze A, Vanderauwera S, Hoeberichts FA, Vandorpe M, Van Gaever T, Van Breusegem F (2012) A subcellular localization compendium of hydrogen peroxide-induced proteins. Plant Cell Environ 35:308–320

Ito J, Batth TS, Petzold CJ, Redding-Johanson AM, Mukhopadhyay A, Verboom R, Meyer EH, Millar AH, Heazlewood JL (2011) Analysis of the Arabidopsis cytosolic proteome highlights subcellular partitioning of central plant metabolism. J Proteome Res 10:1571–1582

Jiang JQ, Wu M (2012) Predicting multiplex subcellular localization of proteins using protein-protein interaction network: a comparative study. BMC Bioinformatics 13(Suppl 10):S20

Jiang ZH, Zhou XP, Tao M, Yuan F, Liu LL, Wu FH, Wu XM, Xiang Y, Niu Y, Liu F, Li CJ, Ye R, Byeon B, Xue Y, Zhao HY, Wang HN, Crawford BM, Johnson DM, Hu CX, Pei C, Zhou W, Swift GB, Zhang H, Vo-Dinh T, Hu ZL, Siedow JN, Pei ZM (2019) Plant cell-surface GIPC sphingolipids sense salt to trigger Ca2+ influx. Nature 572:341

Joshi HJ, Hirsch-Hoffmann M, Baerenfaller K, Gruissem W, Baginsky S, Schmidt R, Schulze WX, Sun Q, van Wijk KJ, Egelhofer V, Wienkoop S, Weckwerth W, Bruley C, Rolland N, Toyoda T, Nakagami H, Jones AM, Briggs SP, Castleden I, Tanz SK, Millar AH, Heazlewood JL (2011) MASCP Gator: an aggregation portal for the visualization of Arabidopsis proteomics data. Plant Physiol 155:259–270

Kamal AH, Cho K, Komatsu S, Uozumi N, Choi JS, Woo SH (2012) Towards an understanding of wheat chloroplasts: a methodical investigation of thylakoid proteome. Mol Biol Rep 39:5069–5083

Kersey PJ, Allen JE, Allot A, Barba M, Boddu S, Bolt BJ, Carvalho-Silva D, Christensen M, Davis P, Grabmueller C, Kumar N, Liu Z, Maurel T, Moore B,

McDowall MD, Maheswari U, Naamati G, Newman V, Ong CK, Paulini M, Pedro H, Perry E, Russell M, Sparrow H, Tapanari E, Taylor K, Vullo A, Williams G, Zadissia A, Olson A, Stein J, Wei S, Tello-Ruiz M, Ware D, Luciani A, Potter S, Finn RD, Urban M, Hammond-Kosack KE, Bolser DM, De Silva N, Howe KL, Langridge N, Maslen G, Staines DM, Yates A (2018) Ensembl Genomes 2018: an integrated omics infrastructure for non-vertebrate species. Nucleic Acids Res 46:D802–D808

Komatsu S, Wang X, Yin X, Nanjo Y, Ohyanagi H, Sakata K (2017) Integration of gel-based and gel-free proteomic data for functional analysis of proteins through Soybean Proteome Database. J Proteome 163:52–66

Koroleva OA, Tomlinson ML, Leader D, Shaw P, Doonan JH (2005) High-throughput protein localization in Arabidopsis using Agrobacterium-mediated transient expression of GFP-ORF fusions. Plant J 41:162–174

Kruft V, Eubel H, Jansch L, Werhahn W, Braun HP (2001) Proteomic approach to identify novel mitochondrial proteins in Arabidopsis. Plant Physiol 127:1694–1710

Lamesch P, Berardini TZ, Li D, Swarbreck D, Wilks C, Sasidharan R, Muller R, Dreher K, Alexander DL, Garcia-Hernandez M, Karthikeyan AS, Lee CH, Nelson WD, Ploetz L, Singh S, Wensel A, Huala E (2012) The Arabidopsis Information Resource (TAIR): improved gene annotation and new tools. Nucleic Acids Res 40:D1202–D1210

Lao J, Oikawa A, Bromley JR, McInerney P, Suttangkakul A, Smith-Moritz AM, Plahar H, Chiu TY, Gonzalez Fernandez-Nino SM, Ebert B, Yang F, Christiansen KM, Hansen SF, Stonebloom S, Adams PD, Ronald PC, Hillson NJ, Hadi MZ, Vega-Sanchez ME, Loque D, Scheller HV, Heazlewood JL (2014) The plant glycosyltransferase clone collection for functional genomics. Plant J 79:517–529

Larrainzar E, Wienkoop S (2017) A proteomic view on the role of legume symbiotic interactions. Front Plant Sci 8:1267

Lau BYC, Othman A, Ramli US (2018) Application of proteomics technologies in oil palm research. Protein J 37:473–499

Lee J, Lee H, Kim J, Lee S, Kim DH, Kim S, Hwang I (2011) Both the hydrophobicity and a positively charged region flanking the C-terminal region of the transmembrane domain of signal-anchored proteins play critical roles in determining their targeting specificity to the endoplasmic reticulum or endosymbiotic organelles in Arabidopsis cells. Plant Cell 23:1588–1607

Li L, Nelson CJ, Solheim C, Whelan J, Millar AH (2012) Determining degradation and synthesis rates of arabidopsis proteins using the kinetics of progressive 15N labeling of two-dimensional gel-separated protein spots. Mol Cell Proteomics 11(M111):010025

Li L, Nelson CJ, Trosch J, Castleden I, Huang S, Millar AH (2017) Protein degradation rate in arabidopsis thaliana leaf growth and development. Plant Cell 29:207–228

Lin TH, Murphy RF, Bar-Joseph Z (2011) Discriminative motif finding for predicting protein subcellular localization. IEEE/ACM Trans Comput Biol Bioinformatics 8:441–451

Lunn JE, Delorge I, Figueroa CM, Van Dijck P, Stitt M (2014) Trehalose metabolism in plants. Plant J 79:544–567

Mann GW, Calley PC, Joshi HJ and Heazlewood JL (2013) MASCP gator: an overview of the Arabidopsis proteomic aggregation portal. Front. Plant Sci. 4:411. https://doi.org/10.3389/fpls.2013.00411

Majeran W, Friso G, Asakura Y, Qu X, Huang M, Ponnala L, Watkins KP, Barkan A, van Wijk KJ (2012) Nucleoid-enriched proteomes in developing plastids and chloroplasts from maize leaves: a new conceptual framework for nucleoid functions. Plant Physiol 158:156–189

Matamoros MA, Kim A, Penuelas M, Ihling C, Griesser E, Hoffmann R, Fedorova M, Frolov A, Becana M (2018) Protein carbonylation and glycation in legume nodules. Plant Physiol 177:1510–1528

Matsuda S, Vert JP, Saigo H, Ueda N, Toh H, Akutsu T (2005) A novel representation of protein sequences for prediction of subcellular location using support vector machines. Protein Sci 14:2804–2813

McKenzie SD, Ibrahim IM, Aryal UK, Puthiyaveetil S (2020) Stoichiometry of protein complexes in plant photosynthetic membranes. BBA-Bioenergetics 1861:148141

de Michele R, McFarlane HE, Parsons HT, Meents MJ, Lao J, Gonzalez Fernandez-Nino SM, Petzold CJ, Frommer WB, Samuels AL, Heazlewood JL (2016) Free-flow electrophoresis of plasma membrane vesicles enriched by two-phase partitioning enhances the quality of the proteome from arabidopsis seedlings. J Proteome Res 15:900–913

Millar AH, Sweetlove LJ, Giege P, Leaver CJ (2001) Analysis of the Arabidopsis mitochondrial proteome. Plant Physiol 127:1711–1727

Millar AH, Carrie C, Pogson B, Whelan J (2009) Exploring the function-location nexus: using multiple lines of evidence in defining the subcellular location of plant proteins. Plant Cell 21:1625–1631

Mintz-Oron S, Meir S, Malitsky S, Ruppin E, Aharoni A, Shlomi T (2012) Reconstruction of Arabidopsis metabolic network models accounting for subcellular compartmentalization and tissue-specificity. Proc Natl Acad Sci U S A 109:339–344

Mitschke J, Fuss J, Blum T, Hoglund A, Reski R, Kohlbacher O, Rensing SA (2009) Prediction of dual protein targeting to plant organelles. New Phytol 183:224–235

Monaco MK, Stein J, Naithani S, Wei S, Dharmawardhana P, Kumari S, Amarasinghe V, Youens-Clark K, Thomason J, Preece J, Pasternak S, Olson A, Jiao Y, Lu Z, Bolser D, Kerhornou A, Staines D, Walts B, Wu G, D'Eustachio P, Haw R, Croft D, Kersey PJ, Stein L, Jaiswal P, Ware D (2014) Gramene 2013: comparative plant genomics resources. Nucleic Acids Res 42:D1193–D1199

Mueller SJ, Lang D, Hoernstein SN, Lang EG, Schuessele C, Schmidt A, Fluck M, Leisibach D, Niegl C, Zimmer AD, Schlosser A, Reski R (2014) Quantitative analysis of the mitochondrial and plastid proteomes of the moss Physcomitrella patens reveals protein macrocompartmentation and microcompartmentation. Plant Physiol 164:2081–2095

Mulvey CM, Breckels LM, Geladaki A, Britovsek NK, Nightingale DJH, Christoforou A, Elzek M, Deery MJ, Gatto L, Lilley KS (2017) Using hyperLOPIT to perform high-resolution mapping of the spatial proteome. Nat Protoc 12:1110–1135

Narsai R, Devenish J, Castleden I, Narsai K, Xu L, Shou H, Whelan J (2013) Rice DB: an Oryza Information Portal linking annotation, subcellular location, function, expression, regulation, and evolutionary information for rice and Arabidopsis. Plant J 76:1057–1073

Natera SH, Ford KL, Cassin AM, Patterson JH, Newbigin EJ, Bacic A (2008) Analysis of the Oryza sativa plasma membrane proteome using combined protein and peptide fractionation approaches in conjunction with mass spectrometry. J Proteome Res 7:1159–1187

Nelson BK, Cai X, Nebenfuhr A (2007) A multicolored set of in vivo organelle markers for co-localization studies in Arabidopsis and other plants. Plant J 51:1126–1136

Neuberger G, Maurer-Stroh S, Eisenhaber B, Hartig A, Eisenhaber F (2003) Prediction of peroxisomal targeting signal 1 containing proteins from amino acid sequence. J Mol Biol 328:581–592

Nguyen-Kim H, San Clemente H, Balliau T, Zivy M, Dunand C, Albenne C, Jamet E (2016) Arabidopsis thaliana root cell wall proteomics: increasing the proteome coverage using a combinatorial peptide ligand library and description of unexpected Hyp in peroxidase amino acid sequences. Proteomics 16:491–503

Nikolovski N, Rubtsov D, Segura MP, Miles GP, Stevens TJ, Dunkley TP, Munro S, Lilley KS, Dupree P (2012) Putative glycosyltransferases and other plant Golgi apparatus proteins are revealed by LOPIT proteomics. Plant Physiol 160:1037–1051

Nikolovski N, Shliaha PV, Gatto L, Dupree P, Lilley KS (2014) Label-free protein quantification for plant Golgi protein localization and abundance. Plant Physiol 166:1033–1043

Niu B, Jin YH, Feng KY, Lu WC, Cai YD, Li GZ (2008) Using AdaBoost for the prediction of subcellular location of prokaryotic and eukaryotic proteins. Mol Divers 12:41–45

One Thousand Plant Transcriptomes, I (2019) One thousand plant transcriptomes and the phylogenomics of green plants. Nature 574:679–685

Otto TD, Guimaraes AC, Degrave WM, de Miranda AB (2008) AnEnPi: identification and annotation of analogous enzymes. BMC Bioinformatics 9:544

Parsons HT, Christiansen K, Knierim B, Carroll A, Ito J, Batth TS, Smith-Moritz AM, Morrison S, McInerney P, Hadi MZ, Auer M, Mukhopadhyay A, Petzold CJ, Scheller HV, Loque D, Heazlewood JL (2012) Isolation and proteomic characterization of the arabidopsis golgi defines functional and novel components involved in plant cell wall biosynthesis. Plant Physiol 159:12

Peltier JB, Ytterberg J, Liberles DA, Roepstorff P, van Wijk KJ (2001) Identification of a 350-kDa ClpP protease complex with 10 different Clp isoforms in chloroplasts of Arabidopsis thaliana. J Biol Chem 276:16318–16327

Petsalaki EI, Bagos PG, Litou ZI, Hamodrakas SJ (2006) PredSL: a tool for the N-terminal sequence-based prediction of protein subcellular localization. Genom Proteom Bioinformatics 4:48–55

Pierleoni A, Martelli PL, Fariselli P, Casadio R (2006) BaCelLo: a balanced subcellular localization predictor. Bioinformatics 22:e408–e416

Piques M, Schulze WX, Hohne M, Usadel B, Gibon Y, Rohwer J, Stitt M (2009) Ribosome and transcript copy numbers, polysome occupancy and enzyme dynamics in Arabidopsis. Mol Syst Biol 5:314

Ploscher M, Reisinger V, Eichacker LA (2011) Proteomic comparison of etioplast and chloroplast protein complexes. J Proteome 74:1256–1265

Powers SK, Holehouse AS, Korasick DA, Schreiber KH, Clark NM, Jing HW, Emenecker R, Han S, Tycksen E, Hwang I, Sozzani R, Jez JM, Pappu RV, Strader LC (2019) Nucleo-cytoplasmic partitioning of ARF proteins controls auxin responses in Arabidopsis thaliana. Mol Cell 76:177

Prokisch H, Scharfe C, Camp DG II, Xiao W, David L, Andreoli C, Monroe ME, Moore RJ, Gritsenko MA, Kozany C, Hixson KK, Mottaz HM, Zischka H, Ueffing M, Herman ZS, Davis RW, Meitinger T, Oefner PJ, Smith RD, Steinmetz LM (2004) Integrative analysis of the mitochondrial proteome in yeast. PLoS Biol 2:e160

Reiland S, Grossmann J, Baerenfaller K, Gehrig P, Nunes-Nesi A, Fernie AR, Gruissem W, Baginsky S (2011) Integrated proteome and metabolite analysis of the de-etiolation process in plastids from rice (Oryza sativa L.). Proteomics 11:1751–1763

Reumann S, Quan S, Aung K, Yang P, Manandhar-Shrestha K, Holbrook D, Linka N, Switzenberg R, Wilkerson CG, Weber AP, Olsen LJ, Hu J (2009) In-depth proteome analysis of Arabidopsis leaf peroxisomes combined with in vivo subcellular targeting verification indicates novel metabolic and regulatory functions of peroxisomes. Plant Physiol 150:125–143

Reynoso MA, Juntawong P, Lancia M, Blanco FA, Bailey-Serres J, Zanetti ME (2015) Translating Ribosome Affinity Purification (TRAP) followed by RNA sequencing technology (TRAP-SEQ) for quantitative assessment of plant translatomes. Methods Mol Biol 1284:185–207

Salvato F, Havelund JF, Chen M, Rao RS, Rogowska-Wrzesinska A, Jensen ON, Gang DR, Thelen JJ, Moller IM (2014) The potato tuber mitochondrial proteome. Plant Physiol 164:637–653

Scheibe R, Backhausen JE, Emmerlich V, Holtgrefe S (2005) Strategies to maintain redox homeostasis during photosynthesis under changing conditions. J Exp Bot 56:1481–1489

Schein AI, Kissinger JC, Ungar LH (2001) Chloroplast transit peptide prediction: a peek inside the black box. Nucleic Acids Res 29:E82

Schenck CA, Last RL (2020) Location, location! cellular relocalization primes specialized metabolic diversification. FEBS J 287:1359–1368

Schneider S, Harant D, Bachmann G, Nagele T, Lang I, Wienkoop S (2019) Subcellular phenotyping: using proteomics to quantitatively link subcellular leaf protein and organelle distribution analyses of pisum sativum cultivars. Front Plant Sci 10:638

Schubert M, Petersson UA, Haas BJ, Funk C, Schroder WP, Kieselbach T (2002) Proteome map of the chloroplast lumen of Arabidopsis thaliana. J Biol Chem 277:8354–8365

Schwacke R, Ponce-Soto GY, Krause K, Bolger AM, Arsova B, Hallab A, Gruden K, Stitt M, Bolger ME, Usadel B (2019) MapMan4: a refined protein classification and annotation framework applicable to multi-omics data analysis. Mol Plant 12:879–892

Seaver SM, Henry CS, Hanson AD (2012) Frontiers in metabolic reconstruction and modeling of plant genomes. J Exp Biol 63:2247–2258

Senkler J, Senkler M, Eubel H, Hildebrandt T, Lengwenus C, Schertl P, Schwarzlander M, Wagner S, Wittig I, Braun HP (2017) The mitochondrial complexome of Arabidopsis thaliana. Plant J 89:1079–1092

Shah P, Powell AL, Orlando R, Bergmann C, Gutierrez-Sanchez G (2012) Proteomic analysis of ripening tomato fruit infected by Botrytis cinerea. J Proteome Res 11:2178–2192

Shen HB, Yang J, Chou KC (2007) Euk-PLoc: an ensemble classifier for large-scale eukaryotic protein subcellular location prediction. Amino Acids 33:57–67

Shen BR, Wang LM, Lin XL, Yao Z, Xu HW, Zhu CH, Teng HY, Cui LL, Liu EE, Zhang JJ, He ZH, Peng XX (2019) Engineering a new chloroplastic photorespiratory bypass to increase photosynthetic efficiency and productivity in rice. Mol Plant 12:199–214

Shi H, Schwender J (2016) Mathematical models of plant metabolism. Curr Opin Biotechnol 37:143–152

Shin CJ, Wong S, Davis MJ, Ragan MA (2009) Protein-protein interaction as a predictor of subcellular location. BMC Syst Biol 3:28

Shinano T, Yoshimura T, Watanabe T, Unno Y, Osaki M, Nanjo Y, Komatsu S (2013) Effect of phosphorus levels on the protein profiles of secreted protein and root surface protein of rice. J Proteome Res 12:4748–4756

Small I, Peeters N, Legeai F, Lurin C (2004) Predotar: a tool for rapidly screening proteomes for N-terminal targeting sequences. Proteomics 4:1581–1590

Sperschneider J, Catanzariti AM, DeBoer K, Petre B, Gardiner DM, Singh KB, Dodds PN, Taylor JM (2017) LOCALIZER: subcellular localization prediction of both plant and effector proteins in the plant cell. Sci Rep 7:44598

Stuart JM, Segal E, Koller D, Kim SK (2003) A gene-coexpression network for global discovery of conserved genetic modules. Science 302:249–255

Suliman M, Chateigner-Boutin AL, Francin-Allami M, Partier A, Bouchet B, Salse J, Pont C, Marion J, Rogniaux H, Tessier D, Guillon F, Larre C (2013) Identification of glycosyltransferases involved in cell wall synthesis of wheat endosperm. J Proteome 78:508–521

Sun Q, Zybailov B, Majeran W, Friso G, Olinares PD, van Wijk KJ (2009) PPDB, the plant proteomics database at Cornell. Nucleic Acids Res 37:D969–D974

Tabbita F, Pearce S, Barneix AJ (2017) Breeding for increased grain protein and micronutrient content in wheat: ten years of the GPC-B1 gene. J Cereal Sci 73:183–191

Tamura T, Akutsu T (2007) Subcellular location prediction of proteins using support vector machines with alignment of block sequences utilizing amino acid composition. BMC Bioinformatics 8:466

Tang H, Bowers JE, Wang X, Ming R, Alam M, Paterson AH (2008) Synteny and collinearity in plant genomes. Science 320:486–488

Tanz SK, Small I (2011) In silico methods for identifying organellar and suborganellar targeting peptides in Arabidopsis chloroplast proteins and for predicting the topology of membrane proteins. Methods Mol Biol 774:243–280

Tanz SK, Castleden I, Hooper CM, Vacher M, Small I, Millar HA (2013a) SUBA3: a database for integrating experimentation and prediction to define the SUBcellular location of proteins in Arabidopsis. Nucleic Acids Res 41:D1185–D1191

Tanz SK, Castleden I, Small ID, Millar AH (2013b) Fluorescent protein tagging as a tool to define the subcellular distribution of proteins in plants. Front Plant Sci 4:214

Taylor NL, Millar AH (2017) Isolation of plant organelles and structures: methods and protocols. In: Methods in molecular biology: Springer protocols. Springer, New York, NY

Taylor NL, Fenske R, Castleden I, Tomaz T, Nelson CJ, Millar AH (2014) Selected reaction monitoring to determine protein abundance in Arabidopsis using the Arabidopsis proteotypic predictor. Plant Physiol 164:525–536

Tello-Ruiz MK, Naithani S, Stein JC, Gupta P, Campbell M, Olson A, Wei S, Preece J, Geniza MJ, Jiao Y, Lee YK, Wang B, Mulvaney J, Chougule K, Elser J, Al-Bader N, Kumari S, Thomason J, Kumar V, Bolser DM, Naamati G, Tapanari E, Fonseca N, Huerta L, Iqbal H, Keays M, Munoz-Pomer Fuentes A, Tang A, Fabregat A, D'Eustachio P, Weiser J, Stein LD, Petryszak R, Papatheodorou I, Kersey PJ, Lockhart P, Taylor C, Jaiswal P, Ware D (2018) Gramene 2018: unifying comparative genomics and pathway resources for plant research. Nucleic Acids Res 46:D1181–D1189

Terasawa Y, Ito M, Tabiki T, Nagasawa K, Hatta K, Nishio Z (2016) Mapping of a major QTL associated with protein content on chromosome 2B in hard red winter wheat (Triticum aestivum L.). Breed Sci 66:471–480

Thompson A, Schafer J, Kuhn K, Kienle S, Schwarz J, Schmidt G, Neumann T, Johnstone R, Mohammed AK, Hamon C (2003) Tandem mass tags: a novel quantification strategy for comparative analysis of complex protein mixtures by MS/MS. Anal Chem 75:1895–1904

Uberegui E, Hall M, Lorenzo O, Schroder WP, Balsera M (2015) An Arabidopsis soluble chloroplast proteomic analysis reveals the participation of the Executer pathway in response to increased light conditions. J Exp Bot 66:2067–2077

Vinga S, Neves AR, Santos H, Brandt BW, Kooijman SA (2010) Subcellular metabolic organization in the context of dynamic energy budget and biochemical systems theories. Phil Trans R Soc Lond B Biol Sci 365:3429–3442

Waese J, Fan J, Pasha A, Yu H, Fucile G, Shi R, Cumming M, Kelley LA, Sternberg MJ, Krishnakumar V, Ferlanti E, Miller J, Town C, Stuerzlinger W, Provart NJ (2017) ePlant: visualizing and exploring multiple levels of data for hypothesis generation in plant biology. Plant Cell 29:1806–1821

Wang M, Weiss M, Simonovic M, Haertinger G, Schrimpf SP, Hengartner MO, von Mering C (2012) PaxDb, a database of protein abundance averages across all three domains of life. Mol Cell Proteomics 11:492–500

Wang M, Herrmann CJ, Simonovic M, Szklarczyk D, von Mering C (2015) Version 4.0 of PaxDb: protein abundance data, integrated across model organisms, tissues, and cell-lines. Proteomics 15:3163–3168

Weckwerth W, Baginsky S, van Wijk K, Heazlewood JL, Millar H (2008) The multinational Arabidopsis steering subcommittee for proteomics assembles the largest proteome database resource for plant systems biology. J Proteome Res 7:4209–4210

Werhahn W, Niemeyer A, Jansch L, Kruft V, Schmitz UK, Braun H (2001) Purification and characterization of the preprotein translocase of the outer mitochondrial membrane from Arabidopsis. Identification of multiple forms of TOM20. Plant Physiol 125:943–954

Wu S, Schalk M, Clark A, Miles RB, Coates R, Chappell J (2006) Redirection of cytosolic or plastidic isoprenoid precursors elevates terpene production in plants. Nat Biotechnol 24:1441–1447

Xing S, Meng X, Zhou L, Mujahid H, Zhao C, Zhang Y, Wang C, Peng Z (2016) Proteome profile of starch granules purified from rice (Oryza sativa) endosperm. PLoS One 11:e0168467

Xu L, Carrie C, Law SR, Murcha MW, Whelan J (2013) Acquisition, conservation, and loss of dual-targeted proteins in land plants. Plant Physiol 161:644–662

Yin L, Vener AV, Spetea C (2015) The membrane proteome of stroma thylakoids from Arabidopsis thaliana studied by successive in-solution and in-gel digestion. Physiol Plant 154:433–446

Yu NY, Wagner JR, Laird MR, Melli G, Rey S, Lo R, Dao P, Sahinalp SC, Ester M, Foster LJ, Brinkman FS (2010) PSORTb 3.0: improved protein subcellular localization prediction with refined localization subcategories and predictive capabilities for all prokaryotes. Bioinformatics 26:1608–1615

Zentella R, Zhang ZL, Park M, Thomas SG, Endo A, Murase K, Fleet CM, Jikumaru Y, Nambara E, Kamiya Y, Sun TP (2007) Global analysis of della direct targets in early gibberellin signaling in Arabidopsis. Plant Cell 19:3037–3057

Zybailov B, Rutschow H, Friso G, Rudella A, Emanuelsson O, Sun Q, van Wijk KJ (2008) Sorting signals, N-terminal modifications and abundance of the chloroplast proteome. PLoS One 3:e1994

The Contribution of Metabolomics to Systems Biology: Current Applications Bridging Genotype and Phenotype in Plant Science

5

Marina C. M. Martins, Valeria Mafra, Carolina C. Monte-Bello, and Camila Caldana

Abstract

Metabolomics is a valuable approach used to acquire comprehensive information about the set of metabolites in a cell or tissue, enabling a functional screen of the cellular activities in biological systems. Although metabolomics provides a more immediate and dynamic picture of phenotypes in comparison to the other omics, it is also the most complicated to measure because no single analytical technology can capture the extraordinary complexity of metabolite diversity in terms of structure and physical properties. Metabolomics has been extensively employed for a wide range of applications in plant science, which will be described in detail in this chapter. Among them, metabolomics is used for discriminating patterns of plant responses to genetic and environmental perturbations, as diagnostics and prediction tool to elucidate the function of genes for important and complex agronomic traits in crop species, and flux measurements are used to dissect the structure and regulatory properties of metabolic networks.

Keywords

Networks · Flux · Metabolites · Phenotype · Metabolism

5.1 Introduction and Overview of Plant Metabolomics

Metabolism is a complex, dynamic and highly integrated network of pathways driving the processes of assimilation, transport and chemical modification of small molecules. Its ultimate function is to maximize growth, survival and reproduction. Metabolites are organic compounds with low molecular weight (<1500 Da), and their properties and functionality dictate the chemistry of life (Fiehn et al. 2000; Fiehn 2002;

M. C. M. Martins (✉)
Departamento de Botânica, Instituto de Biociências, Universidade de São Paulo, São Paulo, Brazil
e-mail: martins.marina@usp.br

V. Mafra
Instituto Federal de Educação, Ciência e Tecnologia do Norte de Minas Gerais, Januária, Brazil

C. C. Monte-Bello
Universidade Estadual de Campinas, Campinas, Brazil

Laboratório Nacional de Ciência e Tecnologia do Bioetanol, Centro Nacional de Pesquisa em Energia e Materiais, Campinas, Brazil

Max Planck Institute of Molecular Plant Physiology, Potsdam, Germany
e-mail: cassano@mpimp-golm.mpg.de

C. Caldana
Max Planck Institute of Molecular Plant Physiology, Potsdam, Germany
e-mail: caldana@mpimp-golm.mpg.de

F. V. Winck (ed.), *Advances in Plant Omics and Systems Biology Approaches*, Advances in Experimental Medicine and Biology 1346, https://doi.org/10.1007/978-3-030-80352-0_5

Bino et al. 2004; Hall 2006). In the plant kingdom, more than 200,000 metabolites have been estimated (Dixon and Strack 2003; Afendi et al. 2012). These molecules are extremely diverse in their chemical structure and physical properties (e.g. polarity, volatility, size, and stability) and have a wide range of relative concentrations (Bino et al. 2004; Saito and Matsuda 2010; Jorge et al. 2016). In addition, plants possess a remarkable degree of compartmentation within their cells (Lunn 2007; Sweetlove and Fernie 2013), and the physical separation of metabolic pathways enables incompatible reactions to occur simultaneously within one cell and also prevents metabolic imbalances. Altogether, these features make the study of plant metabolism particularly challenging (Fiehn et al. 2000; Fiehn 2002; Saito and Matsuda 2010).

Traditionally, metabolites were grouped as 'primary' and 'secondary'. Primary metabolites are compounds that play essential roles in basic cell metabolism (e.g. amino acids, nucleotides, sugars, and lipids) and are required for proper growth and development. These substances are directly involved in the processes of photosynthesis, respiration, nutrient assimilation, and synthesis of macromolecules (Sulpice and McKeown 2015). Besides, plants have the ability to synthesize other compounds termed as 'secondary' or specialized metabolites, which were initially thought to be functionless end products of metabolism with no major significance for plant life (Pichersky and Gang 2000; Bourgaud et al. 2001). Also, unlike primary metabolites, which are found spread throughout the plant kingdom, secondary metabolites are often restricted to particular plant groups (Moore et al. 2014). However, many studies have demonstrated that secondary metabolites play crucial protective roles in the adaptation and survival of plants in different ecological niches and environmental conditions such as defence against herbivores, pests and pathogens (Nakabayashi and Saito 2015; Tenenboim and Brotman 2016). Plant secondary metabolites are extremely heterogeneous and diverse in functions, but can be divided into three major groups according to their biosynthetic pathways: terpenes and steroids, phenolics, and alkaloids (Harborne 1999). Although many metabolites were initially discovered through the study of discrete pathways, the metabolism operates as a systemic integrated network (Sweetlove et al. 2008; Stitt et al. 2010a) and efforts have been directed to increase the understanding of metabolic networks at a systems level.

The functions of metabolites are also correlated with their abundance, which reflects the balance between their rates of synthesis and degradation (Last et al. 2007). Examples include metabolites such as sucrose that is the major product of photosynthesis, the systemic form of transport sugar, and a signal molecule that responds to internal and external environmental cues altering development and stress acclimation (Rolland et al. 2006; Ruan 2014), compared to intermediates of the glycolytic pathway with low steady-state concentration and fast turnover rates (Lunn et al. 2014). Interestingly, some metabolites found in extremely low concentrations also play roles as critical signalling molecules modulating plant growth, development, and physiology. The levels of the sugar trehalose-6-phosphate (Tre6P), the intermediate of trehalose synthesis, act as a sensor of sucrose availability (Lunn et al. 2006, 2014; Yadav et al. 2014) impacting embryogenesis, leaf growth and flowering (Lunn et al. 2014).

Metabolite levels not only fluctuate according to the diel cycle, developmental stage, and environmental stimuli but also diverge between different organs, tissues and cells in a multicellular organism. It has been assumed that higher plants have about 40 distinct cell-types (e.g. trichomes, guard, xylem fibres, and parenchyma cells) (Martin et al. 2001; Misra et al. 2014) that exhibit different morphologies and play specialized functions in different plant organs. For example, a single leaf contains up to 15 different cell types (Martin et al. 2001), and each of them may display highly contrasting metabolite pools posing a difficult task to unbiased identify and quantify these small molecules (Bino et al. 2004; Hall and Hardy 2012).

Unlike DNA, transcripts, and proteins, metabolites are at the endpoint of the flow of genetic information and are considered as the nearest molecular readout of the phenotype–genotype relationship in a biological system (Fiehn 2002; Hall 2006). With the advent of post-genomic era, functional genomic studies have been greatly accelerated with technological advances in data generation from multiple levels (i.e. DNA sequences, transcripts, proteins and metabolites) using new 'omics' tools. While the well-established technologies for high-throughput DNA sequencing, gene expression analysis (transcriptomics), and protein profiling (proteomics) have been routinely adopted and explored in the last decades, metabolomics has developed into a powerful and complementary analytical technology for plant functional genomics in both basic and applied research (Fiehn 2002; Bino et al. 2004; Hall 2006; Roessner-Tunali 2007).

By definition, metabolomics aims an unbiased identification and quantification of all metabolites in a complex biological sample in a particular experimental condition (Fiehn et al. 2000; Fiehn 2002; Bino et al. 2004; Hall 2006). This well-accepted definition has some relevant questions. First, similar to other omics, metabolomics could be assayed in any level of complexity, such as a whole organism, a specific organ, tissue, a suspension cultured cell lines or even a single cell or cellular compartment depending on the biological question in investigation (Fiehn 2002). This aspect reveals the great potential of plant metabolomics to dissect the physiological state in a specific biological context, exploring all the richness and complexity of these small molecules (Hall 2006). Also, compared to transcriptomics and proteomics, metabolomics has the great advantage to be assayed independently from the availability of previous genome or transcriptome data, opening new perspectives to use plant metabolomics to gain more insights about the regulation of metabolism on complex traits in non-model and crops species (Hall 2006; Watanabe et al. 2018). Furthermore, metabolomics is the only instrumental tool allowing the true measurement of fluxes, essential to a comprehensive understanding of metabolism (Nikoloski et al. 2015; Allen 2016; Freund and Hegeman 2017).

Second, due to the dynamic nature of the metabolism, metabolomics experiments must be carefully planned and designed with respect to harvesting (e.g. sample type and size, pooling or not, replication, time scale) and sample preparation in order to obtain reliable and biologically relevant results. Some excellent literature are available to guide beginners in the field (Jenkins et al. 2004; Fiehn et al. 2007; Biais et al. 2012; Gibon and Rolin 2012; Hall and Hardy 2012; Lu et al. 2017).

Third, metabolomics is predicted to take a real picture of the total metabolic activities taking place in plants in a given experimental condition. However, practically, due to the myriad of metabolites with different physico-chemical properties, particularly for plants, no analytical method is able to simultaneously cover all the metabolites from a single extract (Fiehn et al. 2000; Bino et al. 2004; Hall 2006; Roessner-Tunali 2007; Hall and Hardy 2012). In this context, the field of plant metabolomics has greatly advanced with the development of multiple analytical approaches in parallel with the continuous increase in the plant metabolite databases to facilitate metabolite identification.

Basically, there are two general approaches to assess the overall metabolome: targeted and untargeted analyses (Fiehn 2002; Fernie 2003; Goodacre et al. 2004; Last et al. 2007). Targeted metabolomics allows an unbiased detection and quantification of a predefined set of known metabolites, usually applied to screen for selected compounds belonging to specific metabolic classes (Sawada et al. 2009; Cajka and Fiehn 2014). In contrast, untargeted metabolomics combines comprehensiveness with robustness to detect and quantify known and unidentified components. Untargeted analysis has been applied to large-scale profiling studies aiming at the identification of metabolite patterns or 'fingerprints' to discriminate different plant species/cultivars and in response to perturbations without the need of a formal metabolite identification (Keurentjes et al. 2006; Steinfath et al. 2010). In both targeted and untargeted strategies, coverage of detected metabolites and their obtained level of structural information and quantification (e.g. absolute, relative or semi-quantitative) rely on the purpose of the study that will influence the choice of the most appropriate metabolite extraction and analytical platform (Hall 2006; Saito and Matsuda 2010; Lei et al. 2011). A combination of targeted and untargeted metabolomics methods has been often used in the last years as a complementary strategy to address different biological questions in the same study (Last et al. 2007; Farag et al. 2012; Cajka and Fiehn 2014).

Due to the inherent complexity of plant metabolism, no 'silver bullet' or single technology is currently available to cover the full metabolome from a single sample. Technological developments in analytical methods to analyse highly complex mixtures have led to the establishment of two leading platforms applied to plant metabolomics research, namely nuclear magnetic resonance (NMR) spectroscopy and mass-spectrometry (MS), which can be coupled to gas (GC-MS, GC-NMR) or liquid (LC-MS or LC-NMR) chromatographic separation methods to improve resolution (Last et al. 2007; Kim et al. 2011; Tenenboim and Brotman 2016; Jorge et al. 2016). NMR-based metabolomics enables accurate quantification of abundant metabolites and resolution of chemical structures with a high reproducibility and relatively short time (Verpoorte et al. 2007; Kim and Verpoorte 2010; Schripsema 2010; Kim et al. 2011; Markley et al. 2017). Also, a great advantage is its simple sample preparation as metabolites can be measured directly from crude plant extracts or in vivo (Markley et al. 2017). However, the major drawbacks in NMR reside in its poor sensitivity and dynamic range of detection compared to MS, as well as problems related with superimposed spectrum signals that hamper the structural elucidation process, limiting the numbers of metabolites truly resolved (Kim et al. 2011; Markley et al. 2017).

Unlike NMR, MS is by far the primary detection method of metabolomics, due to its higher sensitivity, accuracy, and speed to detect and identify a wide range of metabolites (Last et al. 2007; Gika et al. 2014; Aretz and Meierhofer 2016; Haggarty and Burgess 2017). GC-MS has emerged as the gold-standard MS-based method for plant metabolite analysis due to numerous advantages compared to other analytical instruments such as robust quantification of hundreds of naturally volatile metabolites (e.g. alcohols, esters and monoterpenes) as well as non-volatile and polar metabolites (mainly primary metabolites), which can be converted into volatile and thermally stable compounds through derivatization (Hall 2006). Furthermore, GC-MS has a superior reproducibility and high chromatographic resolution over other analytical instruments (Fernie 2003; Jorge et al. 2016) and allows the development of metabolite libraries (Schauer et al. 2005; Kopka et al. 2005; Kind et al. 2009). Compared to GC-MS, LC-MS is a most versatile technique able to detect a broader range of compounds, being the preferred method of choice for targeted and untargeted analysis of secondary metabolites (Allwood and Goodacre 2010) or specific metabolite classes like phosphorylated compounds, which are less stable during the derivatization process required for GC-MS analysis (Hall 2006). Dedicated literature concerning pros and cons for each technology is available (Ward et al. 2007; Gika et al. 2014; Engskog et al. 2016; Aretz and Meierhofer 2016; Haggarty and Burgess 2017; Lu et al. 2017).

The choice of the analytical platform is a compromise and will be highly dependent on the biological question and availability of instruments or methods. However, metabolome coverage has greatly benefited from multiple analytical approaches (Marshall and Powers 2017) in parallel with continuous increasing in the plant metabolite databases to facilitate metabolite identification.

5.2 Applications of Metabolomics in Plant Sciences

Throughout the substantial advances in metabolomics, this technology has been extensively used as a cornerstone in systems biology to elucidate the link between genotype–phenotype in plants (Aretz and Meierhofer 2016). Deciphering biosynthetic pathways, their regulation and interactions are essential for understanding how plants respond to different sorts of perturbations (developmental, genetic or environmental). This is crucial for functional genomics, metabolic engineering, and synthetic biology approaches aiming at the accumulation of specific products (e.g. pharmacologically relevant metabolites) as well as plants with higher vigour and biomass for food and fuels. In this section, we will illustrate some of the broad potential applicability of metabolomics in plant science.

5.2.1 Pattern Recognition and Discrimination

Due to their autotrophic nature, plants are dependent on the light period to perform photosynthesis and usually accumulate carbon reserves to support growth and metabolic activity during the night (Smith and Stitt 2007; Stitt et al. 2010b). Time-resolved measurements of the metabolome along the diurnal cycle have been investigated in several species from algae to higher plants (Bénard et al. 2015; Hirth et al. 2017), showing that the amplitude and timing of metabolic changes vary. Primary metabolite and lipid profiling in synchronized growing cells of *Chlamydomonas reinhardtii* revealed interesting patterns along light and dark cycle: (1) most amino acids peak after 4 h of light coinciding with the commitment point of the cell cycle and (2) the turnover of membrane lipids (MGDG, SQDG and DGTS) is very distinctive from storage lipids (TAG) (Jüppner et al. 2017). In addition, these authors identified some new lipid species for this model microalgae and pinpointed metabolic signatures that can be used as biomarkers for several phases of the cell cycle. Metabolic profiling in the CAM species Agave indicated some differences along the diel cycle in comparison to *Arabidopsis*, not only in malate and fumarate, organic acids related to the nocturnal CO_2 fixation in CAM, but also in ascorbic acid known to play a role in redox signalling (Abraham et al. 2016).

Plants synthesize a plethora of value-added natural products with multiple applications to pharmaceutical, cosmetic, food, and agrochemical industries. Considering these bioactive molecules, the diversity and characterization of compound classes have been explored with metabolomics not only in model species (Li et al. 2016), but also in citrus (Wang et al. 2017), peach (Monti et al. 2016), yam (Price et al. 2017), pine (Meijón et al. 2016), wild grassland plants (French et al. 2018) and medicinal species (for review see Rai et al. 2017). In an outstanding study, the analysis of 17-hydroxygeranyllinalool diterpene glycosides in 35 solanaceous species identified 105 novel metabolites restricted to genera *Nicotiana*, *Capsicum*, and *Lycium*, indicating the potential of metabolomics to differentiate among species (Heiling et al. 2016). This work give evidence that MS metabolomics can be employed to evaluate phylogenetic occurrence of many secondary metabolic pathways.

With respect to the production of renewable fuels, algal biodiesel holds considerable promise to meet future energy demands. Microalgae have much faster growth rates than crops and are able to accumulate enormous amounts of lipids (from 20% to 40% of dry weight), mainly in the form of TAGs (Scranton et al. 2015; Wase et al. 2017). A vast collection of recent literature using metabolomics to identify lipid species in microalgae and evaluate factors influencing lipid accumulation is available (Yao et al. 2015; Bromke et al. 2015; Chen et al. 2017; Matich et al. 2018; Piligaev et al. 2018; Yang et al. 2018). A current challenge is to promote TAG accumulation and storage without penalties on biomass. GC-MS analysis of lipids and primary metabolites was utilized to test the effect of selected molecules from a high-throughput chemical genetics screening aiming to identify lipid-activating compounds in *C. reinhardtii* (Wase et al. 2017). These authors verified distinct metabolic response to five compounds that promoted TAG accumulation, four of them without decreasing galactolipids and their efficacy was also proved in three other algal species.

This example illustrates the value of metabolomics in assessing the response of plants to chemicals.

Metabolomics is also incredibly useful to recognize a wide range of other patterns that were not mentioned in this section, such as metabolic responses along plant development (Wang et al. 2016; Czedik-Eysenberg et al. 2016; Watanabe et al. 2018) and under stressful conditions that restrict growth (Obata and Fernie 2012; Arbona and Gomez-Cadenas 2016; Jorge et al. 2016). The latter has a huge impact on agriculture due to the identification of markers for increased stress tolerance.

5.2.2 Functional Genomics

Mutants and transgenic lines are excellent tools to determine gene function in plant morphology, biochemistry, and physiology. Metabolomics is very powerful to distinguish among genotypes even in the absence of growth phenotypes (Fukushima et al. 2014b), boosting functional readouts in comparison to classical chemical or genetic screens evaluating growth responses. Therefore, it is routinely employed for characterizing mutants and genetically modified (GM) lines.

Arabidopsis thaliana was the first plant genome to be completely sequenced, and although there are vast genetic resources for this species, only about 12% of gene function assignments were based on in vivo characterization (Rhee and Mutwil 2014). T-DNA sequence-indexed mutant collections have enabled allele coverage for most *Arabidopsis* genes (O'Malley et al. 2015), serving as basis for both forward and reverse genetic strategies. Metabolomics has been employed to determine the metabolomes of several lines containing T-DNA insertions in genes of unknown functions. A combination of various analytical platforms (including LC-MS, CE-MS, UHPLC-QTOF-MS, and GC–TOF-MS) was used to analyse 69 mutants, ensuring detection of important metabolic alterations and creation of a public database (Quanbeck et al. 2012). In a recent work, Monne et al. (2018) have biochemically characterized the properties of recombinant mitochondrial carriers previously thought to be uncoupling proteins 1 and 2, and detected their ability to transport amino acids. GC-MS metabolite profiling in T-DNA insertion mutants confirmed massive changes in organic and amino acids, enabling to assign a new function for these proteins as aspartate and glutamate transporters.

The combination of multiple analytical platforms revealed minimal or no clear metabolic differences between conventional and GM lines of tomato (Kusano et al. 2011) and soybean (Kusano et al. 2015), respectively, showing that metabolomics is also valuable to analyse risk assessment of GM crops.

5.2.3 Metabolomics as a Prediction Tool

Improving crop productivity has been a major issue concerning growing world population and climate change (White et al. 2016; van der Kooi et al. 2016; Shih et al. 2016; Altieri and Nicholls 2017; Frieler et al. 2017). As the composite of metabolic reactions represent the outcome of determinant genes generating the phenotype, metabolomics has contributed to improve the understanding of the genetic architecture and the key elements underlying biological functions and agronomic traits (Kumar et al. 2017). Attributes such as quality, shelf life, biomass production, yield, and resistance to diseases are controlled by multiple genes, and their genomic regions are known as quantitative trait loci (QTLs) (Collard et al. 2005). QTL mapping reveals the localization of loci, enabling the identification of coregulated compounds in naturally variable phenotypes (Keurentjes et al. 2006), with specific impact on crop breeding. However, many traits are controlled by a large number of QTLs (Bernardo 2008; Xu and Crouch 2008), which also have strong interactions with the environment. Metabolomics has greatly assisted genetic analyses to clarify the relationship between genetic and biochemical bases of plant metabolism (Fernie and Tohge 2017), serving as a tool to increase breeding efficiency. The pioneer works on metabolite-based QTL (mQTL) were performed with *Arabidopsis* (Meyer et al. 2007, 2010; Lisec et al. 2008, 2009) and tomato (Schauer et al. 2006), aiming to predict biomass production. These works opened new perspectives for using metabolites as biomarkers for accurate estimation of plant performance based on parental information (for review see Fernandez et al. 2016), and since then, several studies in rice (Matsuda et al. 2012; Dan et al. 2016), potato (Sprenger et al. 2017), tomato (Quadrana et al. 2014; Toubiana et al. 2015), wheat (Hill et al. 2015), and other crops have been conducted. In general, those works provide hints on heritable mechanisms affecting the levels of metabolites, show that various mQTLs have a strong influence on metabolite levels and pinpoint mQTL hotspots, suggesting that modification of small genomic regions could control the metabolic status. Depending on the density of the genetic map, it is even possible to identify candidate genes involved in particular pathways. Gong et al. (2013) successfully assigned the function of genes to many mQTLs related to flavonoid metabolism and other mQTLs of unknown functions in rice. Moreover, they performed functional characterization of three candidate genes confirming their relationship to the accumulation of the corresponding metabolites and could also reconstruct some metabolic pathways.

High-throughput genotyping technologies have revolutionized genome-wide association studies (GWAS), another method suitable for mapping the loci responsible for natural variations in a phenotype of interest. GWAS focus on the identification of significantly associated genetic polymorphisms in a large population and has some advantages in comparison to traditional QTL mapping (Korte and Farlow 2013). Metabolomics has also been combined with GWAS originating high-resolution maps of genomic regions related with metabolite variation (Luo 2015; Fernie and Tohge 2017). A comprehensive study of maize kernel metabolism combined metabolomics analysis by LC-MS/MS and GWAS in an association panel in different locations (Wen et al. 2014). The results made it possible to verify and update the annotation of many maize genes through the identification of novel metabolites and genes involved in the formation of phenolamides and flavonoids, and also to explore biomarkers for kernel weight. Other few recent examples are (1) evaluation of metabolites in maize roots and identification and validation of a terpene synthase gene that plays a role in antifungal defence (Ding et al. 2017) and (2) discovery of candidate genes contributing to steroidal glycoalkaloid and flavonoid metabolism in tomato fruit along domestication, with some of the genes annotated and characterized (Zhu et al. 2018).

Metabolomics has also been employed solely to investigate the relationship between biochemical characteristics and geographic origins, genotypic characteristics and morphological traits in seeds of 100 cultivars of *japonica* and *indica* rice (Hu et al. 2014). Non-targeted UHPLC-MS/MS and GC-MS revealed opposite abundance of some metabolites (e.g. asparagine and alanine) between *japonica* and *indica* cultivars, suggesting different strategies for nitrogen utilization in rice seeds. Few significantly different metabolite and morphological trait correlations between the two subgroups indicated that they tend to be subspecies-specific (Hu et al. 2014). Another study in a panel of sorghum breeding lines determined associations between metabolites in leaves and morpho-physiological traits, revealing that chlorogenic and shikimic acids are related to photosynthesis, initial plant growth, and final biomass (Turner et al. 2016). Together, the above-mentioned studies are examples of the building bases for ameliorating agronomic traits in crops.

5.2.4 Flux Analysis

Although steady-state measurements of metabolites are very valuable for giving a general overview of metabolic alterations in response to a defined perturbation, they do not provide detailed information about flux distributions. Therefore, conventional metabolomics and flux analysis are complementary approaches for characterizing the plant metabolic network. Metabolic reactions are catalysed by enzymes and depend on the concentration of substrate and end products. On another hand, metabolites can regulate enzyme activity at several levels, from allosteric to transcriptional regulation (Wegner et al. 2015). A large number of metabolites are intermediates of branched and circular metabolic pathways, and frequently metabolite levels and enzyme activities have only poor correlations with transcripts or proteins (Gibon et al. 2004; Piques et al. 2009; Stitt and Gibon 2014), which also do not correlate with fluxes (Fernie and Stitt 2012; Schwender et al. 2014). Those findings place posttranslational modifications of enzymes as regulatory events integrating signalling, gene expression, and metabolism (Grabsztunowicz et al. 2017; O'Leary and Plaxton 2017).

Fluxes are challenging to determine because no simple methodology is able to follow the dynamic rate of metabolite interconversions or the intracellular activity of multiple enzymes (Kruger and Ratcliffe 2015). Flux analyses make it possible to determine metabolic pathways that are actively operating and how their activity is coordinated with additional pathways to establish a balanced network (Nikoloski et al. 2015). This information can be used to estimate optimal configuration for a network and fluxes for the production of interesting end-products (Farre et al. 2014). The measurement of metabolome-wide fluxes is an emerging field contributing to a more integrated output of cellular function (Salon et al. 2017).

The use of isotope labelling with radioactive or stable isotopes is a classical biochemical technique for measuring intracellular fluxes (Freund and Hegeman 2017) and is known as metabolic flux analysis (MFA). Briefly, MFA consists of monitoring the redistribution of the labelled compound in a large number of metabolites using MS or NMR, building a model of the network, fitting the model to the MS or NMR data in order to obtain a set of fluxes, and extensive statistics to evaluate the reliability of the estimated flux (Kruger et al. 2012; Kruger and Ratcliffe 2015; Allen et al. 2015; Salon et al. 2017). MS enables resolving fragments or complete isotopic composition of a metabolite, whereas NMR allows to measure positional labelling information. It is worthwhile mentioning that some elements must be taken into consideration when performing this sort of experiment, involving the labelling magnitude of the precursor substrate molecule through the system, the size of the metabolite pool and the conversion rate of the precursor substrate into the metabolite (Nikoloski et al. 2015). In the last years, various protocols to perform MFA in plants have been described (Cocuron and Alonso 2014; Heise et al. 2014; Tivendale et al. 2016; Dethloff et al. 2017; Obata et al. 2017; Acket et al. 2017). Stable isotope-labelling experiments with ^{13}C-pyruvate, ^{13}C-glutamate and ^{15}N-ammonium were used to evaluate a switch of the tricarboxylic acid cycle to a noncyclic operation mode under hypoxia in soybean (António et al. 2016). The monitoring of label redistribution with GC-TOF-MS showed that metabolic alterations were independent from the supply of isotope-labelled substrate and accumulation of alanine, GABA, and succinate occur due to activation of alanine metabolism and GABA shunt.

The other approach typically used to estimate fluxes is a constraint model combining genomic information and biochemical data to predict metabolic fluxes through the network, namely flux balance analysis (FBA). As FBA demands fewer measurements, it is often easier to implement than MFA (Kruger and Ratcliffe 2015). FBA is frequently employed to predict fluxes to maximize biomass production or minimize energy consumption (Colombie et al. 2015; Yuan et al. 2016), and substantial progress in plant metabolic modelling has been achieved in recent years (Shi and Schwender 2016). The power of FBA prediction was confirmed comparing flux profiles between guard and mesophyll leaf cells. Modelling predicted a C4-like metabolism in guard cells (due to higher anaplerotic CO_2 fixation into oxaloacetate) and higher fluxes through sucrose synthesis as a result of a futile cycle, which could be confirmed with a ^{13}C-labelling experiment using isolated mesophyll and guard cells (Robaina-Estévez et al. 2017). This study demonstrates the application of FBA to investigate different cellular types.

5.2.5 Integration with Other Omics

The integration of metabolomics with other high-throughput technologies permits a more holistic view of biological phenomena, as exemplified by the mQTL and GWAS studies above mentioned. Another case is the investigation of transcripts and metabolites in duckweed, the smallest and fastest growing aquatic flowering plants, aimed at elucidating the phenotype of starch accumulation under nitrogen starvation. Duckweeds are able to accumulate impressive amounts of starch, evidencing their potential for bioethanol production (Xu et al. 2011; Cui and Cheng 2015; Fujita et al. 2016). RNASeq analysis hypothesized more partitioning into starch due to the up-regulation of enzymes involved in gluconeogenesis and down-regulation of glycolysis, as well as alterations in genes coding for enzymes of starch and sucrose synthesis (Yu et al. 2017). Metabolite profiling by LC-MS/MS confirmed higher ADP-glucose and lower UDP-glucose amounts, substrates for starch and sucrose synthesis, respectively, and enzymatic activity of the enzymes producing these substrates was also in agreement with transcript and metabolic data. Only due to the integration of the different information levels, it was possible to confirm that the increased starch content was a consequence of increased output from gluconeogenesis and TCA pathways (Yu et al. 2017).

By combining photosynthetic rate, measurements of metabolites, transcripts and proteins, polysome loading and growth analysis, it was possible to achieve a systemic response of metabolism and growth after a shift to higher irradiance in the non-saturating range for photosynthesis in the algal *C. reinhardtii* (Mettler et al. 2014). This temporal analysis revealed an initial increase in photosynthesis prior to stimulation of growth to match increased carbon fixation, and higher metabolic fluxes leading to accumulation of metabolic intermediates and starch. Transcriptional and posttranscriptional regulation were found to be important after primary changes in metabolites, leading to alterations in the abundance of particular proteins, which also brought about subsequently changes in the levels of metabolites. This is an outstanding work showing that the different levels of information present very distinct temporal kinetics, and are orchestrated to ensure fast readjustment of metabolism in a fluctuating light environment.

Usually, the integration of data from two system-levels is primarily made on simple correlations methods (Rajasundaram and Selbig 2016). However, several statistical methods and tools are available for network visualization, pathway analyses, genome-scale metabolic reconstruction and integration of multidimensional data (Rohn et al. 2012; Bartel et al. 2013; Fukushima et al. 2014a; Villaveces et al. 2015; Bersanelli et al. 2016; Sajitz-Hermstein et al. 2016; Schwahn et al. 2017; Therrien-Laperrière et al. 2017; Robaina-Estevez and Nikoloski 2017; Basu et al. 2017).

The use of biological networks for integrative analysis offers new directions to identify how large networks are coregulated. More recently, integrative approaches were shown to provide systemic views of plant defence against insects (Barah and Bones 2015), secondary wall formation (Li et al. 2016), structure and regulation of metabolic pathways (Tohge et al. 2015), hormone signalling (Yoshida et al. 2015), and single cells (Colomé-Tatché and Theis 2018). The integration of multi-omics data has expanded the mechanistic comprehension of plant metabolism and function.

5.3 Final Considerations and Future Perspectives

Since the appearance of metabolomics almost two decades ago, higher resolution analytical platforms and their use in combination have enabled the detection of hundreds of metabolic features within a complex biological sample. However, a significant portion of these detected peaks usually cannot be identified, hindering the accomplishment of a complete metabolome. The elucidation of new metabolites is still very laborious and remains an enormous challenge. Serial combination of columns in tandem and column switching are means to improve metabolome coverage. In addition to the technological advances, efforts in sharing reference compounds and organization of metabolite spectral signatures in public libraries, as well as standardization of protocols to report metabolite data will definitely increase identification confidence and take a leap forward in the use of metabolomics as discovery tool.

Another bottleneck in metabolomics is highly compartmentalization of plant metabolism with a range of biochemical steps in a single pathway taking place in different cellular organelles and/or being catalysed by isoforms of enzymes at different subcellular locations. Strategies to track spatial distribution of metabolites and proteins include isolation or organelles, fractionation techniques, immunohistochemistry and the powerful flux analyses, which has increased the understanding about how metabolic pathways are integrated. These approaches together with natural variation might unravel crucial metabolic modules contributing for efficient manipulation of plant metabolism via metabolic engineering.

There is a growing interest in using metabolomics for a wide range of biological targets, and although it has still some limitations, metabolomics use alone or combined with other omics technologies is revolutionizing plant biology and crop breeding providing new insights into genetic regulation of metabolism, cellular function and the structure of metabolic networks.

References

Abraham PE, Yin H, Borland AM et al (2016) Transcript, protein and metabolite temporal dynamics in the CAM plant Agave. Nat Plants 2:16178

Acket S, Degournay A, Merlier F, Thomasset B (2017) 13C labeling analysis of sugars by high resolution-mass spectrometry for metabolic flux analysis. Anal Biochem 527:45–48

Afendi FM, Okada T, Yamazaki M et al (2012) KNApSAcK family databases: integrated metabolite–plant species databases for multifaceted plant research. Plant Cell Physiol 53:e1

Allen DK (2016) Quantifying plant phenotypes with isotopic labeling & metabolic flux analysis. Curr Opin Biotechnol 37:45–52

Allen DK, Bates PD, Tjellstrom H (2015) Tracking the metabolic pulse of plant lipid production with isotopic labeling and flux analyses: past, present and future. Prog Lipid Res 58:97–120

Allwood JW, Goodacre R (2010) An introduction to liquid chromatography mass spectrometry instrumentation applied in plant metabolomic analyses. Phytochem Anal 21:33–47

Altieri MA, Nicholls CI (2017) The adaptation and mitigation potential of traditional agriculture in a changing climate. Clim Chang 140:33–45

António C, Päpke C, Rocha M et al (2016) Regulation of primary metabolism in response to low oxygen availability as revealed by carbon and nitrogen isotope redistribution. Plant Physiol 170:43–56

Arbona V, Gomez-Cadenas A (2016) Metabolomics of disease resistance in crops. Curr Issues Mol Biol 19:13–30

Aretz I, Meierhofer D (2016) Advantages and pitfalls of mass spectrometry based metabolome profiling in systems biology. Int J Mol Sci 17:632

Barah P, Bones AM (2015) Multidimensional approaches for studying plant defence against insects: from ecology to omics and synthetic biology. J Exp Bot 66:479–493

Bartel J, Krumsiek J, Theis FJ (2013) Statistical methods for the analysis of high-throughput metabolomics data. Comput Struct Biotechnol J 4:e201301009

Basu S, Duren W, Evans CR et al (2017) Sparse network modeling and metscape-based visualization methods for the analysis of large-scale metabolomics data. Bioinformatics 33:1545–1553

Bénard C, Bernillon S, Biais B et al (2015) Metabolomic profiling in tomato reveals diel compositional changes in fruit affected by source–sink relationships. J Exp Bot 66:3391–3404

Bernardo R (2008) Molecular markers and selection for complex traits in plants: learning from the last 20 years. Crop Sci 48:1649–1664

Bersanelli M, Mosca E, Remondini D et al (2016) Methods for the integration of multi-omics data: mathematical aspects. BMC Bioinformatics 17:S15

Biais B, Bernillon S, Deborde C et al (2012) Precautions for harvest, sampling, storage, and transport of crop plant metabolomics samples. In: Hardy NW, Hall RD (eds) Plant metabolomics. Methods in molecular biology (methods and protocols). Humana Press, Totowa, NJ, pp 51–63

Bino RJ, Hall RD, Fiehn O et al (2004) Potential of metabolomics as a functional genomics tool. Trends Plant Sci 9:418–425

Bourgaud F, Gravot A, Milesi S, Gontier E (2001) Production of plant secondary metabolites: a historical perspective. Plant Sci 161:839–851

Bromke MA, Sabir JS, Alfassi FA et al (2015) Metabolomic profiling of 13 diatom cultures and their adaptation to nitrate-limited growth conditions. PLoS One 10:e0138965

Cajka T, Fiehn O (2014) Comprehensive analysis of lipids in biological systems by liquid chromatography-mass spectrometry. Trends Anal Chem 61:192–206

Chen B, Wan C, Mehmood MA et al (2017) Manipulating environmental stresses and stress tolerance of microal-

gae for enhanced production of lipids and value-added products–a review. Bioresour Technol 244:1198–1206
Cocuron J-C, Alonso AP (2014) Liquid chromatography tandem mass spectrometry for measuring (13)C-labeling in intermediates of the glycolysis and pentose phosphate pathway. Methods Mol Biol 1090:131–142
Collard BCY, Jahufer MZZ, Brouwer JB, Pang ECK (2005) An introduction to markers, quantitative trait loci (QTL) mapping and marker-assisted selection for crop improvement: the basic concepts. Euphytica 142:169–196
Colombie S, Nazaret C, Benard C et al (2015) Modelling central metabolic fluxes by constraint-based optimization reveals metabolic reprogramming of developing Solanum lycopersicum (tomato) fruit. Plant J 81:24–39
Colomé-Tatché M, Theis FJ (2018) Statistical single cell multi-omics integration. Curr Opin Syst Biol 7:54–59
Cui W, Cheng JJ (2015) Growing duckweed for biofuel production: a review. Plant Biol 17:16–23
Czedik-Eysenberg A, Arrivault S, Lohse MA et al (2016) The interplay between carbon availability and growth in different zones of the growing maize leaf. Plant Physiol 172:943–967
Dan Z, Hu J, Zhou W et al (2016) Metabolic prediction of important agronomic traits in hybrid rice (Oryza sativa L.). Sci Rep 6:21732
Dethloff F, Orf I, Kopka J (2017) Rapid in situ (13)C tracing of sucrose utilization in Arabidopsis sink and source leaves. Plant Methods 13:87
Ding Y, Huffaker A, Kollner TG et al (2017) Selinene volatiles are essential precursors for maize defense promoting fungal pathogen resistance. Plant Physiol 175:1455–1468
Dixon RA, Strack D (2003) Phytochemistry meets genome analysis, and beyond. Phytochemistry 62:815–816
Engskog MKR, Haglöf J, Arvidsson T, Pettersson C (2016) LC–MS based global metabolite profiling: the necessity of high data quality. Metabolomics 12:114
Farag MA, Porzel A, Wessjohann LA (2012) Comparative metabolite profiling and fingerprinting of medicinal licorice roots using a multiplex approach of GC-MS, LC-MS and 1D NMR techniques. Phytochemistry 76:60–72
Farre G, Blancquaert D, Capell T et al (2014) Engineering complex metabolic pathways in plants. Annu Rev Plant Biol 65:187–223
Fernandez O, Urrutia M, Bernillon S et al (2016) Fortune telling: metabolic markers of plant performance. Metabolomics 12:158
Fernie AR (2003) Metabolome characterization in plant system analysis. Funct Plant Biol 30:111–120
Fernie AR, Stitt M (2012) On the discordance of metabolomics with proteomics and transcriptomics: coping with increasing complexity in logic, chemistry, and network interactions scientific correspondence. Plant Physiol 158:1139–1145
Fernie AR, Tohge T (2017) The genetics of plant metabolism. Annu Rev Genet 51:287–310
Fiehn O (2002) Functional genomics. In: Functional genomics. Springer Netherlands, Dordrecht, pp 155–171
Fiehn O, Kopka J, Dörmann P et al (2000) Metabolite profiling for plant functional genomics. Nat Biotechnol 18:1157–1161
Fiehn O, Robertson D, Griffin J et al (2007) The metabolomics standards initiative (MSI). Metabolomics 3:175–178. https://doi.org/10.1007/s11306-007-0070-6
French KE, Harvey J, McCullagh JSO (2018) Targeted and untargeted metabolic profiling of wild grassland plants identifies antibiotic and anthelmintic compounds targeting pathogen physiology, metabolism and reproduction. Sci Rep 8:1695
Freund DM, Hegeman AD (2017) Recent advances in stable isotope-enabled mass spectrometry-based plant metabolomics. Curr Opin Biotechnol 43:41–48
Frieler K, Schauberger B, Arneth A et al (2017) Understanding the weather signal in national crop-yield variability. Earths Fut 5:605–616
Fujita T, Nakao E, Takeuchi M et al (2016) Characterization of starch-accumulating duckweeds, Wolffia globosa, as renewable carbon source for bioethanol production. Biocatal Agric Biotechnol 6:123–127
Fukushima A, Kanaya S, Nishida K (2014a) Integrated network analysis and effective tools in plant systems biology. Front Plant Sci 5:598
Fukushima A, Kusano M, Mejia RF et al (2014b) Metabolomic characterization of knockout mutants in Arabidopsis: development of a metabolite profiling database for knockout mutants in Arabidopsis. Plant Physiol 165:948–961
Gibon Y, Rolin D (2012) Aspects of experimental design for plant metabolomics experiments and guidelines for growth of plant material. Methods Mol Biol 860:13–30
Gibon Y, Blaesing OE, Hannemann J et al (2004) A robot-based platform to measure multiple enzyme activities in Arabidopsis using a set of cycling assays: comparison of changes of enzyme activities and transcript levels during diurnal cycles and in prolonged darkness. Plant Cell 16:3304–3325
Gika HG, Wilson ID, Theodoridis GA (2014) LC-MS-based holistic metabolic profiling. Problems, limitations, advantages, and future perspectives. J Chromatogr B, Anal Technol Biomed Life Sci 966:1–6
Gong L, Chen W, Gao Y et al (2013) Genetic analysis of the metabolome exemplified using a rice population. Proc Natl Acad Sci U S A 110:20320–20325
Goodacre R, Vaidyanathan S, Dunn WB et al (2004) Metabolomics by numbers: acquiring and understanding global metabolite data. Trends Biotechnol 22:245–252
Grabsztunowicz M, Koskela MM, Mulo P (2017) Post-translational modifications in regulation of chloroplast function: recent advances. Front Plant Sci 8:240
Haggarty J, Burgess KEV (2017) Recent advances in liquid and gas chromatography methodology for extending coverage of the metabolome. Curr Opin Biotechnol 43:77–85

Hall RD (2006) Plant metabolomics: from holistic hope, to hype, to hot topic. New Phytol 169:453–468
Hall RD, Hardy NW (2012) Practical applications of metabolomics in plant biology. In: Methods in molecular biology. Humana Press, Totowa, NJ, pp 1–10
Harborne JB (1999) Classes and functions of secondary products from plants. In: Chemicals from plants. World Scientific; Imperial College Press, Singapore; London, pp 1–25
Heiling S, Khanal S, Barsch A et al (2016) Using the knowns to discover the unknowns: MS-based dereplication uncovers structural diversity in 17-hydroxygeranyllinalool diterpene glycoside production in the Solanaceae. Plant J 85:561–577
Heise R, Arrivault S, Szecowka M et al (2014) Flux profiling of photosynthetic carbon metabolism in intact plants. Nat Protoc 9:1803–1824
Hill CB, Taylor JD, Edwards J et al (2015) Detection of QTL for metabolic and agronomic traits in wheat with adjustments for variation at genetic loci that affect plant phenology. Plant Sci 233:143–154
Hirth M, Liverani S, Mahlow S et al (2017) Metabolic profiling identifies trehalose as an abundant and diurnally fluctuating metabolite in the microalga Ostreococcus tauri. Metabolomics 13:68
Hu C, Shi J, Quan S et al (2014) Metabolic variation between japonica and indica rice cultivars as revealed by non-targeted metabolomics. Sci Rep 4:5067
Jenkins H, Hardy N, Beckmann M et al (2004) A proposed framework for the description of plant metabolomics experiments and their results. Nat Biotechnol 22:1601–1606
Jorge TF, Rodrigues JA, Caldana C et al (2016) Mass spectrometry-based plant metabolomics: metabolite responses to abiotic stress. Mass Spectrom Rev 35:620–649
Jüppner J, Mubeen U, Leisse A et al (2017) Dynamics of lipids and metabolites during the cell cycle of Chlamydomonas reinhardtii. Plant J 92:331–343
Keurentjes JJB, Fu J, de Vos CHR et al (2006) The genetics of plant metabolism. Nat Genet 38:842–849
Kim HK, Verpoorte R (2010) Sample preparation for plant metabolomics. Phytochem Anal 21:4–13
Kim HK, Choi YH, Verpoorte R (2011) NMR-based plant metabolomics: where do we stand, where do we go? Trends Biotechnol 29:267–275
Kind T, Wohlgemuth G, Lee DY et al (2009) FiehnLib: mass spectral and retention index libraries for metabolomics based on quadrupole and time-of-flight gas chromatography/mass spectrometry. Anal Chem 81:10038–10048
van der Kooi CJ, Reich M, Löw M et al (2016) Growth and yield stimulation under elevated CO2 and drought: a meta-analysis on crops. Environ Exp Bot 122:150–157
Kopka J, Schauer N, Krueger S et al (2005) GMD@CSB.DB: the Golm metabolome database. Bioinformatics 21:1635–1638
Korte A, Farlow A (2013) The advantages and limitations of trait analysis with GWAS: a review. Plant Methods 9:29
Kruger NJ, Ratcliffe RG (2015) Fluxes through plant metabolic networks: measurements, predictions, insights and challenges. Biochem J 465:27–38
Kruger NJ, Masakapalli SK, Ratcliffe RG (2012) Strategies for investigating the plant metabolic network with steady-state metabolic flux analysis: lessons from an Arabidopsis cell culture and other systems. J Exp Bot 63:2309–2323
Kumar R, Bohra A, Pandey AK et al (2017) Metabolomics for plant improvement: status and prospects. Front Plant Sci 8:1302
Kusano M, Redestig H, Hirai T et al (2011) Covering chemical diversity of genetically-modified tomatoes using metabolomics for objective substantial equivalence assessment. PLoS One 6:e16989
Kusano M, Baxter I, Fukushima A et al (2015) Assessing metabolomic and chemical diversity of a soybean lineage representing 35 years of breeding. Metabolomics 11:261–270
Last RL, Jones AD, Shachar-Hill Y (2007) Towards the plant metabolome and beyond. Nat Rev Mol Cell Biol 8:167–174
Lei Z, Huhman DV, Sumner LW (2011) Mass spectrometry strategies in metabolomics. J Biol Chem 286:25435–25442
Li Z, Omranian N, Neumetzler L et al (2016) A transcriptional and metabolic framework for secondary wall formation in Arabidopsis. Plant Physiol 172:1334–1351
Lisec J, Meyer RC, Steinfath M et al (2008) Identification of metabolic and biomass QTL in Arabidopsis thaliana in a parallel analysis of RIL and IL populations. Plant J 53:960–972
Lisec J, Steinfath M, Meyer RC et al (2009) Identification of heterotic metabolite QTL in Arabidopsis thaliana RIL and IL populations. Plant J 59:777–788
Lu W, Su X, Klein MS et al (2017) Metabolite measurement: pitfalls to avoid and practices to follow. Annu Rev Biochem 86:277–304
Lunn JE (2007) Compartmentation in plant metabolism. J Exp Bot 58:35–47
Lunn JE, Feil R, Hendriks JHM et al (2006) Sugar-induced increases in trehalose 6-phosphate are correlated with redox activation of ADPglucose pyrophosphorylase and higher rates of starch synthesis in *Arabidopsis thaliana*. Biochem J 397:139–148
Lunn JE, Delorge I, Figueroa CM et al (2014) Trehalose metabolism in plants. Plant J 79:544–567
Luo J (2015) Metabolite-based genome-wide association studies in plants. Curr Opin Plant Biol 24:31–38
Markley JL, Brüschweiler R, Edison AS et al (2017) The future of NMR-based metabolomics. Curr Opin Biotechnol 43:34–40. https://doi.org/10.1016/J.COPBIO.2016.08.001
Marshall DD, Powers R (2017) Beyond the paradigm: combining mass spectrometry and nuclear magnetic resonance for metabolomics. Prog Nucl Magn Reson Spectrosc 100:1–16
Martin C, Bhatt K, Baumann K (2001) Shaping in plant cells. Curr Opin Plant Biol 4:540–549

Matich EK, Ghafari M, Camgoz E et al (2018) Time-series lipidomic analysis of the oleaginous green microalga species Ettlia oleoabundans under nutrient stress. Biotechnol Biofuels 11:29

Matsuda F, Okazaki Y, Oikawa A et al (2012) Dissection of genotype-phenotype associations in rice grains using metabolome quantitative trait loci analysis. Plant J 70:624–636

Meijón M, Feito I, Oravec M et al (2016) Exploring natural variation of Pinus pinaster Aiton using metabolomics: is it possible to identify the region of origin of a pine from its metabolites? Mol Ecol 25:959–976

Mettler T, Mühlhaus T, Hemme D et al (2014) Systems analysis of the response of photosynthesis, metabolism, and growth to an increase in irradiance in the photosynthetic model organism Chlamydomonas reinhardtii. Plant Cell 26:2310–2350

Meyer RC, Steinfath M, Lisec J et al (2007) The metabolic signature related to high plant growth rate in Arabidopsis thaliana. Proc Natl Acad Sci 104:4759–4764

Meyer RC, Kusterer B, Lisec J et al (2010) QTL analysis of early stage heterosis for biomass in Arabidopsis. Theor Appl Genet 120:227–237

Misra BB, Assmann SM, Chen S (2014) Plant single-cell and single-cell-type metabolomics. Trends Plant Sci 19:637–646

Monne M, Daddabbo L, Gagneul D et al (2018) Uncoupling proteins 1 and 2 (UCP1 and UCP2) from Arabidopsis thaliana are mitochondrial transporters of aspartate, glutamate and dicarboxylates. J Biol Chem 293:4213

Monti LL, Bustamante CA, Osorio S et al (2016) Metabolic profiling of a range of peach fruit varieties reveals high metabolic diversity and commonalities and differences during ripening. Food Chem 190:879–888

Moore BD, Andrew RL, Külheim C, Foley WJ (2014) Explaining intraspecific diversity in plant secondary metabolites in an ecological context. New Phytol 201:733–750

Nakabayashi R, Saito K (2015) Integrated metabolomics for abiotic stress responses in plants. Curr Opin Plant Biol 24:10–16

Nikoloski Z, Perez-Storey R, Sweetlove LJ (2015) Inference and prediction of metabolic network fluxes. Plant Physiol 169:1443–1455

O'Leary BM, Plaxton WC (2017) Mechanisms and functions of post-translational enzyme modifications in the organization and control of plant respiratory metabolism. In: Tcherkez G, Ghashghaie J (eds) Plant respiration: metabolic fluxes and carbon balance. Springer International Publishing, Cham, pp 261–284

O'Malley RC, Barragan CC, Ecker JR (2015) A user's guide to the Arabidopsis T-DNA insertional mutant collections. Methods Mol Biol 1284:323–342

Obata T, Fernie AR (2012) The use of metabolomics to dissect plant responses to abiotic stresses. Cell Mol Life Sci 69:3225–3243

Obata T, Rosado-Souza L, Fernie AR (2017) Coupling radiotracer experiments with chemical fractionation for the estimation of respiratory fluxes. Methods Mol Biol 1670:17–30

Pichersky E, Gang DR (2000) Genetics and biochemistry of secondary metabolites in plants: an evolutionary perspective. Trends Plant Sci 5:439–445

Piligaev AV, Sorokina KN, Shashkov MV, Parmon VN (2018) Screening and comparative metabolic profiling of high lipid content microalgae strains for application in wastewater treatment. Bioresour Technol 250:538–547

Piques M, Schulze WX, Höhne M et al (2009) Ribosome and transcript copy numbers, polysome occupancy and enzyme dynamics in Arabidopsis. Mol Syst Biol 5:314

Price EJ, Bhattacharjee R, Lopez-Montes A, Fraser PD (2017) Metabolite profiling of yam (Dioscorea spp.) accessions for use in crop improvement programmes. Metabolomics 13:144

Quadrana L, Almeida J, Asis R et al (2014) Natural occurring epialleles determine vitamin E accumulation in tomato fruits. Nat Commun 5:3027

Quanbeck SM, Brachova L, Campbell AA et al (2012) Metabolomics as a hypothesis-generating functional genomics tool for the annotation of Arabidopsis thaliana genes of "unknown function". Front Plant Sci 3:15

Rai A, Saito K, Yamazaki M (2017) Integrated omics analysis of specialized metabolism in medicinal plants. Plant J 90:764–787

Rajasundaram D, Selbig J (2016) More effort - more results: recent advances in integrative "omics" data analysis. Curr Opin Plant Biol 30:57–61

Rhee SY, Mutwil M (2014) Towards revealing the functions of all genes in plants. Trends Plant Sci 19:212–221

Robaina-Estevez S, Nikoloski Z (2017) On the effects of alternative optima in context-specific metabolic model predictions. PLoS Comput Biol 13:e1005568

Robaina-Estévez S, Daloso DM, Zhang Y et al (2017) Resolving the central metabolism of Arabidopsis guard cells. Sci Rep 7:8307

Roessner-Tunali U (2007) Uncovering the plant metabolome: current and future challenges. Springer Netherlands, Dordrecht

Rohn H, Junker A, Hartmann A et al (2012) VANTED v2: a framework for systems biology applications. BMC Syst Biol 6:139

Rolland F, Baena-Gonzalez E, Sheen J (2006) Sugar sensing and signaling in plants: conserved and novel m. Annu Rev Plant Biol 57:675–709

Ruan Y-L (2014) Sucrose metabolism: gateway to diverse carbon use and sugar signaling. Annu Rev Plant Biol 65:33–67

Saito K, Matsuda F (2010) Metabolomics for functional genomics, systems biology, and biotechnology. Annu Rev Plant Biol 61:463–489

Sajitz-Hermstein M, Topfer N, Kleessen S et al (2016) iReMet-flux: constraint-based approach for integrat-

ing relative metabolite levels into a stoichiometric metabolic models. Bioinformatics 32:i755–i762
Salon C, Avice J-C, Colombie S et al (2017) Fluxomics links cellular functional analyses to whole-plant phenotyping. J Exp Bot 68:2083–2098
Sawada Y, Akiyama K, Sakata A et al (2009) Widely targeted metabolomics based on large-scale MS/MS data for elucidating metabolite accumulation patterns in plants. Plant Cell Physiol 50:37–47
Schauer N, Steinhauser D, Strelkov S et al (2005) GC-MS libraries for the rapid identification of metabolites in complex biological samples. FEBS Lett 579:1332–1337
Schauer N, Semel Y, Roessner U et al (2006) Comprehensive metabolic profiling and phenotyping of interspecific introgression lines for tomato improvement. Nat Biotechnol 24:447–454
Schripsema J (2010) Application of NMR in plant metabolomics: techniques, problems and prospects. Phytochem Anal 21:14–21
Schwahn K, Beleggia R, Omranian N, Nikoloski Z (2017) Stoichiometric correlation analysis: principles of metabolic functionality from m data. Front Plant Sci 8:2152
Schwender J, König C, Klapperstück M et al (2014) Transcript abundance on its own cannot be used to infer fluxes in central metabolism. Front Plant Sci 5:668
Scranton MA, Ostrand JT, Fields FJ, Mayfield SP (2015) Chlamydomonas as a model for biofuels and bioproducts production. Plant J 82:523–531
Shi H, Schwender J (2016) Mathematical models of plant metabolism. Curr Opin Biotechnol 37:143–152
Shih PM, Liang Y, Loqué D (2016) Biotechnology and synthetic biology approaches for metabolic engineering of bioenergy crops. Plant J 87:103–117
Smith AM, Stitt M (2007) Coordination of carbon supply and plant growth. Plant Cell Environ 30:1126–1149. https://doi.org/10.1111/j.1365-3040.2007.01708.x
Sprenger H, Erban A, Seddig S et al (2017) Metabolite and transcript markers for the prediction of potato drought tolerance. Plant Biotechnol J 16:939
Steinfath M, Strehmel N, Peters R et al (2010) Discovering plant metabolic biomarkers for phenotype prediction using an untargeted approach. Plant Biotechnol J 8:900–911
Stitt M, Gibon Y (2014) Why measure enzyme activities in the era of systems biology? Trends Plant Sci 19:256–265
Stitt M, Sulpice R, Keurentjes J (2010a) Metabolic networks: how to identify key components in the regulation of metabolism and growth. Plant Physiol 152:428–444
Stitt M, Lunn J, Usadel B (2010b) Arabidopsis and primary photosynthetic metabolism - more than the icing on the cake. Plant J 61:1067–1091
Sulpice R, McKeown PC (2015) Moving toward a comprehensive map of central plant metabolism. Annu Rev Plant Biol 66:187–210
Sweetlove LJ, Fernie AR (2013) The spatial organization of metabolism within the plant cell. Annu Rev Plant Biol 64:723–746
Sweetlove LJ, Fell D, Fernie AR (2008) Getting to grips with the plant metabolic network. Biochem J 409:27–41
Tenenboim H, Brotman Y (2016) Omic relief for the biotically stressed: metabolomics of plant biotic interactions. Trends Plant Sci 21:781–791
Therrien-Laperrière S, Cherkaoui S, Boucher G et al (2017) PathQuant: a bioinformatic tool to quantitatively annotate the relationship between genes and metabolites through metabolic pathway mapping. FASEB J 31:769.3–769.3
Tivendale ND, Jewett EM, Hegeman AD, Cohen JD (2016) Extraction, purification, methylation and GC–MS analysis of short-chain carboxylic acids for metabolic flux analysis. J Chromatogr B, Anal Technol Biomed Life Sci 1028:165–174
Tohge T, Scossa F, Fernie AR (2015) Integrative approaches to enhance understanding of plant metabolic pathway structure and regulation. Plant Physiol 169:1499–1511
Toubiana D, Batushansky A, Tzfadia O et al (2015) Combined correlation-based network and mQTL analyses efficiently identified loci for branched-chain amino acid, serine to threonine, and proline metabolism in tomato seeds. Plant J 81:121–133
Turner MF, Heuberger AL, Kirkwood JS et al (2016) Non-targeted metabolomics in diverse sorghum breeding lines indicates primary and secondary metabolite profiles are associated with plant biomass accumulation and photosynthesis. Front Plant Sci 7:953
Verpoorte R, Choi YH, Kim HK (2007) NMR-based metabolomics at work in phytochemistry. Phytochem Rev 6:3–14
Villaveces JM, Koti P, Habermann BH (2015) Tools for visualization and analysis of molecular networks, pathways, and -omics data. Adv Appl Bioinforma Chem 8:11–22
Wang L, Nägele T, Doerfler H et al (2016) System level analysis of cacao seed ripening reveals a sequential interplay of primary and secondary metabolism leading to polyphenol accumulation and preparation of stress resistance. Plant J 87:318–332
Wang S, Yang C, Tu H et al (2017) Characterization and metabolic diversity of flavonoids in Citrus species. Sci Rep 7:10549
Ward JL, Baker JM, Beale MH (2007) Recent applications of NMR spectroscopy in plant metabolomics. FEBS J 274:1126–1131
Wase N, Tu B, Allen JW et al (2017) Identification and metabolite profiling of chemical activators of lipid accumulation in green algae. Plant Physiol 174:2146–2165
Watanabe M, Tohge T, Balazadeh S et al (2018) Comprehensive metabolomics studies of plant developmental senescence. In: Guo Y (ed) Plant senescence: methods and protocols. Springer, New York, NY, pp 339–358

Wegner A, Meiser J, Weindl D, Hiller K (2015) How metabolites modulate metabolic flux. Curr Opin Biotechnol 34:16–22
Wen W, Li D, Li X et al (2014) Metabolome-based genome-wide association study of maize kernel leads to novel biochemical insights. Nat Commun 5:3438
White AC, Rogers A, Rees M, Osborne CP (2016) How can we make plants grow faster? A source–sink perspective on growth rate. J Exp Bot 67:31–45
Xu Y, Crouch JH (2008) Marker-assisted selection in plant breeding: from publications to practice. Crop Sci 48:391–407
Xu J, Cui W, Cheng JJ, Stomp A-M (2011) Production of high-starch duckweed and its conversion to bioethanol. Biosyst Eng 110:67–72
Yadav UP, Ivakov A, Feil R et al (2014) The sucrose–trehalose 6-phosphate (Tre6P) nexus: specificity and mechanisms of sucrose signalling by Tre6P. J Exp Bot 65:1051–1068
Yang L, Chen J, Qin S et al (2018) Growth and lipid accumulation by different nutrients in the microalga Chlamydomonas reinhardtii. Biotechnol Biofuels 11:40
Yao L, Gerde JA, Lee S-L et al (2015) Microalgae lipid characterization. J Agric Food Chem 63:1773–1787
Yoshida T, Mogami J, Yamaguchi-Shinozaki K (2015) Omics approaches toward defining the comprehensive abscisic acid signaling network in plants. Plant Cell Physiol 56:1043–1052
Yu C, Zhao X, Qi G et al (2017) Integrated analysis of transcriptome and metabolites reveals an essential role of metabolic flux in starch accumulation under nitrogen starvation in duckweed. Biotechnol Biofuels 10:167
Yuan H, Cheung CYM, Hilbers PAJ, van Riel NAW (2016) Flux balance analysis of plant metabolism: the effect of biomass composition and model structure on model predictions. Front Plant Sci 7:537
Zhu G, Wang S, Huang Z et al (2018) Rewiring of the fruit metabolome in tomato breeding. Cell 172:249–261.e12

Interactomes: Experimental and In Silico Approaches

6

Luíza Lane de Barros Dantas
and Marcelo Mendes Brandão

Abstract

Any part of the Central Dogma of Molecular Biology (DNA replication, transcription, and translation) is based on an intricate protein–protein interaction. On this chapter, we will navigate over the techniques that enable us to construct or fulfill the gaps on an interactome study, directly using assessment of the biochemical and/or molecular machinery that allow two proteins to interact with each other; or rely on computational biology techniques to gather information on PPI from public available databases and evaluate this interaction.

Keywords

Protein–Protein interactions · Interactome · Proteome · Bioinformatics · Databases

L. L. de Barros Dantas (✉)
John Innes Centre, Norwich Research Park, Norwich, UK

M. M. Brandão
Center for Molecular Biology and Genetic Engineering, State University of Campinas, Campinas, SP, Brazil
e-mail: brandaom@unicamp.br

6.1 Introduction

The intricate machinery that sustains all living forms is built upon a large well lubricated network of interaction among different biochemical entities, specially, relying on the protein–protein interactions (PPI). Proteins on their course of action, hardly ever act as a lone wolf since their functions tend to be regulated by other proteins to properly achieve its goal.

Protein–protein interactions are the central controller to all biological processes and its revelation provide the basis to comprehend biology as an integrated system. Michael Cusick, on his 2005 manuscript entitled "Interactome: gateway into systems biology," states that "the full interactome network is the complete collection of all physical protein–protein interactions that can take place within a cell."

The interactome is the next big step for System Biology, after massive worldwide effort for DNA, RNAs, and proteins sequencing and subsequent gene annotation for many model and non-model organisms. The PPI from a specific cell or organ unravel the roles of each interactor on a signal transduction pathway, improving the discover, quantification and new biochemical targets for biotechnology.

F. V. Winck (ed.), *Advances in Plant Omics and Systems Biology Approaches*, Advances in Experimental Medicine and Biology 1346, https://doi.org/10.1007/978-3-030-80352-0_6

6.2 Molecular Technologies for Protein–Protein Interactions (PPI) Identification

The plant cell requires a tight coordination of protein expression, assembly, modification, aggregation into complexes and subcellular localization, in order to properly function. Therefore, it is important to know how proteins work to fully understand how a plant cell works. In addition, as proteins mostly act gathered in complexes rather than isolated, it is critical to understand how proteins interact inside these complexes. What keep proteins together in the macromolecular complexes are protein–protein interactions (PPI). Such interactions are crucial for the maintenance of the cell as a working unity in every plant tissue. The PPI study also helps to elucidate protein cellular localization, which is also relevant to understand protein function. That is why the PPI study provides insights about cell physiology.

PPI can be investigated through many different technologies, used to discover, to confirm or to characterize PPIs and analyze protein proximity on a molecular level in plants. Some techniques are tailored to investigate protein interactions on a binary level or on a multicomplex level with high accuracy. Others allow PPI investigation by imaging living cells or protein complexes, using organisms, purified proteins and cell lysates. There are techniques better suited for PPI screening, while other methods are convenient for confirming PPIs. Before starting a full set of experiments to analyze a specific PPI, a few issues should be considered to avoid both false positive and false negative interactions. Meticulously experiment planning is critical and it is advised to combine at least two different independent molecular methods in PPI analysis (Braun et al. 2013; Hayes et al. 2016). Besides, for already known PPIs, information regarding the binding affinity of the proteins involved in an interaction is useful (Perkins et al. 2010). It also helps on the experiment design when there is prior knowledge about binding domains of the interacting proteins (Keskin et al. 2016) and about subcellular location where proteins interact (Hayes et al. 2016). To help designing PPI analysis in plants, a brief description of the most well-established molecular technologies used to study protein interactions in plants is shown thereupon, as well as some examples of these technologies applied on plant PPIs analysis.

6.2.1 Yeast Two Hybrid (Y2H)

Y2H might be the most popular technique to investigate PPI and for many scientists this is the starting point for PPI studies. This in vivo method is based on the direct interaction between two proteins fused to halves of a transcription factor inside yeast nucleus, which reconstructs a transcription factor that expresses a reporter gene. This reporter gene is in charge of yeast survival on selective media (Fields and Song 1989). There are many versions of this technique (Bruckner et al. 2009; White 1996), but Y2H general principle is quite simple. A transcription factor split in two halves: one half is a DNA-binding domain (DB), that allows DNA binding (called bait), and the second half is a transcriptional activation domain (AD), that activates the gene reporter expression (called prey). The transcription of the reporter gene allows yeast to grow in a selective media (Ito et al. 2001) only when a given pair of proteins fused to bait and prey halves physically interact. Y2H is often used as a screening method to start searching for PPI. This method is suitable because it is easy to operate and inexpensive, ideal to start screening PPIs (Braun et al. 2013). Y2H has several limitations, like generate false positive PPIs. Because the candidate interacting proteins should be expressed in yeast nucleus, Y2H might detect interactions between proteins unnaturally co-localized. Also, Y2H might fail to identify interactions involving proteins requiring post-translational modifications, proteins with transient interactions or proteins expressed and or active on the membrane (Braun et al. 2009). That is why Y2H is a technique to be applied in combination with other PPI detection technologies. In plants, there are many examples of PPI analysis done using Y2H. The first plant interactome, the *Arabidopsis* Interactome 1 (AI-1), was completed using Y2H and shows around 6200 interactions among 2700 proteins, approximately (Arabidopsis Interactome Mapping 2011). In tomato, Y2H was used to examine the interactions of ABA signaling core components (Chen et al. 2016). In tobacco, Y2H assays showed the role of 14-3-3 isoforms in plant signaling by mapping the interaction between protein 14-3-3 and enzyme sucrose-6-phosphate synthase (SPS) (Bornke 2005; Ferro and Trabalzini 2013).

6.2.2 Pull-Down

Pull-down assays are widely used for PPI detection and/or confirmation. This in vitro technique is based on affinity purification, similarly to co-immunoprecipitation (Co-IP). The difference between them is, while Co-IP uses antibodies fused to known proteins, pull-down uses tags fused to known proteins. In pull-down experiments, a known protein is expressed in cells with a tag (called bait). This fused protein is immobilized to an affinity matrix specifically compatible with this tag. The interacting candidate proteins (called preys) are trapped in a protein complex attached to matrix. After a few purification steps, this protein complex is eluted and ready to analysis on SDS-PAGE and western blot or mass spectrometry (Louche et al. 2017). The bait proteins can come from various sources, such as cell lysate, expression systems or purified. That is also true for the prey proteins, depending on the purpose of the pull-down assay, which could be PPI identification or characterization of a known PPI. A crucial step in a pull-down assay preparation is the choice of a tag. Since the tag is going to act as the link between the specific affinity matrix and the protein complex, aspects such as size and polarity of a tag before expressing the bait fused protein must be considered. The glutathione S-transferase (GST) tag has affinity for glutathione-based matrixes. GST tags are significantly large (26 kDa), expensive, and can interact in a nonspecific fashion. An extensively used tag is the histidine (His) tag. This tag is made of six histidine amino acid residues and has a high affinity for nickel-based resins, such Ni-NTA agarose. This is a small tag (1.1 kDa), unlikely to affect the bait protein folding and it is inexpensive. Even though pull-down is a good method to study PPI in complexes, this might not be the best approach to investigate transient PPIs. An example of pull-down assays use in plants comes from rice RING UB E3 ligase (OsSIRP2), whose gene is upregulated under abiotic stress conditions (i.e., salinity stress). E3 ligase was shown to interact with TRANSKETOLASE 1 (OsTKL1) under salinity conditions and to increase OsTKL1 degradation (Chapagain et al. 2017). Pull-down experiments were also done to confirm interactions between JASMONATE ZIM DOMAIN (JAZ) protein and NOVEL INTERACTOR OF JAZ (NINJA) transcriptional repressor in jasmonate responses (Pauwels et al. 2010).

6.2.3 Co-immunoprecipitation (Co-IP)

This technique is another in vitro method based on affinity purification for PPI analysis in a larger scale on protein complexes. Co-IP is generally used for PPI confirmation and/or characterization (Dwane and Kiely 2011; Hayes et al. 2016; Rao et al. 2014). Similarly to the pull-down mechanism, Co-IP assays are based on a known protein (called bait), with which other proteins in a complex (called prey) interact. The complex is isolated due to the connection between the bait protein and a specific antibody. For Co-IP, whole cell lysates can be used as a starting point, as well as purified proteins. The protein complex detected due to the antibody specific connection can be immobilized in a matrix, isolated, eluted and analyzed by western blot or mass spectrometry. Because this method allows the use of cell lysate, Co-IP is a suitable approach for proteins bearing post-translational modifications and is also indicated to analyze endogenous proteins (Rao et al. 2014). Besides, Co-IP can evaluate proteins PPIs in their native conformation and it is relatively inexpensive. A great disadvantage of Co-IP for plant studies is the fact that it is a technique based on the use of antibodies, since there is little variety of antibodies for plant proteins (Braun et al. 2013). Also, Co-IP produces background and false positives, requiring careful planning and use of negative controls (Braun et al. 2013; Ransone 1995). Transient PPI are challenging to be detected using Co-IP. In *Arabidopsis*, Co-IP assays were used to expose the interactions of EFR receptor kinases triggered by innate immunity responses (Roux et al. 2011). Recently, the interaction between PROTEIN TARGETING TO STARCH (PTST) PTST2 and PTST3 with STARCH SYNTHASE4 (SS4) was shown to be related to starch granule initiation regulation in *Arabidopsis* leaves (Seung et al. 2017).

6.2.4 Tandem Affinity Purification: Mass Spectrometry (TAP-MS)

This is a high throughput method for PPIs identification, designed to investigate them in the cell standard conditions (Rigaut et al. 1999). TAP-MS employs a tag fused to the C- or N-terminus of a known protein, called bait (Kaiser et al. 2008). As the tag used in TAP-MS assays is built as a double tag, with two proteins connected by a protease, this method requires a two-step purification using two immobilized matrixes with affinity for each part of the double tag (Gunzl and Schimanski 2009). The protein complex that interacts with the bait protein is isolated from the initial cell lysate or purified protein solution and subsequently analyzed by mass spectrometry. There are several types of double tags used in TAP-MS. One of them is the combination of a double-protein-A domain connected by a tobacco etch virus (TEV) protease cleavage site to a calmodulin-binding peptide (Rigaut et al. 1999). Another tag is the GS tag, which has a double-protein-G domain and a streptavidin-binding-peptide connected by a protease from TEV or rhinovirus 3C (Braun et al. 2013; Van Leene et al. 2008). TAP-MS is a very efficient method able to detect both transient and stable PPI (Yates et al. 2009). However, due to the necessity of specific equipment, it can be expensive. The fused tag might interfere in the bait protein expression and folding, and the two-round purification might interfere in the final PPI yielding in an initial protein material. In plants, TAP-MS was used to elucidate the TCP4 complex components, helping to regulate the expression of CONSTANS (CO) at the right time of the day (Kubota et al. 2017). A classic example of TAP-MS in plants is the platform for *Arabidopsis* cell suspension cultures created to analyze protein complexes (Van Leene et al. 2011).

6.2.5 Förster Resonance Energy Transfer

The resonance energy transfer methods are proximity-dependent techniques that use recombinant fused proteins to analyze proteins pairs within a distance of 10 nm or less from each other (Kerppola 2006; Piston and Kremers 2007). The interacting proteins pairs are fused to donor-acceptor molecules pairs, either fluorescent or bioluminescent, and the energy of an excited donor molecule is transferred to the acceptor molecule, which emits energy as photons (Lonn and Landegren 2017; Wiens and Campbell 2018). There are two different methods based on the principle of resonance energy transfer, according to the molecular nature of the donor-acceptor pair: Fluorescent Resonance Energy Transfer (FRET) (Piston and Kremers 2007) and Bioluminescent Resonance Energy Transfer (BRET) (Pfleger and Eidne 2006). FRET is based on fluorophores donor-acceptors pairs. An example of widely used donor-acceptor pairs in FRET assays are Cyan Fluorescent Protein (CFP), used as the donor fluorophore, and Yellow Fluorescent Protein (YFP), used as the acceptor fluorophore. Each one of these fluorescent proteins is fused to an interacting protein from a PPI pair and, in case both interacting proteins are brought together in a distance of 10 nm or less, light emission can be imaged using standard confocal microscopy or wide-field microscope, for example (Lonn and Landegren 2017). BRET depends upon an enzyme-catalyzed luminescence reaction. The oxidation reaction of a compatible substrate, such as coelenterazine, by luciferase enzyme causes emission of bioluminescence. In BRET, the luciferase acts as the donor that excites the acceptor fluorophore, if the acceptor-donor pair is within a radius of 10 nm or less. The bioluminescence emission can be captured using a cooled-CCD camera (Lonn and Landegren 2017; Xu et al. 2007, 1999). In both FRET and BRET, PPI can be imaged in situ and in planta. Nonetheless, both techniques require expensive equipment for analysis. BRET assays were efficiently used to image tobacco and *Arabidopsis* tissues (Xu et al. 2007), and also to show the role of interaction between enzymes SUCROSE PHOSPHATE SYNTHASE (SPS) and SUCROSE PHOSPHATE PHOSPHATASE (SPP) in *Arabidopsis* growth (Maloney et al. 2015). FRET assays were applied on experiments to identify interactions between VACUOLAR SORTING RECEPTORS (VSRs) and vacuole-targeted proteins, crucial to target proteins for degradation in the vacuole (Kunzl et al. 2016).

6.2.6 Bimolecular Fluorescence Complementation (BiFC)

This molecular in vivo method for PPI analysis is an established form of protein complementation assay (PCA), based on protein-fragment complementation. In BiFC assays, a fluorescent protein, like GFP or YFP, is split in half and each of these parts is fused to the N- or C-terminal end of a candidate interacting proteins pair. Note that those fluorescent protein parts alone are unfunctional. If the recombinant protein pair interacts, both fluorescent protein halves are linked and the fluorescent protein is restored to its full folded version (Ghosh et al. 2000; Lonn and Landegren 2017). The resultant fluorescence emission can be imaged using live microscopy or confocal microscopy. In plants, BiFC experiments are mostly performed prior to transient protein expression in either *Nicotiana* or *Arabidopsis* (Bracha-Drori et al. 2004; Braun et al. 2013; Citovsky et al. 2008). Even though BiFC is a suitable method for identifying the subcellular cell location where PPI occurs, the recombinant fluorescent fused half-protein might affect protein conformation and location. Another limitation is that BiFC assays might give high background fluorescence because of the fluorescent protein parts spontaneous self-assembling. The spontaneous self-assembling might also generate false positives and, therefore, BiFC experiments need a very careful planning and rigorous control. As an alternative to BiFC, but using the same PCA principle, there is the Bimolecular Luminescent Complementation (BiLC). BiLC uses luciferases from different sources instead of fluorescent proteins for complementation (Buntru et al. 2016; Wiens and Campbell 2018). In plants, BiFC assays were performed to prove the homodimerization of transcription factors LATERAL ORGAN BOUNDARIES DOMAIN/ASYMMETRIC LEAVES2-LIKEs (LBD) LBD16 and LBD18, required for activating lateral root formation in *Arabidopsis* (Lee et al. 2017). In rice, experiments showed the relationship between flowering time and phosphorus homeostasis with help of BiFC experiments confirming the interaction between proteins UBIQUITIN-CONJUGATING E2 ENZYME (OsPHO2) and GIGANTEA (OsGI) (Li et al. 2017). BiLC performed in *Nicotiana* showed PPIs in the Golgi apparatus relevant to xyloglucan biosynthesis (Lund et al. 2015).

6.3 In Silico Approaches for Protein–Protein Interactions (PPI) Identification

6.3.1 Databases

Independently on which molecular technique has been used to identify protein interaction, it is necessary to storage this information in a way that useful information might be gathered from the data set, enabling data comparison, exchange and verification. This storage can be done locally using a correlational database manager, such as MySQL or its fork MariaDB, or on spreadsheet software.

The basic database structure and elements for PPI stowage are presented on Table 6.1.

Table 6.1 Basic PPI database schema elements for storing interaction information

IntA	IntB	Type	Method	Pub	FSW
AT5G47790	AT5G63310	Predicted	Affinity Capture-MS	Mol Syst Biol. 2007;3:89. Epub 2007 Mar 13	0.0410
AT1G09570	AT5G63310	Experimental	Yeast two hybrid	J Biol Chem. 2005 Feb 18;280(7):5740–9. Epub 2004 Nov 23	0.0465

IntA = interactor A; IntB = interactor B; Type = How this interaction was identified; Method = Experimental: This means that the indicated PPI was experimentally demonstrated using the same organism model of study. Predicted: The indicated PPI was proposed based on orthology studies; Pub = indicated reference to the publication of which this interaction was annotated from; FSW = the Functional Similarity Weight. It represents the proportion of interaction partners that two proteins have in common

6.3.2 In Silico PPI Reliability Based on Interaction Topology

Two related mathematical approaches, the Czekanowski-Dice distance (CD-distance) (Brun et al. 2003) and Functional Similarity Weight (FSW) (Chua et al. 2006), have been proposed to assess the reliability of protein interaction data based on the number of common neighbors of two proteins.

The FSW algorithm was originally proposed by Chua et al. (2006) and the functional similarity weight index on a pair of proteins A and B in an interaction graph ($FSW_{A,B}$) is defined as:

$$\mathrm{FSW}_{A,B} = \left(\frac{2|N_A \cap N_B|}{|N_A - N_B| + 2|N_A \cap N_B| + \lambda_{A,B}} \right) \times \left(\frac{2|N_A \cap N_B|}{|N_B - N_A| + 2|N_A \cap N_B| + \lambda_{B,A}} \right),$$

where

N_A = set of interaction partners of A; N_B = set of interaction partners of B; $\lambda_{A,B}$ is a weight to penalize similarity weights between protein pairs when any of the proteins has few interacting partners and is calculated as:

$$\lambda_{A,B} = \max\left(0, N_{avg} - \left(|N_A - N_B| + |N_a \cap N_B|\right)\right),$$

where

N_{avg} = Average of interactions made by each protein on a database.

The Czekanowski-Dice distance between two proteins a and b is given by:

$$D(a,b) = \frac{|N'_a \Delta N'_b|}{|N'_a \cup N'_b| + |N'_a \cap N'_b|}$$

where

N'_a = a set of proteins that contain a and its interaction neighbors; aΔb = symmetric difference between two sets, a and b.

Both algorithms were initially projected to predict protein functions, and lately have been shown to perform well for assessing the reliability of protein interactions (Liu et al. 2009). Wong (2008) has shown that using FSW, which estimates the strength of functional association, to remove unreliable interactions (low FSW) improves the performance of clustering algorithms.

The effectiveness of using FSW as a PPI reliability index was demonstrated using 19,452 interactions in yeast obtained from the GRID database (Breitkreutz et al. 2003). Over 80% of the top 10% of protein interactions ranked by FSW have a common cellular role, and over 90% of them have a common subcellular localization (Chen et al. 2006b, c).

One example of FSW application can be seen on the *Arabidopsis thaliana* protein interaction network database—AtPIN (Brandao et al. 2009). Due to its integrative profile, the reliability index for a reported PPI can be postulated in terms of interaction partners proportion that two proteins have in common, and these pairs of interacting proteins highly ranked by this method are likely to be true positive interactors. Contrariwise, the proteins pairs lowly ranked are likely to be false positives. With the same benchmarking approach indicated above, the top 10% of protein interactions, ranked by FSW in AtPIN (release 9 of AtPINDB), have indicated that 59% of PPIs share the same subcellular compartment, and 83% have the same function or participate in the same cellular process. A decent FSW value threshold starting point is the top 20%, since Chua et al. (2006) and Chen et al. (2006b) have demonstrated that a protein pair having a high FSW value, above this value, is likely to share a common function.

The most interesting feature of the CD-distance and FSW is that they can rank the reliability of an interaction between a pair of proteins using only the topology of the interactions between that pair and their neighbors within a short radius in a graph network (Chen et al. 2006b, c).

6.3.3 PPI Reliability Evaluation Based on Subcellular Localization

An additional reliability checking point for in silico PPI predictions is the Cellular Compartment Classification or C^3. The C^3 value is represented as classes and is calculated using simple mathematical sum of three parameters:

$$C^3 = A + B + C$$

where

A = type of interaction; B = co-localization; C = determination of subcellular localization (experimentally or predicted).

Table 6.2 presents a summary of the possible entering values to calculate C^3.

Considering all possibilities, it is possible to divide the PPIs in a dataset into four classes:

- **Class A** (C^3 = 7): The PPI and subcellular location have seemed to be experimentally demonstrated and both proteins are co-localized.
- **Class B** (C^3 = 5): The PPI and subcellular location have been experimentally shown; however, the proteins were localized to different subcellular compartments.
- **Class C** (C^3 = 3): Same as Class A, but the PPI is based on prediction analyses.
- **Class D** (C^3 = 6): Same as Class A, but subcellular location is based on prediction analyses.

6.3.4 Publicly Available Databases

We are living on the Big Data ages, and several available databases with proteins interactions have arisen over the past decades. Zahiri et al. (2013) present on their manuscript a comprehensive list of the most popular PPI repositories for model organisms. To integrate the major public interaction data providers in a mutual agreement to share data, to develop a distinct set of curation rules for collect data from directly deposited PPI data and/or from peer-reviewed publications, the IMEx was created, acronym for International Molecular Exchange Consortium (Orchard et al. 2012).

IMEx aims to make these interactions material available in an intuitive browsing and search interface on a single website. One of the key points of this concatenated and curated dataset is to provide all the information in standard format, facilitating the usage and incorporation of this data on a variety of bio computational applications.

This sharing standardization is mandatory since each database provider might storage its PPI datasets on a particular format. Currently, the most used standard format for molecular interaction data exchange is the PSI-MI XML (Kerrien et al. 2007), proposed by the Proteomics Standards Initiative, maintained by the Human Proteome Organization (HUPO). Another very popular exchange format is the PSI-mitab, differently from PSI-MI XML, all the molecular interactions are presented on tab-delimited format with up to 42 fields of information. Both formats

Table 6.2 Numeric values for each parameter for C^3 calculation

Parameter	Value	Description
Type of interaction	4	Based on experimental data
	0	No experimental data available (predicted)
Co-localization	2	Same cellular compartment
	0	Different cellular compartment
Determination of subcellular localization	1	Based on experimental analyses
	0	If one or both are predicted

Table 6.3 Largest IMEx partners caretakers of publicly available data

Database	URL	References
Biological General Repository for Interaction Datasets (Biogrid)	http://www.thebiogrid.org	Stark et al. (2006)
Database of interacting proteins (DIP)	http://dip.doe-mbi.ucla.edu	Salwinski et al. (2004), Xenarios et al. (2002)
IntAct	http://www.ebi.ac.uk/intact/	Orchard et al. (2014)
Molecular Interaction Database (MINT DB)	http://mint.bio.uniroma2.it/mint	Licata et al. (2012)

previously cited, and a few other molecular interactions exchange layouts can be found at HUPO GitHub address at https://github.com/HUPO-PSI or at the HUPO-PSI web site (http://www.psidev.info/).

The four most active and cited IMEx partners' datasets are presented on Table 6.3. All of them are focused on model organisms PPIs and sharable information on standards formats previously discussed.

6.3.5 In Silico Predictions

All the members within a protein family are homologous and can be further separated into orthologs, which are genes of different species that evolved from a common ancestral gene by speciation. Generally, orthologs retain the same molecular function during evolution. Researchers rely on these characteristics to predict possible interactions for a non-model organism from a well curated and annotated PPI dataset.

Studies using multiple sequences alignment, from different organisms, have demonstrated that when average amino acid identity is over 50% of correctly aligned residues, we assume that the involved proteins will present the same ancestry, and, therefore, might be considered orthologs (Ogden and Rosenberg 2007; Thompson et al. 1999). There are many ways to detect orthologs genes/proteins;

6.3.5.1 Reciprocal BLAST

The simplest is the reciprocal BLAST analyses. Having two datasets named Species A and Species B, two separated blast datasets so-called A_DB and B_DB are created. Basically, the appropriate BLAST program is run querying Species A on B_DB and Species B on A_DB, using arbitrary threshold for sequence similarity over 80%. A second parameter to evaluate is the *e*-value but keeping in mind that *e*-value depends on the database size, so, the larger the database, the smaller the *e*-value can be. A good cutoff starting point is something between 10^{-5} and 10^{-20}. Moreno-Hagelsieb has summarized few hints on how to choose the best BLAST parameters values for reciprocal BLAST ortholog identification approach (Moreno-Hagelsieb and Latimer 2008). To identify the best reciprocal hit among all sequences on the datasets, the BackBlast Reciprocal Blast script (https://github.com/LeeBergstrand/) can be used to automate the analyses process. The BackBlast algorithm will identify those best reciprocal hits and return a filtered list of most plausible orthologs among Species A and Species B. Now it is possible to transfer PPI information from one dataset to another.

6.3.5.2 OrthoMCL

Its algorithm firstly identify sequence similarities by reciprocal best BLAST, and then, joins proteins into ortholog groups based on normalized BLAST scores between proteins using Markov clustering (Enright et al. 2002; Li et al. 2003). It is also available in the orthoMCL-DB website (Chen et al. 2006a), which contains ortholog groups for most completely sequenced and annotated eukaryotes and for a number of completely sequenced and annotated prokaryotes (http://orthomcl.org/). There is an ample tutorial written by Fischer et al. (2011), encompassing all the steps needed to identify the most plausible orthologs.

6.3.5.3 InParanoid

This program uses the pairwise similarity scores between two datasets, calculated using BLASTP, for assembling orthology groups. These orthology groups are initially composed of two so-called seed orthologs found by reciprocal best hits between two datasets. On second step, more sequences are added to the group if the sequences in the two datasets are closer to the corresponding seed ortholog than to any sequence not present into the ortholog group in question. The orthology group participants are now called inparalogs, and, a confidence value is provided for each of them, representing how closely related it is to its seed ortholog (O'Brien et al. 2005). The Inparanoid DB (Sonnhammer and Ostlund 2015) is an online database for ortholog groups with inparalogs (http://inparanoid.sbc.su.se/).

References

Arabidopsis Interactome Mapping C (2011) Evidence for network evolution in an arabidopsis interactome map. Science 333:601–608. https://doi.org/10.1126/science.1203877

Bornke F (2005) The variable C-terminus of 14-3-3 proteins mediates isoform-specific interaction with sucrose-phosphate synthase in the yeast two-hybrid system. J Plant Physiol 162:161–168. https://doi.org/10.1016/j.jplph.2004.09.006

Bracha-Drori K, Shichrur K, Katz A, Oliva M, Angelovici R, Yalovsky S, Ohad N (2004) Detection of protein-protein interactions in plants using bimolecular fluorescence complementation. Plant J 40:419–427. https://doi.org/10.1111/j.1365-313X.2004.02206.x

Brandao MM, Dantas LL, Silva-Filho MC (2009) AtPIN: Arabidopsis thaliana protein interaction network. BMC Bioinformatics 10:454. https://doi.org/10.1186/1471-2105-10-454

Braun P et al (2009) An experimentally derived confidence score for binary protein-protein interactions. Nat Methods 6:91–97. https://doi.org/10.1038/nmeth.1281

Braun P, Aubourg S, Van Leene J, De Jaeger G, Lurin C (2013) Plant protein interactomes. Annu Rev Plant Biol 64:161–187. https://doi.org/10.1146/annurev-arplant-050312-120140

Breitkreutz BJ, Stark C, Tyers M (2003) The GRID: the general repository for interaction datasets. Genome Biol 4:R23

Bruckner A, Polge C, Lentze N, Auerbach D, Schlattner U (2009) Yeast two-hybrid, a powerful tool for systems biology. Int J Mol Sci 10:2763–2788. https://doi.org/10.3390/ijms10062763

Brun C, Chevenet F, Martin D, Wojcik J, Guenoche A, Jacq B (2003) Functional classification of proteins for the prediction of cellular function from a protein-protein interaction network. Genome Biol 5:R6

Buntru A, Trepte P, Klockmeier K, Schnoegl S, Wanker EE (2016) Current approaches toward quantitative mapping of the interactome. Front Genet 7:74. https://doi.org/10.3389/fgene.2016.00074

Chapagain S, Park YC, Kim JH, Jang CS (2017) Oryza sativa salt-induced RING E3 ligase 2 (OsSIRP2) acts as a positive regulator of transketolase in plant response to salinity and osmotic stress. Planta 247:925. https://doi.org/10.1007/s00425-017-2838-x

Chen F, Mackey AJ, Stoeckert CJ Jr, Roos DS (2006a) OrthoMCL-DB: querying a comprehensive multi-species collection of ortholog groups. Nucleic Acids Res 34:D363–D368. https://doi.org/10.1093/nar/gkj123

Chen J et al (2006b) Increasing confidence of protein-protein interactomes. In: 17th International Conference on Genome Informatics, Yokohama, Japan, pp 284–297

Chen J, Hsu W, Lee ML, Ng SK (2006c) Increasing confidence of protein interactomes using network topological metrics. Bioinformatics 22:1998–2004

Chen P et al (2016) Interactions of ABA signaling core components (SlPYLs, SlPP2Cs, and SlSnRK2s) in tomato (Solanum lycopersicon). J Plant Physiol 205:67–74. https://doi.org/10.1016/j.jplph.2016.07.016

Chua HN, Sung WK, Wong L (2006) Exploiting indirect neighbours and topological weight to predict protein function from protein-protein interactions. Bioinformatics 22:1623–1630

Citovsky V, Gafni Y, Tzfira T (2008) Localizing protein-protein interactions by bimolecular fluorescence complementation in planta. Methods 45:196–206. https://doi.org/10.1016/j.ymeth.2008.06.007

Dwane S, Kiely PA (2011) Tools used to study how protein complexes are assembled in signaling cascades. Bioeng Bugs 2:247–259. https://doi.org/10.4161/bbug.2.5.17844

Enright AJ, Van Dongen S, Ouzounis CA (2002) An efficient algorithm for large-scale detection of protein families. Nucleic Acids Res 30:1575–1584

Ferro E, Trabalzini L (2013) The yeast two-hybrid and related methods as powerful tools to study plant cell signalling. Plant Mol Biol 83:287–301. https://doi.org/10.1007/s11103-013-0094-4

Fields S, Song O-K (1989) A novel genetic system to detect protein-protein interactions. Nature 340:245–246. https://doi.org/10.1038/340245a0

Fischer S et al (2011) Using OrthoMCL to assign proteins to OrthoMCL-DB groups or to cluster proteomes into new ortholog groups. Curr Protoc Bioinformatics Chapter 6:Unit 6.12.11–Unit 6.12.19. https://doi.org/10.1002/0471250953.bi0612s35

Ghosh I, Hamilton AD, Regan L (2000) Antiparallel leucine zipper-directed protein reassembly: application to the green fluorescent protein. J Am Chem Soc 122:5658–5659. https://doi.org/10.1021/ja994421w

Gunzl A, Schimanski B (2009) Tandem affinity purification of proteins. Curr Protoc Prot Sci Chapter 19:Unit.19.19. https://doi.org/10.1002/0471140864.ps1919s55

Hayes S, Malacrida B, Kiely M, Kiely PA (2016) Studying protein-protein interactions: progress, pitfalls and solutions. Biochem Soc Trans 44:994–1004. https://doi.org/10.1042/BST20160092

Ito T, Chiba T, Ozawa R, Yoshida M, Hattori M, Sakaki Y (2001) A comprehensive two-hybrid analysis to explore the yeast protein interactome. Proc Natl Acad Sci U S A 98:4569–4574. https://doi.org/10.1073/pnas.061034498

Kaiser P, Meierhofer D, Wang X, Huang L (2008) Tandem affinity purification combined with mass spectrometry to identify components of protein complexes. In: Starkey M, Elaswarapu R (eds) Methods in molecular biology. Humana Press, Totowa, NJ, pp 309–326. https://doi.org/10.1007/978-1-59745-188-8_21

Kerppola TK (2006) Visualization of molecular interactions by fluorescence complementation. Nat Rev Mol Cell Biol 7:449–456. https://doi.org/10.1038/nrm1929

Kerrien S et al (2007) Broadening the horizon--level 2.5 of the HUPO-PSI format for molecular interactions. BMC Biol 5:44. https://doi.org/10.1186/1741-7007-5-44

Keskin O, Tuncbag N, Gursoy A (2016) Predicting protein-protein interactions from the molecular to the proteome level. Chem Rev 116:4884–4909. https://doi.org/10.1021/acs.chemrev.5b00683

Kubota A et al (2017) TCP4-dependent induction of CONSTANS transcription requires GIGANTEA in photoperiodic flowering in Arabidopsis. PLoS Genet 13:e1006856

Kunzl F, Fruholz S, Fassler F, Li B, Pimpl P (2016) Receptor-mediated sorting of soluble vacuolar proteins ends at the trans-Golgi network/early endosome. Nat Plants 2:16017. https://doi.org/10.1038/nplants.2016.17

Lee HW, Kang NY, Pandey SK, Cho C, Lee SH, Kim J (2017) Dimerization in LBD16 and LBD18 transcription factors is critical for lateral root formation. Plant Physiol 174:301–311. https://doi.org/10.1104/pp.17.00013

Li L, Stoeckert CJ Jr, Roos DS (2003) OrthoMCL: identification of ortholog groups for eukaryotic genomes. Genome Res 13:2178–2189. https://doi.org/10.1101/gr.1224503

Li S, Ying Y, Secco D, Wang C, Narsai R, Whelan J, Shou H (2017) Molecular interaction between PHO2 and GIGANTEA reveals a new crosstalk between flowering time and phosphate homeostasis in Oryza sativa. Plant Cell Environ 40:1487–1499. https://doi.org/10.1111/pce.12945

Licata L et al (2012) MINT, the molecular interaction database: 2012 update. Nucleic Acids Res 40:D857–D861. https://doi.org/10.1093/nar/gkr930

Liu G, Wong L, Chua HN (2009) Complex discovery from weighted PPI networks. Bioinformatics 25:1891–1897

Lonn P, Landegren U (2017) Close encounters - probing proximal proteins in live or fixed cells. Trends Biochem Sci 42:504–515. https://doi.org/10.1016/j.tibs.2017.05.003

Louche A, Salcedo SP, Bigot S (2017) Protein–Protein Interactions: Pull-Down Assays. In: Journet L, Cascales E (eds) Bacterial protein secretion systems. Methods in Molecular Biology, vol 1615. Humana Press, New York, NY. https://doi.org/10.1007/978-1-4939-7033-9_20

Lund CH, Bromley JR, Stenbæk A, Rasmussen RE, Scheller HV (2015) A reversible Renilla luciferase protein complementation assay for rapid identification of protein – protein interactions reveals the existence of an interaction network involved in xyloglucan biosynthesis in the plant Golgi apparatus. J Exp Bot 66:85–97. https://doi.org/10.1093/jxb/eru401

Maloney VJ, Park JY, Unda F, Mansfield SD (2015) Sucrose phosphate synthase and sucrose phosphate phosphatase interact in planta and promote plant growth and biomass accumulation. J Exp Bot 66:4383–4394. https://doi.org/10.1093/jxb/erv101

Moreno-Hagelsieb G, Latimer K (2008) Choosing BLAST options for better detection of orthologs as reciprocal best hits. Bioinformatics 24:319–324. https://doi.org/10.1093/bioinformatics/btm585

O'Brien KP, Remm M, Sonnhammer EL (2005) Inparanoid: a comprehensive database of eukaryotic orthologs. Nucleic Acids Res 33:D476–D480. https://doi.org/10.1093/nar/gki107

Ogden TH, Rosenberg MS (2007) How should gaps be treated in parsimony? A comparison of approaches using simulation. Mol Phylogenet Evol 42:817–826

Orchard S et al (2012) Protein interaction data curation: the International Molecular Exchange (IMEx)

consortium. Nat Methods 9:345–350. https://doi.org/10.1038/nmeth.1931

Orchard S et al (2014) The MIntAct project--IntAct as a common curation platform for 11 molecular interaction databases. Nucleic Acids Res 42:D358–D363. https://doi.org/10.1093/nar/gkt1115

Pauwels L et al (2010) NINJA connects the co-repressor TOPLESS to jasmonate signalling. Nature 464:788–791. https://doi.org/10.1038/nature08854

Perkins JR, Diboun I, Dessailly BH, Lees JG, Orengo C (2010) Review transient protein-protein interactions: structural, functional, and network properties. Struct Fold Des 18:1233–1243. https://doi.org/10.1016/j.str.2010.08.007

Pfleger KDG, Eidne KA (2006) Illuminating insights into protein-protein interactions using bioluminescence resonance energy transfer (BRET). Nat Methods 3:165–174. https://doi.org/10.1038/nmeth841

Piston DW, Kremers G-J (2007) Fluorescent protein FRET: the good, the bad and the ugly. Trends Biochem Sci 32:407–414. https://doi.org/10.1016/j.tibs.2007.08.003

Ransone LJ (1995) Detection of protein-protein interactions by coimmunoprecipitation and dimerization. In: Methods in enzymology, vol 254. Academic Press, New York, NY, pp 491–497. https://doi.org/10.1016/0076-6879(95)54034-2

Rao VS, Srinivas K, Sujini GN, Kumar GNS (2014) Protein-protein interaction detection: methods and analysis. Int J Proteom 2014:147648. https://doi.org/10.1155/2014/147648

Rigaut G, Shevchenko A, Rutz B, Wilm M, Mann M, Séraphin B (1999) A generic protein purification method for protein complex characterization and proteome exploration. Nat Biotechnol 17:1030–1032

Roux M et al (2011) The Arabidopsis leucine-rich repeat receptor – like kinases BAK1/SERK3 and BKK1/SERK4 are required for innate immunity to hemibiotrophic and biotrophic pathogens. Plant Cell 23:2440–2455. https://doi.org/10.1105/tpc.111.084301

Salwinski L, Miller CS, Smith AJ, Pettit FK, Bowie JU, Eisenberg D (2004) The database of interacting proteins: 2004 update. Nucleic Acids Res 32:D449–D451

Seung D et al (2017) Homologs of PROTEIN TARGETING TO STARCH control starch granule initiation in arabidopsis leaves. Plant Cell 29:1657–1677. https://doi.org/10.1105/tpc.17.00222

Sonnhammer EL, Ostlund G (2015) InParanoid 8: orthology analysis between 273 proteomes, mostly eukaryotic. Nucleic Acids Res 43:D234–D239. https://doi.org/10.1093/nar/gku1203

Stark C, Breitkreutz BJ, Reguly T, Boucher L, Breitkreutz A, Tyers M (2006) BioGRID: a general repository for interaction datasets. Nucleic Acids Res 34:D535–D539

Thompson JD, Plewniak F, Poch O (1999) A comprehensive comparison of multiple sequence alignment programs. Nucleic Acids Res 27:2682–2690

Van Leene J, Witters E, Inze D, De Jaeger G (2008) Boosting tandem affinity purification of plant protein complexes. Trends Plant Sci 13:517–520. https://doi.org/10.1093/dnares/dsn008

Van Leene J et al (2011) Isolation of transcription factor complexes from arabidopsis cell suspension cultures by tandem affinity purification. In: Yuan L, Perry SE (eds) Methods in molecular biology. Humana Press, Totowa, NJ, pp 195–218. https://doi.org/10.1007/978-1-61779-154-3_11

White MA (1996) The yeast two-hybrid system: forward and reverse. Proc Natl Acad Sci U S A 93:10001–10003

Wiens MD, Campbell RE (2018) Surveying the landscape of optogenetic methods for detection of protein-protein interactions. Wiley Interdiscip Rev Syst Biol Med 10:e1415. https://doi.org/10.1002/wsbm.1415

Wong L (2008) Constructing more reliable protein-protein interaction maps. In: International Symposium on Computational Biology & Bioinformatics, University of Kerala, 17–19 January 2008, pp 284–297

Xenarios I, Salwinski L, Duan XJ, Higney P, Kim SM, Eisenberg D (2002) DIP, the Database of Interacting Proteins: a research tool for studying cellular networks of protein interactions. Nucleic Acids Res 30:303–305

Xu Y, Piston DW, Johnson CH (1999) A bioluminescence resonance energy transfer (BRET) system: application to interacting circadian clock proteins. Proc Natl Acad Sci U S A 96:151–156

Xu X, Soutto M, Xie Q, Servick S, Subramanian C, von Arnim AG, Johnson CH (2007) Imaging protein interactions with bioluminescence resonance energy transfer (BRET) in plant and mammalian cells and tissues. Proc Natl Acad Sci U S A 104:10264–10269. https://doi.org/10.1073/pnas.0701987104

Yates JR, Ruse CI, Nakorchevsky A (2009) Proteomics by mass spectrometry: approaches, advances, and applications. Annu Rev Biomed Eng 11:49–79. https://doi.org/10.1146/annurev-bioeng-061008-124934

Zahiri J, Bozorgmehr JH, Masoudi-Nejad A (2013) Computational prediction of protein-protein interaction networks: algorithms and resources. Curr Genom 14:397–414. https://doi.org/10.2174/1389202911314060004

Probabilistic Graphical Models Applied to Biological Networks

7

Natalia Faraj Murad and Marcelo Mendes Brandão

Abstract

Biological networks can be defined as a set of molecules and all the interactions among them. Their study can be useful to predict gene function, phenotypes, and regulate molecular patterns. Probabilistic graphical models (PGMs) are being widely used to integrate different data sources with modeled biological networks. The inference of these models applied to large-scale experiments of molecular biology allows us to predict influences of the experimental treatments in the behavior/phenotype of organisms. Here, we introduce the main types of PGMs and their applications in a biological networks context.

Keywords

System biology · Biological networks · Bioinformatics

N. F. Murad · M. M. Brandão (✉)
Center for Molecular Biology and Genetic Engineering, State University of Campinas, Campinas, São Paulo, Brazil
e-mail: brandaom@unicamp.br

7.1 Biological Regulatory Networks

Biological networks are a generic term that embraces protein–protein interaction (PPI) network, gene regulatory networks and metabolic networks. This study can be useful to predict gene function and/or interactions, identify functional associations, detect modular complexes that work together, identify groups of genes responsible for phenotypic characteristics.

The research is based on the central dogma in which the flow of information starts from DNA, it is transcribed in a RNA and can be translated into a protein. This process results in the production of the specific biomolecules (RNAs, proteins) needed for the realization and maintenance of the vital cellular activity. Changes in this flow regulates molecules concentration and ensures that all intermediaries of metabolic pathways will be in adequate quantities and the pathways and processes will be deactivated when not needed.

DNA modifications or compaction can influence which sequences are available for transcription as well as the transcription rate. The amount of protein produced depends on the amount of mRNA synthetized and the rate at which it is degraded. There are also posttranscriptional regulations through alternative splicing, mRNAs transport outside the nucleus and its stabilization, microRNAs pathways. The availability of factors necessary for translation may also regulate the

F. V. Winck (ed.), *Advances in Plant Omics and Systems Biology Approaches*, Advances in Experimental Medicine and Biology 1346, https://doi.org/10.1007/978-3-030-80352-0_7

amount of mRNA that is translated, and noncoding RNAs and proteins can cause posttranslational modifications in translation products, also affecting gene expression.

The biological systems involve several molecular entities connected by direct interactions that might lead to the activation or repression of more complex networks. These networks are characterized by the set of molecular species and their interactions, which together control and affect several cellular processes (Karlebach and Shamir 2008). They can represent the set of interactions among genes, transcription factors, and other biomolecules. All genes depend on one or more biochemical signals as an activator to start the transcription or an inhibitor to prevent or reduce their expression even if in the presence of an appropriated activator (Chen et al. 2001).

Data availability from high-throughput techniques and public molecular biology databases has provided the opportunity to conduct genome-wide studies of biological processes that arise from interactions between genetic entities (Costa et al. 2008) and combine them with other data sources, amplifying our assumptions through the observation of the amount of molecules and the application of mathematical methods and statistical analysis.

7.2 Probabilistic Graphical Models

Modern molecular biology experiments, such as chip-arrays, RNA-seq, mass spectrometry, allowed us to measure the behavior of thousands of genes and/or proteins simultaneously under different conditions. Moreover, data science has provided tools and the ability to integrate many information resources, increasing the analytical power that enhances our capability for new discoveries and better visualizations.

Probabilistic Graphical Models (PGM) are a combination of graph and probability theories (Larrañaga et al. 2012), and they are models of joint distributions in which paradigms of random sampling are assumed (Dobra et al. 2004). It is a powerful tool used to analyze and visualize conditional dependencies among genes or other biological entities. These graphs have nodes that are different variables involved in a problem and the edges are assumptions of conditional dependence, in other words, probabilistic relationships between them (Buntine 1996). When there is no edge it is assumed that the variables are independents (Buntine 1996). The most used types of PGM in biology are Bayesian networks (Pearl 1997; Young et al. 2014; Lan et al. 2016; Su et al. 2013; Nagarajan and Scutari 2013; Vera-Licona et al. 2014; Spirtes et al. 2000; Friedman et al. 2000a; Bansal et al. 2007) and Graphical Gaussian Models (Wu et al. 2003; Schäfer and Strimmer 2005; Werhli et al. 2006; Ma et al. 2007).

These models are useful to describe explicitly our hypothesis about the relationship between two types of data: gene expression and binding sites (Friedman 2004). The learning process can combine evidences from several datasets and more robust conclusions due to the different treatment given to datasets inside only one model (Friedman 2004).

In a Graphical Model, the first-order neighbors of a specific gene *A* are genes that, together, provide the set of predictors of variation of the gene expression in *A*, and they become *A* conditionally independent of all other genes (Dobra et al. 2004). Some genes can be slightly related to *A*, but hugely correlated in terms of partial correlation considering other neighbors and then they will be identified as neighbors of *A* (Dobra et al. 2004).

This kind of relationship between a pair of genes could not be identified by methods based on correlation, and they can be important in a biological pathway (Dobra et al. 2004). Then, another key interest applied on these models is the capacity of expanding our ability to identify genes candidates that can play important role in a specific pathway under investigation (Dobra et al. 2004).

Algorithms for PGM inference can be classified in two types (Su et al. 2013; Scutari and Nagarajan 2011; Berkan Sesen et al. 2013): *Constraint-based methods* that focus on identifying conditional independence relationships between variables using observed data and *Score-based methods*, which consider the number of all possible network structures and assign a score to each one that measures how well it explains the observed set of data. These approaches require high computational effort.

7.3 Data Preparation and Normalization

The first step inferring networks is to make sure the quality of the data is good and all values on datasets are comparable. Working with data from different experiments requires normalization to reduce noise due to the measures and experimental differences.

Noise and its variation across molecular entities are caused by factors, such as: (1) stochastic mechanisms from systems dynamics; (2) sensitivity and precision of the experiment measurement; (3) abundant variations of specific molecules; (4) preferential binding affinities; and (5) experimental artifacts that are an outcome of the estimation process (Nagarajan and Scutari 2013; Okoniewski and Miller 2006; Steen 1992; Welsh et al. 2015).

These variations are caused by uncontrollable errors unrelated to the biological variation, and they are also known as batch effects (Gagnon-Bartsch and Speed 2012). Batch effects are a type of potential latent variable in genomic experiments (Leek et al. 2012). Understanding and identifying the impact of the source of noise on network inference procedure is a critical point to avoid identification of spurious associations (Nagarajan and Scutari 2013) and ensure the reproducibility of the experiments.

Such complex phenotypic traits are characterized by a high level of unpredictability, as both the number and nature of the interactions are difficult to distinguish using conventional methods (Su et al. 2013). It is required the development of tools and methods able to transform all these heterogeneous data into biological knowledge about the underlying mechanism (Larrañaga et al. 2006).

Most probabilistic methods are computationally expensive, thus, sometimes, genes of interest need to be selected with a previous analysis. Some methods also require previous discretization of the data and this process can cause loss of information.

From a statistical point of view, the great problem of biological experiments is that the number of samples in biological large-scale experiments is insufficient to estimate accurate networks and entails a large number of spurious edges in the model (Husmeier 2003). The number of genes (p) is much larger than the number of observations (n) (Wang et al. 2013; Djebbari and Quackenbush 2008; Tamada et al. 2003). In addition, the number of actual regulators for a particular gene is only a small fraction of the number of possible regulators (Young et al. 2014). It is important to select a model or algorithm able to overcome this problem and work with a minimal number of samples.

Making effective use of prior knowledge is also crucial in any inference problem (Jaynes 1984), because it is usually vastly underdetermined due to insufficient data quantity and/or quality. Prior information about the topology of the network can be available from two different sources (Vera-Licona et al. 2014): (1) from previous biological knowledge and (2) from knowledge acquired from the prior use of another inference method, thus applying other method as a "meta-inference method."

The former is available in several public databases, such as National Center for Biotechnology Information—NCBI database (Agarwala et al. 2016), Gene Ontology (Ashburner et al. 2000), Kyoto encyclopedia for genes and genomes—KEGG (Kanehisa and Goto 2000), MetaCyc, EcoCyC (Karp 2000), and Plant Metabolic Pathways Database—PlantCyC, MIPS yeast pathways (Guldener 2006), MEROPS (Rawlings et al. 2016), BioGRID (Stark 2006), Reactome (Joshi-Tope et al. 2005), etc. The latter can be included limiting the space of all possible network topologies. It can be done setting the probability of weak connections to 0, simply imposing a maximum number of regulators per gene (Vera-Licona et al. 2014) or choosing which genes can be regulators in the network.

7.4 Graph Theory and Biological Networks

Graph theory describes the properties that allow the mathematic representation and understanding of the networks (Garroway et al. 2008). The use of graphs allows the probability model to be structurally visualized without losing mathematical details. These approaches use the observations to fill many details of the model and also provide principles to combine local multiple models in a joint global model (Friedman 2004).

A network is composed of a set of random variables $X = \{X_1, X_2, \ldots, X_n\}$ describing the observations and a graph denoted as $G = (V, E)$ in which each node or vertex $v \in V$ is associated with one of the random variables in X. The edges $e \in E$ express the dependence/coexpression relationships among the variables X.

X can be composed of variables with continuous or discrete values. Thus, the set of nodes V is given by $V = \Delta \cup \Gamma$, where Δ and Γ are sets of nodes representing discrete (values that can be counted) and continuous (values obtained from measures) variables (Bøttcher and Dethlefsen 2003a). The set of variables X can be denoted $X = (X_v)_{v \in V} = (I, Y) = ((I_\delta)_{\delta \in \Delta}, (Y_\gamma)_{\gamma \in \Gamma})$, where I and Y are sets of discrete and continuous variables (Bøttcher and Dethlefsen 2003a). For a discrete variable, δ, Γ_δ denotes the set of levels (Bøttcher and Dethlefsen 2003a).

Each edge $e_p = (i, j)$ represents the connection between the vertices i and j. If the edges have direction, the graph is called a *directed* graph, and G is an ordered pair $G = (V, E^{\rightarrow})$, where V is the set of vertices and $E^{\rightarrow}$ is the set of ordered pairs of *arcs*. The edge origins from node i to node j are called regulator and target, respectively. If there is no information about the direction, the graph is *undirected.* In addition, the graph is *signed* when its edges indicate if the relationships between the nodes are positive/induction or negative/repression. When the relationships are not indicated, the graph is *unsigned*.

7.4.1 Concepts and Properties

Betweenness: number of shortest paths between pairs of other vertices that pass through the vertice v (Garroway et al. 2008).

Centrality measures the relative position of a node or an edge in terms of connectivity or facility of node interaction (Albert 2005). It is based on the number of shortest paths that pass through the given vertex. A node with high centrality is thus crucial to efficient communication (He et al. 2007).

Connection density: the ratio of number of edges in the network by the total number of all possible edges (Sharan et al. 2007).

Clustering Coefficient: a measure of the probability of two neighbors of a given node being themselves connected (Garroway et al. 2008). This coefficient is given by the ratio (Costa et al. 2008):

$$cci = \frac{\text{number of edges among neighbors } i}{\text{max .number possible among neighbors}}$$

Degree distribution: the distribution of node degree values of a network, and it is important to measure network topology (Garroway et al. 2008). It represents the probability of a node having a given number of degree (Costa et al. 2008).

Diameter: maximum distance between any two of its vertices (Sharan et al. 2007).

Guilt by Association: genes with similar functional properties will tend to interact (Ballouz et al. 2015) or stay more connected in the network and exhibit similar profiles in network data.

Hubs: nodes with high degree or high centrality (Bullmore and Sporns 2009).

Module: a set of nodes that have strong interactions and a common function (Alon 2003).

Network Motifs: a set of genes or gene products with specific molecular functions arranged together corresponding to patterns of connectivity which are statistically more abundant in real networks than in their respectively random versions (Costa et al. 2008). The most common are the *feed-forward loops*, in which a transcription factor and its downstream target both regulate a third target (Alm and Arkin 2003).

Node degree: the number of direct connections that it establishes in the network (Lesne 2006). When the graph is directed, it can be distinguished as *incoming-degree*, considering the number of edges that connects it to the node and as *outgoing-degree*, considering the number of edges that originates from the node (Morris et al. 2012).

Path Length: minimum number of edges that must be traversed to go from one node to another (Sharan et al. 2007).

Robustness refers either to the structural integrity of the network following deletion of nodes or edges or to the effects of perturbations on local or global network states (Bullmore and Sporns 2009).

7.4.2 Types of Topology

The topology of a network is how its nodes and edges are arranged. It can provide deeper information about the network because it allows the extraction of information about biological significance across the comparisons with patterns and rules already known (Garroway et al. 2008). There are basically three types of topology:

Random: also known as Erdös–Rényi (ER) model. Each pair of nodes has an equal probability, *p*, of being connected (Bollobas 1984). A class of networks characterized by a short path length, the node degree follows a Poisson distribution indicating that most nodes have, approximately, the same number of links (Barabasi and Oltvai 2004), and a small average clustering coefficient (Garroway et al. 2008).

Scale-free: its degree distributions follow a power law (Barabasi and Oltvai 2004). Each new node that is added to the network connects, preferentially, to other nodes that already have high degree. A class of networks characterized by a short path length and a power law degree distribution. The average clustering coefficient can vary. Most nodes have relatively few connections, while a few nodes are highly connected (Barabasi and Oltvai 2004). Because most nodes are not particularly well connected, the random removal of even a high proportion of nodes tends to have little impact on the network's characteristic path length (Bullmore and Sporns 2009). However, the targeted removal of the most connected nodes leads to a rapid increase in the characteristic path length and network fragmentation (Garroway et al. 2008).

Small-world: this property combines high levels of local clustering among nodes of a network and short paths that globally link all nodes of the network. This means that all nodes of a large system are linked through relatively few intermediate steps, despite the fact that most nodes maintain only a few direct connections (Bullmore and Sporns 2009). It presents a large mean clustering coefficient and binomial degree distribution (Garroway et al. 2008). If a node is removed, the high feature of clustering might create alternate paths between nodes and such impact of node removal could be less than on random networks (Garroway et al. 2008).

7.4.3 Bayesian Networks (BNs)

Bayesian networks (BNs) are a type of model that represents causal relationships between the variables (Pearl 1997). They represent the dependency structure between multiple entities that interact locally and are useful to describe process in which the value of each component depends directly on the values of a relatively small number of components (Friedman et al. 2000a). The probabilistic nature of this approach is able to deal with the noise from biological process and experiments, making a robust and reliable inference (Husmeier 2003).

BN consists of two components (Aguilera et al. 2011): (1) A network structure in the form of a directed graph. In this graph, nodes represent the random variables, and directed edges represent stochastic dependencies among variables; (2) a set of conditional probability distributions specified by parameters, one for each variable, characterizing the stochastic dependencies represented by the edges.

The network structure is represented by graph *G*, as described above. For each node *v*, with parents pa(*v*), a local probability distribution, *p*(*xv*|*x*pa(*v*)) is defined (Bøttcher and Dethlefsen 2003a). This set of local probability distribution for all variables in the network is *P*. The Bayesian network for a set of random variables *X* is the pair (*G*, *P*) (Bøttcher and Dethlefsen 2003a). The relationships are originated from a node, called parent node, and reach another, called descendent node.

BN can have both discrete and continuous variables. It is an important tool to integrate different types of datasets. It gives us the opportunity of visualizing which variables are more related to a specific state and which one contributes to the increase/decrease or activation/deactivation of a state.

A rule that characterizes a BN is the Markov condition (Klinke et al. 2014; Friedman et al. 2000b). It states that a node is conditionally independent of its non-descendants given its parents (Werhli et al. 2006). Based on this, the joint probability distribution for the entire set of variables represented by BN can be decomposed into a product of conditional probabilities (Su et al. 2013) following the equation:

$$P(X_1,X_2,\ldots,X_n)=\prod_{i=1}^{n}P(X_i \mid \pi_M D(X_i))$$

where X_i are random variables and π_M (X_i) are parents of a node X_i in the M model.

The base of many search algorithms for BN learning from data is in this rule because it allows the division of the total probability distribution in modules that can interconnect, and the network will show all the significant dependencies without disintegrating the information (Sebastiani et al. 2005). This property reduces the number of necessary parameters to characterize the joint probability distribution of the variables (Sebastiani et al. 2005).

Inferring a model of BN for biological networks requires finding a graph G that better describes the dataset d. It is realized by the choice of a function that measures the score of each possible topology of a network (graph G) in relation to the dataset and then it searches for the graph that maximizes the *score* (Bansal et al. 2007). The metric punctuation can be defined using the Bayes' Theorem:

$$P(G \mid d)=\frac{P(d \mid G)*P(G)}{P(d)}$$

in which $P(G)$ can contain a prior knowledge about the network structure, if available, or can be a non-informative constant prior; $P(d|G)$ is a function to be chosen by the algorithm that evaluates the probability of the data d being generate by the graph G (Bansal et al. 2007).

The most popular methods for score quantification are the *Minimal Description Length* (MDL) and the *Bayesian Dirichlet Equivalence* (BDe). Both methods incorporate a penalty for the complexity to avoid over adjusted data (Wilczyński and Dojer 2009). BDe score corresponds to the posterior probability of a network to a dataset (Wilczyński and Dojer 2009). MDL score is originated from Information Theory and corresponds to the length of the data compressed with the compression model derived from the network structure, and it also has a statistic interpretation with an approach of the posterior probability (Wilczyński and Dojer 2009).

Testing all possible combinations of interactions between the genes and choosing a graph with higher score is a complex problem with high computational cost, which increases exponentially with the number of variables. Thereby heuristic methods are used, such as *Greedy-Hill Climbing*, Monte Carlo Markov Chain (MCMC), or *Simulated Annealing* (Heckerman et al. 1995). Greedy search examines a single random change in the network and keep it if the score is higher or discard if it is slower than the current. Simulated Annealing uses some parameters to do the search. MCMC samples consider a probability distribution.

BNs can be static when derived from steady-state data and allow only acyclic graphs (Bansal et al. 2007). This constraint is disadvantageous since feedback mechanisms are an essential feature of biological systems (Husmeier 2003). BNs can also be dynamic, derived from dynamic and time-series datasets, which can be used to predict cyclic phenomena, overcoming the static BN limitation (Yu et al. 2004).

The nodes in the graph represent the stochastic variables, and the edges directed dependency relationships between the variables, quantified by the conditional probability distributions (Sebastiani et al. 2005). The family of conditional probability distributions and their parameters specify the functional form of the conditional probabilities associated with the edges, that is, they indicate the nature of the interactions and the intensities of these interactions (Werhli et al. 2006).

Figure 7.1 shows an example of a BN. This network was obtained using microarray data from sugarcane (McCormick et al. 2008). Sugarcane leaves were treated with exposition to cold (treated leaf) or normal temperature (control leaf). We can observe, in this example, a gene related with stress response that is a hub in the network. Its expression is activated in the "*Treated Leaf*" and suppressed in the "*Control Leaf*." Moreover, this central gene is related with the expression of several other genes of stress response, and it suppresses a gene of sugar signaling, an important category for this kind of experiment.

There are many software formats for BN inferring. Here, we list some of them:

- *BaNJO* (https://users.cs.duke.edu/~amink/software/banjo/): developed in Java, efficient for large datasets. It can infer static or dynamic networks. It requires high memory when working with many experiments. Low performance with noisy data. Allows multiprocessing and time limit. Accepts only discrete variables.
- *BNFinder* (Dojer et al. 2013): it can infer both static and dynamic networks. It allows configuration of prior information. There are three types of scoring BDe (Bayesian Dirichlet equivalence), MIT (Mutual Information Test) and MDL (Minimum Description Length). The choice of the score depends on the size of the dataset and computational cost. Developed in Python. Memory can be a limiting factor. Allows multiprocessing and continuous or discrete variables.
- *PEBL* (Shah and Woolf 2013): it is a Python library. Allows both discrete or continuous variables and multiprocessing. It uses BDe score and Gibbs Sampler. It has low sensitivity, so it generates network with low density.
- *ScanBMA* (Young et al. 2014): it searches the model space more efficiently and thoroughly than previous algorithms. It infers dynamic networks. It uses Bayesian Information Criterion (BIC) as scoring function. Only dynamic networks.
- *R packages*:
  – *deal* (Bøttcher and Dethlefsen 2003b) and *BNArray* (Chen et al. 2006): *deal* owns proceedings to include prior, estimate parameters, calculate scores. *BNArray* provide features in the context of gene expression. They work together. Used for small networks.
  – *bnlearn* (Scutari and Nagarajan 2011) and *gRain* (Højsgaard 2012): it allows discrete or continuous variables. Possibility of using several learning and classifier meth-

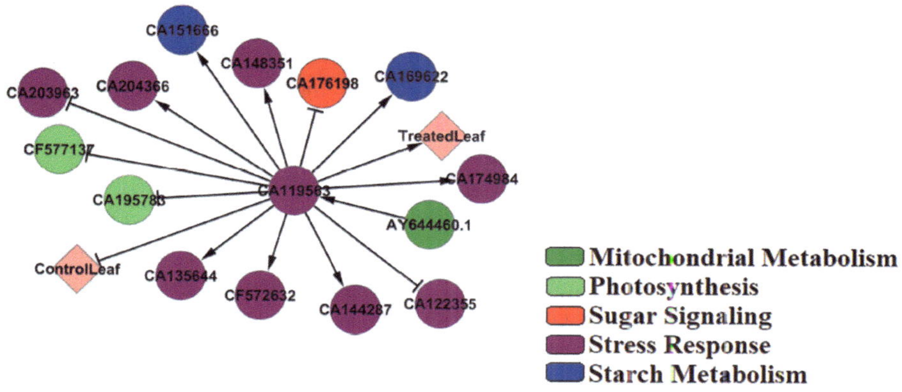

Fig. 7.1 Example of a Bayesian network. Subnetwork learnt from microarray data of leaves of sugarcane exposed to cold and control situation. Circles represent the genes and diamonds represent discrete variables, in this case, the treatments (Murad 2013)

ods for the network. *gRain* creates conditional probability tables.
- *BANFF* (Young et al. 2014): data preprocessing, efficient Bayesian model fitting with diagnostics, quantitatively and graphically summarizing posterior samples of parameters.

7.4.4 Graphical Gaussian Models (GGMs)

GGMs, also known as covariance selection models, are a strategy for undirected graphs G inference. It assumes multivariate Gaussian distribution for underlying data (Wang et al. 2013) and satisfies the pairwise conditional independence restrictions, which are shown in the independence graph of a jointly normal set of random variables (Wu et al. 2003). The inference of GGMs is based on a (stable) estimation of the covariance matrix of this distribution (Werhli 2012). This kind of inference has lower computational cost when compared with BNs.

An edge connects a pair of genes if they are partially correlated (Krämer et al. 2009). Partial correlation only measures the strength of direct interaction and is used to infer conditional independence relationships (Krämer et al. 2009). Once a direct gene association network is complete, the knowledge about indirect gene associations can be easily obtained (Wang et al. 2013). When compared with traditional correlation methods, they allow to distinguish direct and indirect relationships, so the graph is sparser, or less dense.

For two random variables, X_i and X_j, on the remaining variables in the data set, the partial correlation coefficient between X_i and X_j is given by the Pearson correlation of the residuals from both regressions (Krumsiek et al. 2011). The effects of all other variables on X_i and X_j are removed and the remaining signals are compared to test if the variables are still correlated (Krumsiek et al. 2011). Then it is determined if the association of X_i and X_j is direct and not mediated by the other variables (Krumsiek et al. 2011).

For an undirected graph $G = (V, E)$, let $X_v \equiv X$ be a random normal p-vector indexed by $V = \{1, \ldots, p\}$ with probability distribution P_v. For a GGM with graph G, the sufficient statistics are given by the sample mean vector and the sample covariance matrices S_{CC} for $C \in C$, where C is the set of cliques of G. A condition is necessary for the computation of several statistical quantities, such as the maximum likelihood estimates, and the partial correlations are that S_{CC} has full rank for all $C \in C$.

Figure 7.2 represents a subnetwork of a Gaussian Graphical Model for microarray data from *Arabidopsis thaliana* (Ma et al. 2007). Partial correlation was used to infer co-regulation patterns between gene pairs conditioned on the behavior of other genes. They found locally coherent subnetworks mainly related to metabolic functions, and stress responses emerged.

Some R packages used to infer these models are listed below:

GeneNet (Schäfer and Strimmer 2005): it uses time-series expression data. It has functions to estimate partial correlation matrix, assesses significance of the edges, and generates the network.

gRc (Højsgaard and Lauritzen 2007): two types of models: RCON models: selected elements of the concentration matrix (the inverse covariance matrix) are restricted to being identical. RCOR models: the partial correlations are restricted to being equal rather than the concentrations.

FastGGM (Wang et al. 2016): It quickly estimates the precision matrix, partial correlations, as well as p-values and confidence intervals for the graph.

GGMridge (Ha et al. 2015): it implements an approach of three steps. Obtains a penalized estimate of a partial correlation matrix, selects non-zero entries of this matrix and reestimates the partial correlation coefficients at these non-zero entries. It promises to be a method with good accuracy on predictions.

GGMSelect (Giraud et al. 2012): It is a fast algorithm. Procedure in two stages: builds a family of candidate graphs from the data-driven

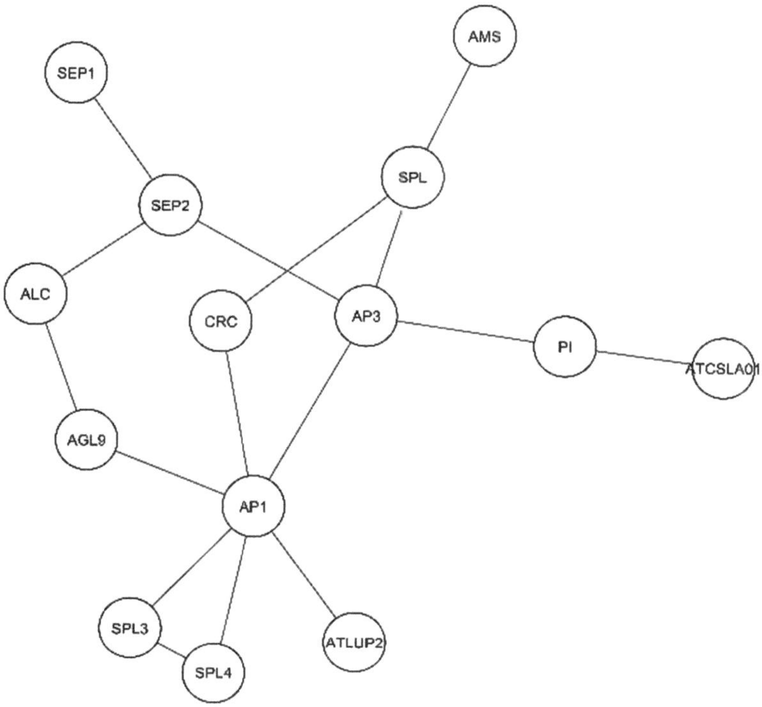

Fig. 7.2 Subnetwork of a Gaussian graphical model obtained from A. thaliana microarray data (Ma et al. 2007). Nodes represent the genes and the edges co-regulation between them

method or some prior knowledge and then selects one graph according to a dedicated criterion based on conditional least-squares. It focuses on data with samples size n smaller than number of variables p showing consistency.

7.5 Network Representation and Visualization Tools

There are four main types of network representation. First, as a graph such as the one described previously (Fig. 7.3a). Second, as an adjacency matrix A with elements a_{ij} that can be 0 for no relationship or equal to 1 when the genes are correlated. If the graph is undirected, this matrix is symmetric, the elements $a_{ij} = a_{ji}$ for any i and j. The third way is the circular plot (Fig. 7.3b). And the last, as the table with three columns, in which the first and third columns are the genes and the one in the middle is the type of interaction between them. This representation can also have more columns representing parameters of the edges.

The most used software formats for network visualization are listed below:

Cytoscape (Shannon et al. 2003): friendly graphical interface, but computationally expensive. It does not support the view of very large networks. It has plugins that provide network analysis and data integration.

Pajek (Batagelj and Marver 1998) and *Medusa* (Pavlopoulos et al. 2011): these software formats are also used in other areas of knowledge. Well-known among all programs, and they use pajek and net format files. They can work fast with large networks.

circlize (Gu et al. 2014) and *igraph* (Csárdi and Nepusz 2006): R packages. igraph is used for graph representation and *circlize* for circular plots also known as chord diagram.

Gephi (Bastian et al. 2009): it is a platform for graph visualization. Intuitive graphical interface. It can work with large networks. Also, it has plugins and shows metrics about the network.

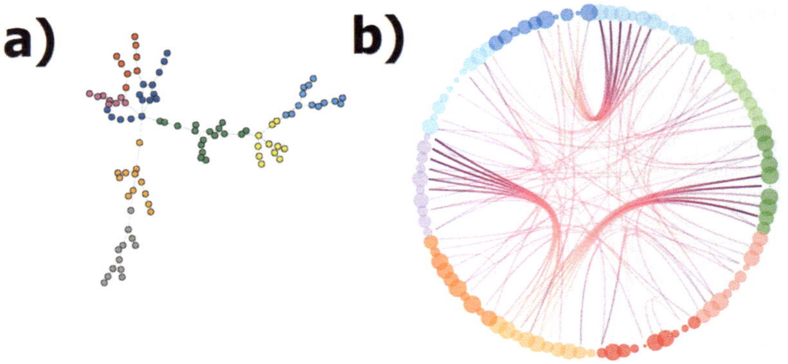

Fig. 7.3 Example of network visualization. (**a**) Visualization as a graph. (**b**) Visualization as a circular plot. (http://www.r-graph-gallery.com)

References

Agarwala R, Barrett T, Beck J, Benson DA, Bollin C, Bolton E et al (2016) Database resources of the national center for biotechnology information. Nucleic Acids Res 44(D1):D7–D19

Aguilera PA, Fernández A, Fernández R, Rumí R, Salmerón A (2011) Bayesian networks in environmental modelling. Environ Model Softw 26(12):1376–1388. https://doi.org/10.1016/j.envsoft.2011.06.004

Albert R (2005) Scale-free networks in cell biology. J Cell Sci 118(21):4947–4957. https://doi.org/10.1242/jcs.02714

Alm E, Arkin AP (2003) Biological networks. Curr Opin Struct Biol 13(2):193–202

Alon U (2003) Biological network: the tinkerer as an engineer. Science (80-) 301(September):1866–1867

Ashburner M, Ball CA, Blake JA, Botstein D, Butler H, Cherry JM et al (2000) Gene ontology: tool for the unification of biology. The Gene Ontology Consortium. Nat Genet 25(1):25–29

Ballouz S, Verleyen W, Gillis J (2015) Guidance for RNA-seq co-expression network construction and analysis: safety in numbers. Bioinformatics 31(13):2123–2130

Bansal M, Belcastro V, Ambesi-Impiombato A, di Bernardo D (2007) How to infer gene networks from expression profiles. Mol Syst Biol 3(78):78. http://www.ncbi.nlm.nih.gov/pubmed/17299415

Barabasi A, Oltvai ZN (2004) Network biology: understanding the cell's functional organization. Nat Rev Genet 5(2):101–113. http://www.ncbi.nlm.nih.gov/pubmed/14735121

Bastian M, Heymann S, Jacomy M (2009) Gephi. An open source software for exploring and manipulating networks. In: Third Int AAAI Conf Weblogs Soc Media, pp 361–362. http://www.aaai.org/ocs/index.php/ICWSM/09/paper/view/154%5Cnpapers2://publication/uuid/CCEBC82E-0D18-4FFC-91EC-6E4A7F1A1972

Batagelj V, Marver A (1998) Pajek – a program for large network analysis. Connections 21:47–57. http://vlado.fmf.uni-lj.si/pub/networks/doc/pajek.pdf

Berkan Sesen M, Nicholson AE, Banares-Alcantara R, Kadir T, Brady M (2013) Bayesian networks for clinical decision support in lung cancer care. PLoS One 8(12):1–13

Bollobas B (1984) The evolution of random graphs. Trans Am Math Soc 286(1):257. http://www.jstor.org/stable/1999405?origin=crossref

Bøttcher SG, Dethlefsen C (2003a) Learning Bayesian networks with R. DSC 2003 working paper

Bøttcher SG, Dethlefsen C (2003b) deal: a package for learning Bayesian networks. J Stat Softw 8(20):1–40. http://www.jstatsoft.org/v08/i20/paper

Bullmore E, Sporns O (2009) Complex brain networks: graph theoretical analysis of structural and functional systems. Nat Rev Neurosci 10(3):186–198

Buntine W (1996) A guide to the literature on learning probabilistic networks from data. IEEE Trans Knowl Data Eng 8(2):195–210

Chen T, Filkov V, Skiena SS (2001) Identifying gene regulatory networks from experimental data. Parallel Comput 27(1–2):141–162

Chen X, Chen M, Ning K (2006) BNArray: an R package for constructing gene regulatory networks from microarray data by using Bayesian network. Bioinformatics 22(23):2952–2954

Costa LF, Rodrigues FA, Cristino AS (2008) Complex networks: the key to systems biology. Genet Mol Biol 31(3):591–601

Csárdi G, Nepusz T (2006) The igraph software package for complex network research. Int J Complex Syst 1695:1–9

Djebbari A, Quackenbush J (2008) Seeded Bayesian Networks: constructing genetic networks from microarray data. BMC Syst Biol 2(1):57. http://www.biomedcentral.com/1752-0509/2/57

Dobra A, Hans C, Jones B, Nevins JR, Yao G, West M (2004) Sparse graphical models for exploring gene expression data. J Multivar Anal 90(1 Spec Issue):196–212

Dojer N, Bednarz P, Podsiadło A, Wilczyński B (2013) BNFinder2: faster Bayesian network learning and Bayesian classification. Bioinformatics 29(16):2068–2070

Friedman N (2004) Inferring cellular networks using probabilistic graphical models. Science (80-) 303(5659):799–805

Friedman N, Linial M, Nachman I, Pe'er D (2000a) Using Bayesian networks to analyze expression data. J Comput Biol 7:127–135. http://dl.acm.org/citation.cfm?id=332306.332355

Friedman N, Linial M, Nachman I, Pe'er D (2000b) Using Bayesian networks to analyze expression data. J Comput Biol 7(3–4):601–620

Gagnon-Bartsch JA, Speed TP (2012) Using control genes to correct for unwanted variation in microarray data. Biostatistics 13(3):539–552

Garroway CJ, Bowman J, Carr D, Wilson PJ (2008) Applications of graph theory to landscape genetics. Evol Appl 1:620–630. https://doi.org/10.1111/j.1752-4571.2008.00047.x

Giraud C, Huet S, Verzelen N (2012) Graph selection with GGMselect. Stat Appl Genet Mol Biol 11(3):3

Gu Z, Gu L, Eils R, Schlesner M, Brors B (2014) Circlize implements and enhances circular visualization in R. Bioinformatics 30(19):2811–2812

Guldener U (2006) MPact: the MIPS protein interaction resource on yeast. Nucleic Acids Res 34(90001):D436–D441. https://doi.org/10.1093/nar/gkj003

Ha MJ, Carolina N, Sun W, Carolina N (2015) Partial correlation matrix estimation using ridge penalty followed by thresholding and reestimation. Biometrics 70(3):762–770

He Y, Chen ZJ, Evans AC (2007) Small-world anatomical networks in the human brain revealed by cortical thickness from MRI. Cereb Cortex 17(10):2407–2419

Heckerman D, Geiger D, Chickering DM (1995) Learning {B}ayesian networks: the combination of knowledge and statistical data. Mach Learn 20(3):197–243. https://doi.org/10.1023/A:1022623210503

Højsgaard S (2012) Graphical independence networks with the gRain Package for R. J Stat Softw 46(10):1–26. http://www.jstatsoft.org/index.php/jss/article/view/v046i10/v46i10.pdf

Højsgaard S, Lauritzen SL (2007) Inference in graphical Gaussian models with edge and vertex symmetries with the gRc package for R. J Stat Softw 23(6):1–26

Husmeier D (2003) Sensitivity and specificity of inferring genetic regulatory interactions from microarray experiments with dynamic Bayesian networks. Bioinformatics 19(17):2271–2282

Jaynes E (1984) Prior information and ambiguity in inverse problems. In: Inverse problems, vol 14, pp 151–166. http://bayes.wustl.edu/etj/articles/ambiguity.pdf

Joshi-Tope G, Gillespie M, Vastrik I, D'Eustachio P, Schmidt E, de Bono B et al (2005) Reactome: a knowledgebase of biological pathways. Nucleic Acids Res 33(Database Issue):428–432

Kanehisa M, Goto S (2000) KEGG: Kyoto encyclopedia of genes and genomes. Nucleic Acids Res 28(1):27–30

Karlebach G, Shamir R (2008) Modelling and analysis of gene regulatory networks. Nat Rev Mol Cell Biol 9(10):770–780. http://www.ncbi.nlm.nih.gov/pubmed/18797474

Karp PD (2000) The EcoCyc and MetaCyc databases. Nucleic Acids Res 28(1):56–59. https://doi.org/10.1093/nar/28.1.56

Klinke D, Barnett J, Cuff C, et al (2014) Using Bayesian networks to identify control topography between cancer processes and immune responses via metagene constructs. Jacob Kaiser Thesis submitted to the College of Medicine at West Virginia University in partial fulfillment of the requirements

Krämer N, Schäfer J, Boulesteix A-L (2009) Regularized estimation of large-scale gene association networks using graphical Gaussian models. BMC Bioinformatics 10(1):384. https://doi.org/10.1186/1471-2105-10-384

Krumsiek J, Suhre K, Illig T, Adamski J, Theis FJ (2011) Gaussian graphical modeling reconstructs pathway reactions from high-throughput metabolomics data. BMC Syst Biol 5(1):21. http://www.biomedcentral.com/1752-0509/5/21

Lan Z, Zhao Y, Kang J, Yu T (2016) Bayesian network feature finder (BANFF): an R package for gene network feature selection. Bioinformatics 32(23):3685–3687

Larrañaga P, Calvo B, Santana R, Bielza C, Galdiano J, Inza I et al (2006) Machine learning in bioinformatics. Brief Bioinform 7(1):86–112

Larrañaga P, Karshenas H, Bielza C, Santana R (2012) A review on probabilistic graphical models in evolutionary computation. J Heuristics 18(5):795–819

Leek JT, Johnson WE, Parker HS, Jaffe AE, Storey JD (2012) The SVA package for removing batch effects and other unwanted variation in high-throughput experiments. Bioinformatics 28(6):882–883

Lesne A (2006) Complex networks: from graph theory to biology. Lett Math Phys 78(3):235–262

Ma S, Gong Q, Bohnert HJ (2007) An Arabidopsis gene network based on the graphical Gaussian model. Genome Res 17:1514–1625

McCormick AJ, Cramer MD, Watt DA (2008) Differential expression of genes in the leaves of sugarcane in response to sugar accumulation. Trop Plant Biol 1(2):142–158. https://doi.org/10.1007/s12042-008-9013-2

Morris JS, Kuchinsky A, Pico A, Institutes G (2012) Analysis and visualization of biological networks with cytoscape. UCSF, p 65. http://www.cgl.ucsf.edu/Outreach/Workshops/NIH-Oct-2012/Cytoscape/Analysis%20and%20Visualization%20of%20Biological%20Networks%20with%20Cytoscape%20v6.pdf

Murad NF (2013) REDES DE REGULAÇÃO GÊNICA DO METABOLISMO DE SACAROSE EM CANA-DE-Açúcar Utilizando Bayesianas, Redes. p 21

Nagarajan R, Scutari M (2013) Impact of noise on molecular network inference. PLoS One 8(12):e80735

Okoniewski MJ, Miller CJ (2006) Hybridization interactions between probesets in short oligo microarrays lead to spurious correlations. BMC Bioinformatics 7:1–14

Pavlopoulos GA, Hooper SD, Sifrim A, Schneider R, Medusa AJ (2011) A tool for exploring and clustering biological networks. BMC Res Notes 4(1):384. http://www.biomedcentral.com/1756-0500/4/384

Pearl J (1997) Bayesian networks. Tech Rep R-246 (Rev II). In: The MIT encyclopedia of the cognitive sciences, pp 3–6

Rawlings ND, Barrett AJ, Finn R (2016) Twenty years of the MEROPS database of proteolytic enzymes, their substrates and inhibitors. Nucleic Acids Res 44(D1):D343–D350

Schäfer J, Strimmer K (2005) Learning large-scale graphical Gaussian models from genomic data. Proc Natl Acad Sci U S A 776:263

Scutari M, Nagarajan R (2011) On identifying significant edges in graphical models of molecular networks. ArXiv. http://arxiv.org/abs/1104.0896

Sebastiani P, Abad M, Ramoni M (2005) Bayesian networks for genomic analysis. In: Genomic signal processing, pp 1–38. http://128.197.153.21/sebas/pdf-papers/gsp.pdf

Shah A, Woolf P (2013) Python environment for Bayesian learning: inferring the structure of Bayesian networks from knowledge and data. J Mach Learn Res 10:159–162

Shannon P, Markiel A, Ozier O, Baliga NS, Wang JT, Ramage D et al (2003) Cytoscape: a software environment for integrated models of biomolecular interaction networks. Genome Res 13:2498–2504

Sharan R, Ulitsky I, Shamir R (2007) Network-based prediction of protein function. Mol Syst Biol 3(88):1–13

Spirtes P, Glymour C, Scheines R, Kauffman S (2000) Constructing Bayesian network models of gene expression networks from microarray data. https://citeseerx.ist.psu.edu/viewdoc/download?doi=10.1.1.645.1959&rep=rep1&type=pdf

Stark C (2006) BioGRID: a general repository for interaction datasets. Nucleic Acids Res 34(90001):D535–D539. https://doi.org/10.1093/nar/gkj109

Steen HB (1992) Noise, sensitivity, and resolution of flow cytometers. Cytometry 13(8):822–830

Su C, Andrew A, Karagas MR, Borsuk ME (2013) Using Bayesian networks to discover relations between genes, environment, and disease. BioData Min 6(1):6. https://doi.org/10.1186/1756-0381-6-6

Tamada Y, Kim S, Bannai H, Imoto S, Tashiro K, Kuhara S et al (2003) Estimating gene networks from gene expression data by combining Bayesian network model with promoter element detection. Bioinformatics 19(Suppl 2):227–236

Vera-Licona P, Jarrah A, Garcia-Puente LD, McGee J, Laubenbacher R (2014) An algebra-based method for inferring gene regulatory networks. BMC Syst Biol 8:37. http://www.pubmedcentral.nih.gov/articlerender.fcgi?artid=4022379&tool=pmcentrez&rendertype=abstract

Wang Z, Xu W, Lucas FAS, Liu Y (2013) Incorporating prior knowledge into Gene Network Study. Bioinformatics 29(20):2633–2640

Wang T, Ren Z, Ding Y, Fang Z, Sun Z, MacDonald ML et al (2016) FastGGM: an efficient algorithm for the inference of Gaussian graphical model in biological networks. PLoS Comput Biol 12(2):1–16

Welsh IC, Kwak H, Chen FL, Werner M, Shopland LS, Danko CG et al (2015) Chromatin architecture of the Pitx2 locus requires CTCF- and Pitx2-dependent asymmetry that mirrors embryonic gut laterality. Cell Rep 13(2):337–349. https://doi.org/10.1016/j.celrep.2015.08.075

Werhli AV (2012) Comparing the reconstruction of regulatory pathways with distinct Bayesian networks inference methods. BMC Genomics 13(Suppl 5):S2. http://www.ncbi.nlm.nih.gov/pubmed/23095805

Werhli AV, Grzegorczyk M, Husmeier D (2006) Comparative evaluation of reverse engineering gene regulatory networks with relevance networks, graphical Gaussian models and Bayesian networks. Bioinformatics 22(20):2523–2531

Wilczyński B, Dojer N (2009) BNFinder: exact and efficient method for learning Bayesian networks. Bioinformatics 25(2):286

Wu X, Ye Y, Subramanian KR (2003) Interactive analysis of gene interactions using graphical gaussian model. In: BIOKDD03: 3rd ACM SIGKDD Workshop on Data Mining in Bioinformatics, pp 1–7

Young WC, Raftery AE, Yeung KY (2014) Fast Bayesian inference for gene regulatory networks using ScanBMA. BMC Syst Biol 8(1):47. http://www.ncbi.nlm.nih.gov/pubmed/24742092

Yu J, Smith VA, Wang PP, Hartemink AJ, Jarvis ED (2004) Advances to Bayesian network inference for generating causal networks from observational biological data. Bioinformatics 20(18):3594–3603

8 Cataloging Posttranslational Modifications in Plant Histones

Ericka Zacarias and J. Armando Casas-Mollano

Abstract

Eukaryotic DNA exist in the nuclei in the form of a complex with proteins called chromatin. Access to the information encoded in the DNA requires the opening of the chromatin. Modulation of the chromatin structure is therefore an important layer of regulation for DNA-templated processes. The basic unit of the chromatin is the nucleosome, which contains DNA wrapped around an octamer of histones, H2A, H2B, H3, and H4. Because histones are a structural part of the nucleosome, its modification can lead to changes in chromatin structure. Amino acid residues in histones could be modified with at least 20 different types of functional groups leading to a vast number of modified residues. Here, an overview of the histone modifications found in plants is provided. We focus mainly in proteomic-based studies either aimed to identify PTMs on purified histones or proteome-wide analysis of particular modifications. The strategies used for cataloging modifications in plants are also described. Profiling of histone modifications is important to begin to understand their functions as mediators of gene regulation in plant biological systems.

Keywords

Chromatin · Core histones · Histone modifications · Proteomics · Mass spectrometry · Acetylation · Methylation · Acylation

E. Zacarias
School of Biological Sciences and Engineering,
Yachay Tech University,
San Miguel de Urcuquí, Ecuador

J. A. Casas-Mollano (✉)
School of Biological Sciences and Engineering,
Yachay Tech University,
San Miguel de Urcuquí, Ecuador

The BioTechnology Institute, College of Biological Sciences, University of Minnesota,
Saint Paul, MN, USA

8.1 Introduction

Eukaryotic DNA is organized inside the nucleus by histones and other proteins in the polymer called chromatin. The basic structure of the chromatin has remained unchanged in all eukaryotic lineages with the nucleosome as its basic unit. The nucleosome consist of ~150 bp of DNA wrapped around an octamer containing two copies of the core histones, H2A, H2B, H3, and H4. In the chromatin, nucleosomes are organized into linear arrays that may be further arranged into higher-order structures with the help of the linker histone H1 (Luger et al. 1997).

The presence of chromatin is restrictive to transcription and other DNA-templated pro-

F. V. Winck (ed.), *Advances in Plant Omics and Systems Biology Approaches*, Advances in Experimental Medicine and Biology 1346, https://doi.org/10.1007/978-3-030-80352-0_8

cesses. Indeed, changes in chromatin structure are required for the proper expression of a gene to be reached (Struhl 1993). Consequently, mechanisms modulating accessibility to chromatin may have a higher hierarchy in the regulation of the gene expression programs encoded in the genome. Regulation of chromatin structure involves mechanisms for the stabilization and assembling of nucleosomes into higher-order structures, but also for the disruption of nucleosome condensation. Some of these mechanisms are based on the modification of DNA and histones, ATP-dependent remodeling of the chromatin and the post-replicative incorporation of histone isoforms (variants) into nucleosomes (Feng et al. 2010; Tessarz and Kouzarides 2014; Swygert and Peterson 2014).

As structural components of the nucleosome, histones, and the DNA itself are poised to become targets of mechanisms modulating chromatin structure. Regulation is usually achieved by modifying DNA with methyl groups and histones with several functional groups. Indeed, histones are subject to a myriad of posttranslational modifications (PTMs) which include phosphorylation, acetylation, methylation (monomethylation, dimethylation, and trimethylation), propionylation, butyrylation, crotonylation, 2-hydroxylisobutyrylation, β-hydroxybutyrylation, malonylation, benzoylation, succinylation, glutarylation, formylation, hydroxylation, ubiquitination, SUMOylation, O-GlcNAcylation, ADP-ribosylation, proline isomerization, and citrullination (Huang et al. 2015, 2014, 2018; Sabari et al. 2017). These modifications occur in over 120 amino acid residues, specially lysine, arginine, serine, threonine, tyrosine, and alanine, in the core, H2B, H2A, H3, and H4, and also the H1 linker histones (Tan et al. 2011b; Huang et al. 2014). PTMs of histones are usually clustered at their N-terminal tails, and the C-terminal tail of histone H2A, but there is evidence of them occurring in the globular domains as well (Mersfelder and Parthun 2006). Modifying histones with different functional groups may affect nucleosome structure in at least two ways. First, modifications, especially those changing the polarity of the residues involved, may function by disrupting histone-DNA and histone-histone interactions, leading to changes in chromatin structure. Second, histone PTMs may result in recruiting and/or eviction of binding modules, and their effector complexes, that regulate chromatin structure (Tessarz and Kouzarides 2014; Berger 2007; Kouzarides 2007; Taverna et al. 2007a). While binding modules recognize histone PTMs acting as chromatin "readers," other proteins enzymatically add or cleave modifications at specific amino acid residues becoming chromatin "writers" and "erasers," respectively (Kouzarides 2007; Tessarz and Kouzarides 2014; Taverna et al. 2007a; Berger 2007). "Writers", "readers," and "erasers" carefully regulate the dynamics of histone modifications, which in turn influence all the DNA-templated process including transcription, replication, and DNA repair among others (Kouzarides 2007; Tessarz and Kouzarides 2014).

Histone modifications, alone or in combination, may index a specific chromatin domain that like a "histone code" can be translated by effector complexes into a specific biological readout (Jenuwein and Allis 2001; Strahl and Allis 2000). However, histone modifications appear to be more complex than just a code input producing a distinct output. Instead, individual histone modifications may have different meanings that may only become clear in the context of other modifications, or even genomic regions (Berger 2007; Kouzarides 2007; Sims and Reinberg 2008; Cerutti and Casas-Mollano 2009). In addition, the functional meaning of some histone modifications may be different between lineages indicating that the histone code may not be universal (Cerutti and Casas-Mollano 2009; Feng and Jacobsen 2011; Fuchs et al. 2006).

Many histone modifications are conserved between plant and animals, even their functional meaning appear to be the same. For other histone PTMs, however, differences in their biological connotation are seemingly clear suggesting that novel functions, and also modifications, may have emerged during plant and animal evolution (Fuchs et al. 2006; Feng and Jacobsen 2011; Cerutti and Casas-Mollano 2009). Thus, elucidating the plant "dialect" of the histone language will require an initial inventory of all the histone modifications and their subsequent characterization in several plant species. This chapter provides information on the histone modifications identified in different plant species by several proteomic-based approaches. Examples of functional clues deduced from these studies are also described in order to provide examples of how these techniques may serve beyond merely cataloging histone modifications. An overview of the methods used to profile posttranslational modifications in plants is also given. Completing a catalog of histone modifications will not only be the first step into their characterization but also these same techniques may be used to obtain functional information. Considering the importance of chromatin structure in the regulation of gene expression, a complete characterization of histone modifications will be necessary for the understanding of plant biological systems.

8.2 Profiling of Histone Modifications in Plants

Initial studies of histone PTMs and their functions in plants have been carried out using modification-specific antibodies raised against a modified amino acid residue in a particular position. The use of these antibodies has allowed the characterization of function, dynamic and genome distribution of a few histone modifications. However, the availability of robust, very specific, commercially available antibodies, restrict severely and create bias in the type and number of histone modifications that could be analyzed. To overcome the problems inherent to the use of antibodies, proteomic approaches based on the use of mass spectrometry can help us to identify the site and type of modification present in a particular histone (Wu et al. 2009; Johnson et al. 2004). Together with antibody-based strategies, proteomic approaches have allowed for the identification of an increasing number of histone modification types present in many histone residues in mammalian cells (Huang et al. 2014).

Attempts to use mass spectrometry to catalog PTMs have been carried out in a variety of tissues and a number of plant species, including Arabidopsis, cauliflower, soybean, rice, sugarcane, papaya, wheat, Brachypodium, tobacco, tomato, and Physcomitrella. In addition, while some of these studies were directed to a single type of modification in whole proteomes, others concentrate on finding several PTMs in core or even single histone types (Table 8.1). With relatively few proteomic studies, profiling of histone modifications in plants is running behind that of other organisms. Indeed, from the 20 types of histone PTMs that had been reported, only nine, phosphorylation, acetylation, methylation (monomethylation, dimethylation, trimethylation), succinylation, crotonylation, butyrylation, malonylation, ubiquitination, 2-hydroxyisobutyrylation have been identified in plants so far (Table 8.1).

In terms of cataloging histone modifications plants, initial studies were first carried out in *Arabidopsis thaliana*. In this plant, Zhang et al. (2007) analyzed methylation (at lysine and arginine), acetylation (at lysine), phosphorylation (at serine, threonine, and tyrosine) and ubiquitination (at lysine) profiles in the four core histones. Mass spectrometry analysis of histone modifications from Arabidopsis suggested conservation of modification sites between plants, yeast, and animals. The conserved nature of some histone modifications, especially acetylation and methylation, was already demonstrated with the use of modification-specific antibodies and have since then confirmed in other plant species for many amino acid residues by mass spectrometry or immunoblot analysis. For instance, the residues Lysine 9 (K9), K14, K18, K23, K27, K36 of histone H3 and K5, K8, K12, K16, K20 of H4 are all commonly acetylated in Arabidopsis, sugarcane, and animals (Mahrez et al. 2016; Moraes et al. 2015; Huang et al. 2014; Berr et al. 2011). The presence of similarly modified residues between plants and animals is not very surprising given the highly conserved nature of amino acid sequence of the core histones, especially histone H3 and H4 (Malik and Henikoff 2003).

Besides conserved PTMs, Zhang et al. (2007) also found several modified residues that were

Table 8.1 Profiling of histone modifications in plants

Species	Plant material used	Targeted proteins	Modifications analyzed/discovered[a]	References
Arabidopsis thaliana	Inflorescences	H3	Methylation (K); acetylation (K)	Johnson et al. (2004)
	Above ground plant tissues	H2A, H2B, H3, H4	Methylation (K); acetylation (K); phosphorylation (S); ubiquitination (K)	Zhang et al. (2007)
	Cultured cells	H2B	Methylation (K, A); acetylation (K); ubiquitination (K)	Bergmuller et al. (2007)
	Leaves	Whole proteome	Acetylation (K)	Hartl et al. (2017)
Oryza sativa	Cell suspension and protoplasts	H3	Methylation (K); acetylation (K)	Tan et al. (2011a)
	Whole plants, developing seeds, embryo, leaf blades	Whole proteome	Acetylation (K)	Meng et al. (2018), Xiong et al. (2016), Xue et al. (2018), He et al. (2016)
	Leaves	Whole proteome	Ubiquitination (K)	Xie et al. (2015)
	Seedlings	H2A, H2B, H3, H4	Crotonylation (K); butyrylation (K)	Lu et al. (2018)
	Leaves	Whole proteome	Crotonylation (K)	Liu et al. (2018c)
	Developing seeds	Whole proteome	Malonylation (K)	Mujahid et al. (2018)
	Seedling leaves	Whole proteome	Acetylation (K); succinylation (K)	Zhou et al. (2018)
	Developing seeds, seedlings	Whole proteome	2-hydroxyisobutyrylation (K)	Meng et al. (2017), Xue et al. (2020)
Glycine max	Leaves	H3, H4	Methylation (K); acetylation (K)	Wu et al. (2009)
	Developing seeds	Whole proteome	Acetylation (K)	Smith-Hammond et al. (2014)
Brassica oleracea	Cauliflower heads	H3	Methylation (K); acetylation (K)	Mahrez et al. (2016)
Saccharum sp.	Leaf rolls	H3, H4	Methylation (K, R); acetylation (K, S, T, Y)	Moraes et al. (2015)
Triticum aestivum	leaves	Whole proteome	Ubiquitination (K)	Zhang et al. (2017)
	Seedlings	Whole proteome	Malonylation	Liu et al. (2018a)
Lycopersicon esculentum	Mixture of roots, stems, and leaves	Whole proteome	Succinylation (K)	Jin and Wu (2016)
Nicotiana tabacum	Seedling leaves	Whole proteome	Crotonylation (K)	Sun et al. (2017)
Carica papaya	Fruits	Whole proteome	Crotonylation (K)	Liu et al. (2018b)
Brachypodium distachyon	Seedling leaves	Whole proteome	Acetylation (K); succinylation (K)	Zhen et al. (2016)
Physcomitrella patens	Whole plants	Whole proteome	2-hydroxyisobutyrylation (K)	Yu et al. (2017)

K, lysine; S, serine; A, adenine; R, arginine; T, threonine; and Y, tyrosine
[a]The letters within parenthesis refer to the modified amino acid

considered unique to plants. Among these were H3K20 acetylation and the absence of H3K79 methylation in Arabidopsis. However, shortly after H3K20 acetylation was shown to exist in yeast and later found in animals (Garcia et al. 2007a; Zheng et al. 2013). H3K79 methylation a conserved modification found in mammals, Drosophila and yeast (Feng et al. 2002), but not in Arabidopsis, was later detected in soybean (Wu et al. 2009). Although, to our knowledge the occurrence of methyl H3K79 has not been reported in any other plant, its presence in soybean suggests that this residue may indeed be methylated in other plant species. The apparent absence of certain modified residues may reflect differences in the abundance of these modifications between different organisms and even tissues. Furthermore, histone isolation methods, sensitivity of the mass spectrometry platforms used and limitations in the protein sequences available, may all contribute to the differences in the histone modifications observed.

Not all histone modifications previously considered unique to plants have been detected in other organisms suggesting they may be truly plant-specific. In addition, mass spectrometry-based proteomic approaches have the advantage of allowing the identification of histone isoforms, or variants, that are also part of the chromatin together with the canonical core histones. While some variants are conserved between animals and plants, some others are plant-specific and contain non-conserved residues that could harbor modifications. In fact, some of the plant-specific modifications reported in Arabidopsis including acetylation of K144 and phosphorylation of serine 145 (S145) in H2A.W.6 and H2A.W.7, and phosphorylation of S141 and S129 in H2A.W.6 are all found in the C-terminal part of H2A.W type of plant-specific variants (Zhang et al. 2007). In Arabidopsis, the three H2A.W variants were shown to promote chromatin condensation by enhancing fiber–fiber interactions via their C-terminal motif KSPKK, whereas H2A.W.7 is involved the response to DNA damage in constitutive heterochromatin (Yelagandula et al. 2014; Lorkovic et al. 2017). Interestingly, acetylation at K144 and phosphorylation of S145 localize to the C-terminal KSPKK motif suggesting that these modifications may have a mechanistic significance in the functions attributed to the H2A.W variants. However, up to now the role of these modifications during heterochromatin formation and DNA damage response remains unexplored.

Another study in Arabidopsis focused on a histone that has been less well studied, the H2B or HTB (for histone H Two B). Bergmuller et al. (2007) used purified histone H2B from Arabidopsis to identify histone modifications specific to this protein form and its variants. From the 11 putative isoforms identified in the Arabidopsis genome, five different H2B proteins, HTB1, HTB2, HTB4, HTB9, and HTB11, were detected in histones isolated from cell suspension cultures. Interestingly, these same five isoforms were those whose expression was the highest in cell suspension cultures indicating these were the most abundant proteins owing to their gene expression levels (Bergmuller et al. 2007). Considering that the five histones identified in cell suspensions are highly expressed in the shoot apex, a highly dividing tissue, it is likely that these H2B isoforms correspond to canonical core histones characterized by their deposition in chromatin during the S-phase of the cell cycle. In contrast, the remaining *HTB* genes are expressed in a tissue/organ-specific or developmentally regulated manner. For instance, *HTB8* is specifically expressed in seeds and the pollen sperm cell whereas *HTB3* transcripts are enriched in mature leaves (Jiang et al. 2020). Similarly, expression of *HTB7* and *HTB10* is restricted to reproductive tissues whereas *HTB5* and *HTB6* display preferential expression in sperm cells (Jiang et al. 2020). The specific expression patterns exhibit by the latter group of *HTB* genes suggest they may encode specialized variants. Indeed, it has recently been shown that *HTB8* encode a novel seed-specific variant present in other angiosperms and HTB3 is a replacement histone deposited in chromatin outside the S-phase (Jiang et al. 2020).

Among the modifications identified in the Arabidopsis H2B isoforms, lysine acetylation was the most numerous. K5, K10, K15, K27, K33, and K34 were found acetylated in histone

HTB2, whereas similar residues were acetylated in HTB9 and HTB11 (Bergmuller et al. 2007). The localization of acetylated lysine residues in Arabidopsis HTB isoforms suggest that this modification is limited to the N-terminal tail domain, which in the isoforms HTB2, HTB9, and HTB11 extends beyond the 45th residue. Lysine methylation, in contrast, was less extensive being dimethylation at K11 of HTB2 and monomethylation at K3 of HTB11 the only sites detected. One ubiquitination site was also found corresponding to K145 of HTB9 (Bergmuller et al. 2007). However, because the tryptic peptide "AVTKFTSS" in which this modification was found is common to most of the Arabidopsis H2B isoforms, it was not possible to determine which isoforms, if not all, carry this modification. Notably, H2B ubiquitination is the only modification in plant H2B for which a function is known. During photomorphogenesis in Arabidopsis, gene activation is usually characterized by increased H2B ubiquitination levels along their transcribed regions (Bourbousse et al. 2012). In addition, genome-wide analysis of multiple histone modifications in Arabidopsis showed that H2B ubiquitination tends to be associated with highly expressed genes together with other activating marks such as, H3K56ac (Histone H3 lysine 56 acetylation), H3K4me3 (H3K4 trimethylation), and H3K36me3 (Roudier et al. 2011). Thus, H2B ubiquitination (in HTB9K145 and equivalent residues) represents a modification associated with gene activity. In addition, its distribution across transcribed regions of expressed genes indicates this modification is likely linked to transcriptional elongation (Bourbousse et al. 2012; Roudier et al. 2011).

Analysis of histone H2B from Arabidopsis also identified different methylation states of the N-terminal alanine residue. Mono-, di-, and trimethylation of the N-terminal alanine was found in HTB1, HTB4, HTB9, and HTB11. Due to the similarity in sequence at the N-terminal end, it was not possible to determine the exact isoform to which each modification corresponds, but it was shown that they occur at least in the HTB9/HTB1 and HTB4/HTB11 pairs (Bergmuller et al. 2007). N-terminal methylation of histone H2B have also been observed in other organisms. In Tetrahymena, the N-terminal alanine of histone H2B was found to be trimethylated (Nomoto et al. 1982). Similarly, H2B, H2A, and H4 in the trypanosomatid *Trypanosoma brucei*, are methylated at the N-terminal alanine, even though; only monomethylation was observed (Janzen et al. 2006; Mandava et al. 2007; Picchi et al. 2017). N-terminal proline has been also found methylated, although only to dimethyl proline, in the histone H2B of Drosophila and several invertebrates (Webb et al. 2010; Desrosiers and Tanguay 1988). The functional roles of N-terminal methylation of histones remain unknown, but clues are emerging from recent studies in other chromatin-associated proteins. In human DDB2 (DNA damage-binding protein 2), a protein that participates in global genome nucleotide excision repair, the α-amino group of its N-terminal alanine residue is also mono-, di-, and trimethylated like in H2B (Cai et al. 2014). More important, N-terminal methylation is required for nuclear localization and promotes recruitment of DDB2 to UV light-induced DNA lesions such as cyclobutane pyrimidine dimers (Cai et al. 2014). In a similar way, chromatin association of RCC1 (Regulator of chromosome condensation 1), a protein involved in nuclear transport to the cytoplasm, nuclear envelop assembly and mitosis, requires N-terminal methylation of serine or proline (Chen et al. 2007). Also, human CENP-B (Centromere protein B) is trimethylated at its N-terminal glycine, a modification that facilitates binding to their target sequence, the CENP-B box, in the centromere (Dai et al. 2013). Interestingly, it was proposed that while monomethylation of the α amine group will have a small effect, the addition of two and three methyl groups will produce a permanent positive charge at the N-terminal amino acid (Stock et al. 1987). The resulting structure may have an increased affinity for the negatively charged phosphates of the DNA (Cai et al. 2014; Chen et al. 2007). Thus, it is tempting to speculate that N-terminal alanine methylation of histone H2B in Arabidopsis may result in altered interactions between H2B and the other histones, or with the DNA, leading to changes in chromatin structure.

Recent studies aimed to catalog PTMs in other plant species have updated their searches to include newly discovered modifications resulting in an expanded repertoire of histone marks. The discovery of serine, threonine and tyrosine O-acetylation, a novel H3 modification present in several organisms including yeast, Tetrahymena and metazoans (Britton et al. 2013), prompted the search for similar modifications in sugarcane histones. Indeed, peptides containing O-acetylation in serine 10 (S10), threonine 22 (T22), S28, tyrosine 41 (Y41), Y54 of histone H3, and T30 of histone H4 were detected in a mass spectrometry analysis of histones from sugarcane (Moraes et al. 2015). In humans, histone H3 serine 10 acetylation (H3S10ac) is enriched during the S-phase of the cell cycle, suggesting it may play a role in DNA replication and/or histone deposition (Britton et al. 2013). H3S10ac was also enriched in induced pluripotent stem cells indicating that it may also be involved in the maintenance of the pluripotent state (Britton et al. 2013). However, further evidence of the functional significance of H3S10ac or any other serine, threonine or tyrosine acetylation is lacking for in any organism.

8.2.1 Novel Lysine Acylations Discovered in Plants

Recent reports suggest that besides acetylation, lysine could be modified with a number of other acyl groups including, propionylation, butyrylation, crotonylation, 2-hydroxyisobutyrylation, β-hydroxybutyrylation, benzoylation, malonylation, glutarylation, and succinylation (Sabari et al. 2017; Huang et al. 2018, 2014). Five of these novel lysine acylations, succinylation, malonylation, butyrylation, crotonylation, and 2-hydroxyisobutyrylation, have been proven to exist in histones from several plant species (Table 8.1, Fig. 8.1).

In rice, enrichment of lysine succinylated peptides using an anti-succinyllysine pan antibody lead to the discovery of lysine succinylation in 2593 proteins (Zhou et al. 2018). K36 and K79 of histone H3 (isoforms H3.3 and H3.1, respectively); K13, K23, K67, K86, K137, and K145 of H2B (isoform H2B.7); K114 and K128 of H2A (isoform H2A.3); and K31 and K91 of Histone H4 were all found succinylated in rice. In another study, succinylation of K56 of H3 and K108, K137, and K145 of H2B (isoform H2B.1) were found in Brachypodium (Zhen et al. 2016). Also in tomato, antibody-based enrichment of lysine succinylated peptides lead to the discovery of lysine succinylation in 202 proteins (Jin and Wu 2016). From the peptides analyzed, it was deduced that K79 of histone H3 and K22 of the H2B isoform, H2B.3, were succinylated in tomato. From all the succinylated residues found in plants, succinyl H3K56 and H4K91 were also found in human, mouse and Drosophila, whereas succinyl H3K79 and H4K31 were found in the same organisms, but also in yeast cells (Xie et al. 2012). Even though H2B and H2A are less conserved than H3 and H4 between animals and plants, similar succinylated residues were also observed. H2BK108 and H2BK116, corresponding to rice H2BK137 and H2BK145, respectively, are succinylated in mammals and plants. However, other conserved residues found succinylated in plants were not modified in mammalian or yeast cells. H3K36, H2BK86 (equivalent to mammalian H2BK57) and H2AK114 (equivalent to mammalian H2AK99) are all succinylated in plants but not in mammals. Yet, for other residues, conservation of succinylation is not possible to stablish. For instance, the succinylation of the residues equivalent to the plant H2BK13, H2BK23, H2BK67, and H2B.3K22 is not possible to stablish since the N-terminal tail of plant histone H2B is not very well conserved with that of mammals. Taken together these observations suggest that even though lysine succinylation is conserved between plant and animals, there is also a number of plant-specific residues succinylated in plants. Intriguingly, lysine succinylation is rare in the H2B N-terminal tail of animals (Xie et al. 2012; Huang et al. 2014) indicating that this could be a specific feature of plant histone H2B.

Lysine crotonylation sites have been comprehensively determined in rice histones (Lu et al. 2018). By using mass spectrometry to analyze isolated histones, Lu et al. (2018) were able to

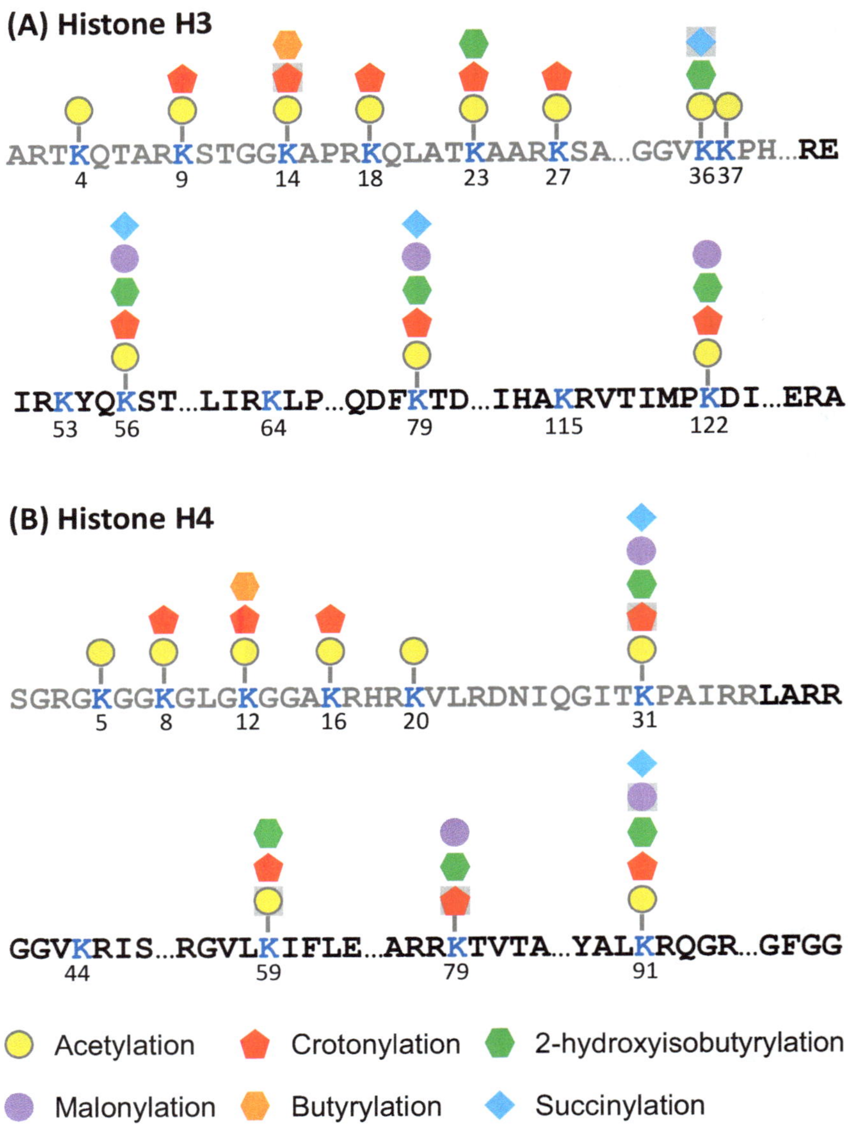

Fig. 8.1 Lysine acylation sites identified in histone H3 (**a**) and H4 (**b**) from plants. The amino acid sequence of the histones shown correspond to H3.1 and H4.1 from Arabidopsis. Gray and black letters indicate residues corresponding to the N-terminal tail and globular domains, respectively. Lysine residues are indicated in blue and their position is indicated by the numbers bellow. All lysine acylation sites indicated correspond to residues modified in at least one plant species. Modifications non-conserved with animals are highlighted with a gray square

identify 45 lysine crotonylation sites in H3, H2A, H2B, and H4. Few of these sites were independently identified in another study aimed at the identification of global protein crotonylation in rice (Liu et al. 2018c). K9, K14, K18, K23, K27, K56, K79 and K122 of H3 and K8, K12, K16, K31, K79, and K91 of H4 were shown to be crotonylated in rice (Lu et al. 2018). The remaining 31 crotonylation sites were found distributed between the lysines of the histones H2A, H2B, and their variants (Lu et al. 2018). Lysine crotonylation has also been studied in tobacco, this time using affinity enrichment of crotonylated peptides in the whole proteome with an anti-crotonyllysine pan antibody. Although this study was not targeted at histones, lysine crotonylation was found in the four core histones (Sun et al. 2017). K56, K79 and K122 of histone H3; K59 and K91 of histone H4; K127 of histone H2A, isoform H2A.1; and K18, K55, K60, K102, K125, and K133 of histone H2B, isoform H2B.2, were found to be crotonylated in tobacco (Sun et al. 2017). Similarly, a proteomic study of crotonylation in *Carica papaya* fruits identified lysine crotonylation in three sites in the histone H3 and five sites in H2B (Liu et al. 2018b). All these sites, with the exception of H2B.11K99, were previously identified as crotonylated in either rice or tobacco. The majority of the sites identified in the histones H3 (K9, K18, K23, K27, K56, K79, and K122) and H4 (K8, K12, K16, K59, and K91) of rice, tobacco or papaya were previously shown to be crotonylated in mammalian cells (Tan et al. 2011b; Huang et al. 2014). Similarly, residues crotonylated in H2A and H2B in the above mentioned plants were also found to contain the same modification in humans. For instance, the equivalent residue to tobacco H2AK127, human H2AK118, and the residues corresponding to tobacco H2BK125 and H2BK133 (human H2BK108 and H2B116, respectively) were previously demonstrated to be crotonylated in mammals (Tan et al. 2011b; Huang et al. 2014). However, crotonylation of other conserved residues in rice and tobacco core histones appears to be unique to plants. For instance, crotonylation of rice H3K14, H4K31, H4K79; and tobacco H2BK60 and H2BK102 (corresponding to human H2BK43 and H2BK85, respectively) has not been observed in humans so far. Furthermore, lysine residues, not conserved with mammalian histones, are also crotonylated in rice and tobacco. The residue crotonylated in the N-terminal tail of tobacco, H2B.2K18, was also modified in another H2B isoform, H2B.1K24. The counterpart of these residues does not exist human H2B indicating that crotonyl H2B.2K18 and H2B.1K24 are plant-specific modifications. Thus, it is apparent that while crotonylation of some lysine residues is conserved between animals and plants, other residues are specifically modified in plants.

Proteome-wide analyses using pan anti-malonyllysine antibodies allowed the discovery of several malonyl lysine residues in the histones of wheat and rice (Mujahid et al. 2018; Liu et al. 2018a). In wheat, 14 malonylation sites were identified in the four core histones (Liu et al. 2018a). K56, K79, K122, and K31, K79, K91 were shown to be malonylated in the histones H3 and H4, respectively (Liu et al. 2018a). H2AK16, H2AK130, H2BK13, H2BK16; H2B.2K35, H2B.2K111, H2B.2K134, and H2BK142 were also found malonylated in wheat (Liu et al. 2018a). In rice, proteomic analysis found a single site malonylated in one of the histone H2B isoforms, H2B.11K61 (Mujahid et al. 2018). Notably, some of these malonylation sites also occur in human and mouse histones. Wheat H3K56, H3K79, H3K122, H4K31, H4K79, H2B.2K142 (equivalent of human H2BK116) and H2B.11K61 (equivalent to human H2BK46) are all also malonylated in mammalian histones (Mujahid et al. 2018; Colak et al. 2015; Xie et al. 2012). The remaining malonylated residues found in wheat histones are either conserved but not modified in human cells (H4K91, H2AK130, H2B.2K35, H2B.2K111, and H2B.2K134) or correspond to plant-specific residues (H2AK16, H2BK13, and H2BK16).

Lysine butyrylation have been determined in isolated histones from rice (Lu et al. 2018). Four butyrylation sites, H3K14, H4K12, H2B.7K42, and H2B.7K114, were detected by mass spectrometry in rice seedlings. From these sites H3K14 and H4K12 are also butyrylated in human histones whereas H2B.7K42 an H2BK114 are plant-specific residues (Kebede et al. 2017). Thus, similar to other modifications, butyrylation may occur in the same residues in animals and plants but also in non-conserved lysines.

2-hydroxyisobutyrylation of lysines have been characterized in the proteomes of two plant species, rice and the moss *Physcomitrella patens* (Yu et al. 2017; Meng et al. 2017; Xue et al. 2020). In rice, proteome-wide analysis of 2-hydroxyisobutyrylation found 16 modified lysines in core histones; three residues in histone

H3, four in histone H4, three in histone H2A and six in histone H2B (Meng et al. 2017). Similarly, in Physcomitrella 19 lysine residues in core histones were found to have this modification, including; five sites in histone H3, three sites in histone H4, four sites in histone H2A and seven sites in histone H2B (Yu et al. 2017). From the modification sites found in both species, 2-hydroxyisobutyrylation of all of the lysine residues in histones H3 and H4, H3K23, H3K26, H3K56, H3K79, H3K122, H4K31, H4K59, H4K79 and H4K91, was also found in mammalian cells (Meng et al. 2017; Yu et al. 2017; Huang et al. 2014; Dai et al. 2014). In addition, the plant residues corresponding to the human H2A and H2B lysines, H2AK95, H2BK43, H2BK46, H2BK57, H2BK85, H2BK108 and H2BK116, were also 2-hydroxyisobutyrylated in either rice, Physcomitrella or both (Meng et al. 2017; Yu et al. 2017; Huang et al. 2014; Dai et al. 2014). Interestingly, the equivalent of the human H2AK99 residue, Physcomitrella H2AK114 and rice H2AK14, is modified only in the last two organisms suggesting these may be plant-specific modifications. Furthermore, several residues unique to plant H2A and H2B (H2AK22, H2AK43, H2AK136, and H2BK107), were also 2-hydroxyisobutyrylated. Interestingly, rice H2B.9K107, a lysine without a corresponding residue in mammalian H2B, was shown to be 2-hydroxyisobutyrylated in three different isoforms, H2B.9, H2B.2, and H2B.10 (Meng et al. 2017). Thus, even though lysine 2-hydroxyisobutyrylation is conserved between animals and plants, especially at residues corresponding to histones H3 and H4, there are still some plant-specific modifications localized in the more divergent histones H2A and H2B.

With the exception of acetylation, the number of crotonyl and 2-hydroxyisobutyryl lysine residues is larger than any other acylation studied in plants (Fig. 8.1). In mammalian cells, 2-hydroxyisobutyrylation was shown to be present in 63 lysine residues, which surpasses that of other lysine acylations, including acetylation (Dai et al. 2014). In plants, after acetylation, the number of crotonylated lysine residues is the largest among all histone acylations. However, while lineage-specific differences in the number of acylations may have a biological connotation, we should take in account that detection of lysine crotonylation, and also butyrylation, was carried out in isolated histones whereas the other acyl modifications were discovered in whole-proteome studies, not specifically directed to histones. Indeed, the apparent clustering of succinyl, malonyl and 2-hydroxyisobutyryl lysine residues in the histone globular domain (Fig. 8.1) could be due in part to the use of trypsin in proteome-wide studies that may lead to the generation of very small, hydrophilic peptides from the lysine/arginine-rich histone tails (see Sect. 8.3 for discussion). Future studies targeted to the identification of acyl lysine and other PTMs in purified histones will shed more light into the true extent of these modifications in plant histones.

Lysine acetylation is likely the most studied histone modification and one of the first PTMs to be discovered in the histones of any eukaryote (Allfrey et al. 1964). Conversely, for other lysine acylations, which were only recently identified, there is limited information on their functional significance. Nevertheless, it was postulated since the beginning that because lysine acylation will require the corresponding acyl-CoA as groups donors, these modifications might serve as a way to modulate chromatin, and gene expression, in response to metabolic changes (Dai et al. 2014; Dutta et al. 2016). Since their discovery, few non-acetyl lysine acylations have been characterized. In human somatic cells and during mouse spermatogenesis, histone lysine crotonylation was shown associated with active promoters and enhancers suggesting a role for this PTM in gene activation (Tan et al. 2011b). In addition, the "writers", "erasers," and "readers" of histone crotonylation have been identified in mammals. Histone crotonylation is deposited by the histone acetyltransferase p300/CBP (CREB-binding protein), but could also be removed by histone deacetylates from the "Sirtuin" family, Sirt1, Sirt2, and Sirt3 (Sabari et al. 2018; Bao et al. 2014). The YEATS (Yaf9, ENL, AF9, Taf14, Sas5) domain was identified as a chromatin binding module with a preference for crotonylated lysines (Zhao et al. 2016; Li et al. 2016; Andrews et al. 2016). More important, YEATS domain-containing proteins, such as AF9, are able to stimulate transcription linking histone crotonylation to gene expression (Li et al. 2016). In the same way as crotonylation, lysine butyrylation and 2-hydroxyisobutyrylation were also linked to active gene transcription in liver cells and during male germ cell differentiation, respectively (Dai et al. 2014; Kebede et al. 2017). In rice, histone lysine butyrylation and crotonylation have been also associated with actively transcribed chromatin and were shown to co-localize with acetylation suggesting they may function similarly to their animal counterparts (Lu et al. 2018; Liu et al. 2019, 2018c). Association with highly transcribed genes is not restricted to crotonylation, butyrylation and 2-hydroxyisobutyrylation since several other histone lysine acylations are associated with the promoters of active genes in a pattern that resembles that of lysine acetylation (Dutta et al. 2016). Thus, it was proposed by Dutta et al. (2016) that transcription could be stimulated using alternative strategies (i.e., different lysine acylations marks) that reflect the balance of metabolites in the cell.

The presence of conserved acyl lysine residues, including succinylation, malonylation, crotonylation, butyrylation and 2-hydroxyisobutyrylation, in the histones of animals and plants indicates these modifications may have been present in the last common ancestor of these organisms. Furthermore, together with the PTMs one may expect that their "readers", "writers," and "erasers" have also been conserved throughout evolution. Indeed, homologs of the enzymes and effector proteins associated to histone lysine crotonylation, the acetyltransferase p300/CBP, Sirtuin deacetylases, and YEATS domain-containing proteins, are present in plants (Pandey et al. 2002; Zacharaki et al. 2012). Interestingly, several of these lysine acylations in plant histones occurs in the same residue (Fig. 8.1) so that in a given loci a lysine may be modified with either acylation. In animals, competition between acyl modifications in the same residues at promoter regions appears to constitute an important regulatory mechanism (Goudarzi et al. 2016).

8.2.2 Beyond Identification of Histone Modifications

Mass spectrometry-based proteomic analysis of histones is not only a mean for the identification of PTMs but also may offer additional information that could be used to begin inferring the function of certain modifications. Histone PTMs do not occur alone and a single histone may contain combinations of modifications that index a particular locus. For example, a histone that contains a modification associated with gene silencing may contain additional PTMs also associated with gene silencing, a "guilt by association" argument. This same histone may also lack PTMs indexing the opposite gene state, i.e., gene activity. For instance, the mark associated with transcriptional activity, H3K4me3, usually occur together with multiple acetylated sites at H3K9 and H3K14, H3K18, H3K23, and H3K27 (Taverna et al. 2007b; Hazzalin and Mahadevan 2005). In contrast, H3K4me3 do not coexist in the same tail with the euchromatic silencing mark, H3K27me3, because methylation of H3K27 is inhibited by either H3K4me3 or H3K36me3 (Voigt et al. 2012). Acetylation at H3K9, H3K14, and H3K27 also co-occurs with H3K4me1 indexing active enhancers, while H3K4me1 and low levels of H3K14 acetylation marks inactive or poised enhancers (Karmodiya et al. 2012). Thus, variations in the combination of different PTMs, with similar or even opposite functions, may lead to unique functional outcomes.

Due to its ability to determine more than one modification in the same peptide, mass spectrometry is able to provide initial information on the mutual exclusion or co-occurrence of modifications. This is especially true for the heavily modified histone tails in which neighboring residues are usually modified. In Arabidopsis and sugarcane, analysis of the peptide "KSTGGKAPR," corresponding to residues 9–17 of the histone H3, revealed the presence of acetylation and methylation at both H3K9 and H3K14. From these modifications, mono- and dimethylation were the predominant marks at K9, whereas K14 was typically acetylated. However, in spite of their abundance methylation at K9 and acetyl K14 were rarely found in the same peptide implying their roles may be incompatible or even antagonistic (Johnson et al. 2004; Moraes et al. 2015). Indeed, both modifications index opposite chromatin states in plants; H3K9me2 is a histone mark associated with the transcriptionally silent heterochromatin, and transposable elements and repeats in euchromatic regions, whereas H3K14ac is associated with active genes in euchromatin (Earley et al. 2006; Bernatavichute et al. 2008; Chen et al. 2010). In Arabidopsis, the promoters of active rRNA genes are associated with H3K4me3 and acetylation of several residues including H3K14. Conversely, silent RNA loci lose their acetylation marks and become associated with H3K9me2 and DNA methylation (Earley et al. 2006). In the same way, induction of the abscisic acid (ABA)-responsive genes, *ABA INSENSITIVE 1* (*ABI1*), *ABI2*, and *RESPONSIVE TO DESSICATION 29B* (*RD29B*), and the salt stress-responsive genes, *DRE-BINDING PROTEIN 2A* (*DREB2A*), *RD29A*, and *RD29B* correlates with increased H3K14ac, and H3K9ac, but also decreased H3K9me2 (Chen et al. 2010). None of these modifications have been studied in sugarcane, but the similarities in occurrence with Arabidopsis may be an indication that these PTMs may have similar functions in this crop (Moraes et al. 2015).

Recently, O-acetylation of serine, threonine and tyrosine (S/T/Y) was described as a novel modification found in several eukaryotes including yeast, Tetrahymena, metazoans, and plants (Moraes et al. 2015; Britton et al. 2013). In sugarcane, when peptides containing S/T/Y acetylation are found with other PTMs, these are frequently lysine and arginine methylation, but rarely acetylation. This observation suggests that S/T/Y acetylation could coexist with silencing marks such as H3K9me1/3 and H3K27me3, but not at all with lysine acetylation, PTMs usually associated with gene expression. The function of these modifications is currently unknown, but it was proposed that acetylation of H3S10 might inhibit phosphorylation of the same residue, blocking the formation of H3S10ph (Britton et al. 2013). H3S10ph can modulate the binding of effector proteins to the neighboring methyl H3K9 forming a methyl/phospho or binary switch (Fischle et al. 2003). Considering all the S/T/Y acetylation sites found in histones H3 and H4 of sugarcane are adjacent to a lysine residue, and in one case to an arginine, it is possible to assume that S/T/Y acetylation may help to fine tune the interactions between methylation and phosphorylation in these binary switches. Future research will provide more information on the interplay between lysine methylation, and acetylation and phosphorylation of S/T/Y.

Under the right conditions, quantitative data on the histone PTMs in different samples could be obtained with mass spectrometry. When histone derivatization followed by trypsin digestion is used (see Sect. 8.3.2), it leads to the generation of defined peptides (Johnson et al. 2004). Thus, if all the modified versions of a particular peptide (isoforms) can be accounted for, after mass spectrometry, the relative abundance of the modifications can be determined by dividing the peak area (peak integration of the ion chromatogram) of a particular isoform by the sum of the areas of all the isoforms (Johnson et al. 2004; Moraes et al. 2015). Using this approach, the relative levels of H3K9me2 in histone H3 were measured at ~10% and ~40% in Arabidopsis and sugarcane, respectively (Moraes et al. 2015; Johnson et al. 2004). These estimates are consistent with the content of heterochromatin in Arabidopsis (~10%) and sugarcane (~50%) suggesting that the relative level H3K9me2, a mark for constitutive heterochromatin in plants, may correlate with the heterochromatic content in plant genomes.

Another interesting observation comes from the analysis of the peptide corresponding to residues 27–40 of the histone H3. This region allows distinguishing between H3 variants, H3.1 (KSAPATGGVKKPHR) and H3.3 (KSAPTTGGVKKPHR), due to an A versus T change at amino acid 31 (Johnson et al. 2004). Comparative analysis of the relative levels of modifications in these two peptides demonstrate that H3.1 was enriched in the silencing mark, H3K27me3, whereas H3.3 contains higher levels of H3K36me2/3 a mark associated with transcriptionally active genes (Johnson et al. 2004;

Moraes et al. 2015). These observations agree with the distributions of these histones in the Arabidopsis genome. H3.1 is deposited into chromatin during replication and is associated to transcriptionally silent regions enriched in methyl H3K27, methyl H3K9, and DNA methylation. In contrast, H3.3 deposition correlates with transcriptional activity and is distributed in gene bodies and a subset of promoters (Shu et al. 2014; Stroud et al. 2012). Thus, the determination of the relative abundance of modifications lends support for the involvement of H3.3 in active transcription and H3.1 in silent chromatin.

Mass spectrometry could be carried out in a quantitative way to compare the levels of PTMs in different samples. In this way, quantitative differences in histone modifications may be determined between different tissues, organs, genotypes, treatments, and even species, allowing for the identification of particular modifications associated to the biological differences among samples. There have been at least two reports using mass spectrometry to compare histone modifications in two different samples. In Arabidopsis, mass spectrometry was used to determine quantitative differences in lysine methylation of histone H3 between wild-type and a mutant in *KRYPTONITE* (*KYP*), a gene encoding a methyltransferase specific for H3K9 (Johnson et al. 2004). In this work, stable isotopes were used to label peptides coming from the wild-type and the *kyp* mutant in order to differentiate them during mass spectrometry. With this method, Johnson et al. (2004) found that the levels of H3K9me2/3 were reduced in the *kyp* mutant when compared to the wild-type, confirming previous observation stating that KRYPTONITE is indeed a methyltransferase responsible for H3K9me2 in vivo.

The development of newer-generation mass spectrometry equipment made possible novel quantitative approaches based on the targeted analysis of peptides (Liebler and Zimmerman 2013). One of these approaches, LC/MRM-MS (liquid chromatography/multiple reaction monitoring-mass spectrometry) was used to quantify differences in histone PTMs between wild-type *Arabidopsis* and the null-mutant *clf-29* (Chen et al. 2015). CLF (CURLY LEAF) is one of three histone methyltransferases in Arabidopsis responsible for the deposition of H3K27me3 at its target genes (Schubert et al. 2006; Jiang et al. 2008). Chen et al. (2015) shown that indeed the *clf-29* mutant contains significantly reduced amounts of H3K27me3 when compared to wild-type plants. These results were comparable to independent determinations of H3K27me3 using immunoblots with an anti-H3K27me3 antibody. Quantification of a dozen other modified lysines did not reveal any other changes with the exception of a 12% increase in H3K36me3 detected in the *clf-29* mutant. Thus, LC/MRM-MS not only confirmed that H3K27me3 is dependent on CLF, but also hint to the possibility of a crosstalk between H3K27me3 and H3K36me3 (Chen et al. 2015). Analysis of such large number of modifications between two different genotypes will be cumbersome and time-consuming using immunoblots with specific antibodies.

8.3 Strategies for the Analysis of Plant Histone Modifications

Several approaches have been used to catalog PTMs in plant histones. They may be divided in two depending on whether a study is focused on identifying a particular type of modification occurring in the whole-proteome or multiple modifications occurring in core histones or even a histone type in particular (Fig. 8.2). These approaches have also been combined in order to obtain information on a single type of modification, usually occurring at very low levels in histones.

8.3.1 The Whole-Proteome Approach

The whole-proteome approach has been applied in several plant species and using a variety of organs and tissues (Table 8.1). This methodology is used when the target is a modification in particular and not a specific family of proteins so

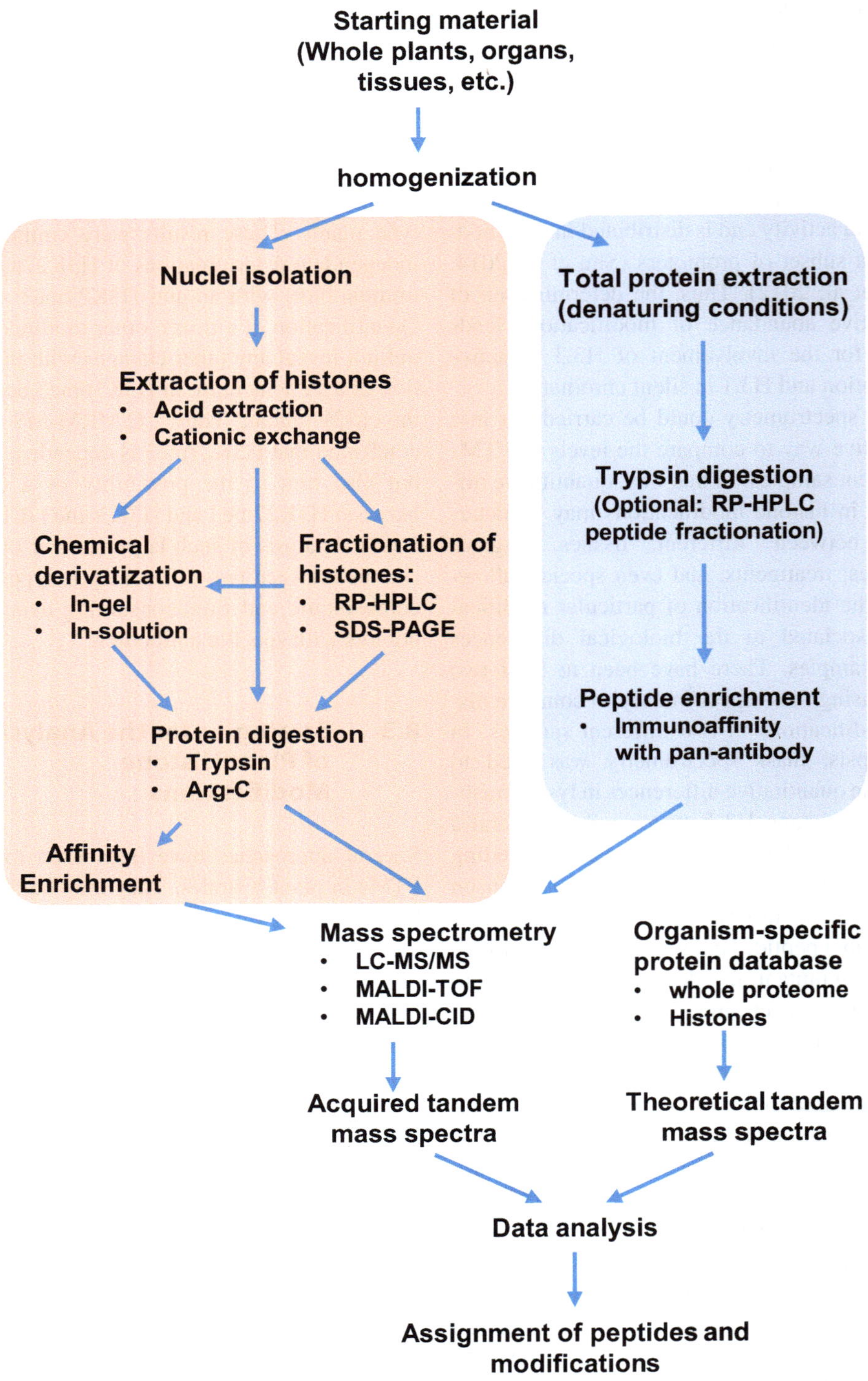

Fig. 8.2 Schematic overview of the proteomic approaches used to characterize histone modifications in plants. All the strategies, and their variations, used in plants are integrated in the diagram. The unique steps involved in the analysis of single modifications in whole proteomes or multiple modifications in isolated histones are enclosed in a light red rectangle in the left and a light blue rectangle in the right, respectively

that the whole proteome is explored and not only histones or a specific family of proteins. The modifications analyzed using this approach include acetylation, ubiquitination, malonylation, succinylation, crotonylation, and 2-hydroxyisobutyrylation specifically on lysines (Table 8.1). Even though these studies usually report the presence of the target modification in a wide range of proteins, we could mine the data reported in order to find PTMs in the proteins of interest, histones in this case.

The process of identifying modifications in a whole-proteome analysis (Fig. 8.2) starts by grinding the tissues of interest with liquid nitrogen and resuspending the homogenized tissues in a denaturing buffer usually containing urea as a chaotropic agent, protease inhibitor cocktails, and, when available, inhibitors for the enzymes that may otherwise remove the modifications. Then, the amount of proteins in the homogenate is quantified and subjected to trypsin digestion. At this point, the complexity of the peptide mixture produced after trypsin digestion may be reduced by fractionating the samples using reverse phase-high performance liquid chromatography (RP-HPLC). Because RP-HPLC fractioning led to an increased number of samples to be handled this step is usually skipped and after trypsin digestion peptides are usually precipitated and resuspended in a buffer suitable for affinity enrichment. In all the studies described here, affinity enrichment has been carried out using a pan antibody recognizing a modified amino acid regardless of the residues surrounding it. Many of these pan antibodies are commercially available, facilitating the study of several PTMs. Once the fractions are enriched for peptides containing the target modification, they are analyzed using a mass spectrometry platform. The platform of choice for this type of whole-proteome analysis is LC-MS/MS (liquid chromatography-tandem mass spectrometry) in which an *online* reverse phase column separate the peptides before injecting them in the tandem mass spectrometer.

After mass spectrometry, the MS/MS spectrum obtained for each peptide analyzed is converted into searchable data (peptide mass, fragment ions peaks and intensities). For database search, all the protein entries corresponding to the organism under study are digested in silico. Then, the masses of the intact peptides and their expected fragment ions are calculated. After that, the experimental and calculated MS/MS spectra are compared, first for matches between the total mass of the peptides and then for matching of the masses of the fragment ions (Cottrell 2011). Because we are interested in the modifications present in the peptides, the mass of the modification added to the peptide is taken into consideration during the database search, as it is the expected mass of the fragment ions resulting from the fragmentation of the peptide with the modification present at a particular residue. Protein identification software like Mascot and SEQUEST are often used for identifying the PTMs of interest. The search engine will finally provide with a list of peptides found containing the modification under study indicating the position of the modified site and protein entry to which it corresponds. However, manual validation of the results is always required (Garcia et al. 2007b).

8.3.2 Approaches Based on Histone Purification

The second approach based on the analysis of core histones involves the extraction of histones from the initial plant material (Fig. 8.2). The first step in the purification of histones involves the isolation of intact nuclei from plant cells. Just by performing this step, the complexity of the sample is reduced because virtually all the histones are forming part of the chromatin inside the nucleus and, in addition, ~40% of the mammalian diploid nucleus composition correspond to histones (Dumortier and Muller 2007). In order to release the nuclei from the cells, the plant material needs to be homogenized to disrupt the rigid plant cell walls but leaving the nuclei intact. Blender-type homogenizer works with most plant tissues but also grinding the samples in liquid nitrogen have proven to be effective in homogenizing plant tissues without disintegrating the nuclei (Sikorskaite et al. 2013). During or after homogenization, depending on the use of a blender or liquid nitrogen for cell disruption, a buffer is added to stabilize the nuclei and protect the chromatin from degradation during the isolation process. The nuclear isolation buffer usually includes: chromatin stabilizers ($MgCl_2$, spermine), membrane stabilizers (sucrose, glycerol, hexylene glycol) inorganic salts (KCl, NaCl) to provide proper ionic strength, inhibitor cocktails (for proteases, phosphatases, acetylases, deacetylases) and organic buffers (MOPS, Tris, HEPES, PIPES) to maintain the pH of the solution around 7.0–8.0. After homogenization, the non-ionic detergent Triton X-100 is added to the buffer to reach a concentration high enough to disrupt the membranes of the chloroplast and mitochondria, but not the nuclear envelope (Sikorskaite et al. 2013; Loureiro et al. 2006). After suspension in the buffer, intact cells, large pieces of tissues and debris are filtered out using Miracloth. The filtered solution contains the nuclei, which in virtue of its large size could then be separated from the rest of the solution by low-speed centrifugation. The nuclei isolated in this way may also contain membrane pieces and sometimes starch grains (Sikorskaite et al. 2013), but it is usually suitable for histone extraction.

Histones are composed of basic amino acids that allow them to interact with the phosphate groups of the DNA. This basic property of histones is used to isolate them from crude nuclear preparations. One method homogenizes the nuclei in diluted acid solutions (HCL or H_2SO_4) in which histones are soluble but other nuclear proteins precipitate. Precipitated proteins are removed by high-speed centrifugation and the extracted histones are concentrated by precipitation with strong acids, like TCA, or dialyzed with diluted acetic acid and lyophilized (Shechter et al. 2007; Sidoli et al. 2016). A second approach uses cation exchange chromatography with the resin BioRex-70 to separate histones from other nuclear proteins (Waterborg et al. 1995; Waterborg 2000).

Both methods produce fairly pure bulk histones consisting of the linker histone H1 and the core histones H2A, H2B, H3, and H4 and also their variants. In plants, acid extraction have been successfully used to isolate histones for mass spectrometry analysis from Arabidopsis, soybean, cauliflower and rice (Bergmuller et al. 2007; Wu et al. 2009; Mahrez et al. 2016; Tan et al. 2011a), whereas cation exchange chromatography with BioRex-70 have been carried out in Arabidopsis and sugarcane (Zhang et al. 2007; Moraes et al. 2015).

Bulk histones could be fractionated into individual histones by RP-HPLC. The high resolution of RP-HPLC allows the separation of individual, highly pure histones (Shechter et al. 2007). Both individually separated and bulk histones have been used for mass spectrometry analysis in several plant species (Table 8.1). However, before mass spectrometry histones are converted into small peptides, usually by digesting them with endoproteases such as trypsin. Digestion with trypsin, which cuts at arginine and lysine, is problematic for histones because their basic nature is given by a high content of lysine and arginine residues especially in their tails. Thus, digestion of histones with trypsin will produce very small peptides difficult to analyze by mass spectrometry. To overcome this problem, histones could be digested with limited amounts of trypsin or for limited time so that not all residues are cleaved by the enzyme. This approach was used to determine PTMs in histones H2A, H2B, H3, and H4 from Arabidopsis, and H3 and H4 from soybean (Wu et al. 2009; Bergmuller et al. 2007; Zhang et al. 2007). Another approach is to replace trypsin by Arg-C, an endoprotease that cleaves only at arginine residues producing larger peptides from histones. Arg-C has been used so far to analyze histone H3 from Arabidopsis and Cauliflower (Mahrez et al. 2016; Zhang et al. 2007). Analysis of histones using both approaches have been carried out in order to increase the chance of finding a larger number of modified residues (Mahrez et al. 2016; Zhang et al. 2007).

Even though the methodologies mentioned above have been successfully used to identify histone modifications, they have some drawbacks. For example, the peptides produced by limited trypsin digestion are often irreproducible and Arg-C is apparently less efficient and not as specific as trypsin (Plazas-Mayorca et al. 2009). To avoid these shortcomings, derivatization methods that chemically block lysine residues so that trypsin will only cleave at arginine residues, producing uniform peptides, have been developed. The most used method involves the propionylation of unmodified and monomethyl lysines before trypsin digestion and then of the free amino group in the N-terminus of the digested peptides. Propionylation of histones not only results in reproducible peptide fragments, but also in less hydrophilic peptides that could be better separated in RP-HPLC and improve fragmentation of the peptides producing MS/MS spectra that are easier to interpret (Garcia et al. 2007b). The generation of uniform peptides during tryptic digestion of histones have additional advantages, for example allowing quantification of the relative amounts of some PTMs (see Sect. 8.2.2). Propionylation of histones using propionic anhydride has been applied to find PTMs in the histones H3 and H4 from sugarcane and H3 from Arabidopsis (Johnson et al. 2004; Moraes et al. 2015). A similar propionylation method, using propionic acid *N*-hydroxysuccinimide ester (NHS-propionate), was also used to analyze modifications in Arabidopsis histone H3 (Chen et al. 2015).

After digestion with endoproteases, samples could be enriched for peptides carrying particular types of modifications, especially those present at very low levels, using affinity chromatography or pan antibodies. Zhang et al. (2007) carried out affinity chromatography with titanium dioxide (TiO_2) to selectively enrich for phosphopeptides in histones from Arabidopsis digested with trypsin. Interestingly, in this work the authors were able to find several phosphorylated residues in histones H2A and H2B (Zhang et al. 2007). Another study in Arabidopsis used IMAC (immobilized metal affinity chromatography) also to enrich for phosphopeptides in the histone H3 from Arabidopsis but not phosphorylated residue was reported. In rice, enrichment of crotonylated and butyrylated peptides after trypsin digestion of purified histones was achieved using anti-butyryllysine and anti-crotonyllysine pan antibodies (Lu et al. 2018).

The platform of choice for the analysis of histone peptides and their modifications is LC-MS/MS, but, other methods have also been used including matrix-assisted laser desorption/ionization time-of-flight (MALDI-TOF/TOF) and hybrid linear ion trap Fourier transform ion cyclotron resonance trap (LTQ-FTICR). The advantages and drawbacks of the different mass spectrometers will not be detailed here as it is beyond the scope of this chapter.

Once the experimental MS/MS spectra are obtained, the database search is carried out in the same way as for the whole-proteome approach (see Sect. 8.3.1). However, the generation of the theoretical MS/MS spectra could be done either using all the protein entries in existence for the organism under study or a custom database could be constructed by using just sequences corresponding to the histones from the particular organism. The latter approach has the advantage of greatly reducing the number of proteins to be searched and by curating the protein sequences, we will avoid mispredicted and misannotated entries in the database, especially those present in poorly annotated genomes. In species without a genome database, the proteins sequences of the histones from related species could be used and/or histone sequences could be obtained from contigs made of Expressed Sequence Tags (ESTs) or RNA Sequencing data. Since histones are highly conserved, using the protein sequences from a close relative often works well for histones H3 and H4. However, histone H2A and H2B are usually more divergent and lineage-specific variants may be present on particular species. For the search of modifications, the differential mass of the added groups is considered. However, since not all residues are completely modified and some may be modified with different functional groups, all possible peptide configurations must be considered in the search. Thus, the possible combinations increase exponentially with the addition of a modification to the search. Although, several search engines have been used to identify PTMs in histones, pFind and Mascot, have been shown to produce the most confident results (Yuan et al. 2014). However, manual inspection of the data is always necessary after database search.

The use of either approach for cataloging protein modifications have yield important information which can be used to start the functional analysis of specific PTMs. Whole-proteome analysis have allowed the identification of several modified residues with a number of functional groups, especially acylations. However, because of the high number of lysine and arginine residues in histones, the use of trypsin digestion will produce samples with underrepresented peptides from histones. This bias could be easily observed in Fig. 8.1, in which most of the modifications identified by whole-proteome approaches are clustered in the globular domains of the histone, the region containing less lysine and arginine residues. Other drawback of the whole-proteome analysis is the high complexity of the samples, which may hamper the detection of low abundance modifications. The use of purified histones, on the other hand, has the advantage of focusing on a reduced number of proteins allowing the identification of even low abundance modifications. However, purification procedures may lead to the loss of some modifications, especially those for which inhibitor cocktails for PTM modifying enzymes are not available. With the increased availability of pan antibody recognizing specific modifications regardless of their sequence context, the ability of carrying out enrichment of modified peptides from isolated histones will improve our ability to detect histones modifications.

8.4 Perspective

In recent years, there have been many advances in the discovery and cataloging of PTMs in histones and whole proteomes from several organisms. The biological function of several modifications, including the "readers", "writers," and "erasers" of these modifications, have been also began to be known and understood in plants. This knowledge is, however, limited to a few modifications like acetylation, methylation, phosphorylation, ubiquitination, crotonylation, and butyrylation. The discovery of new functional groups that can be added to amino acid residues and for which no function is currently known have added more complexity and make us realize that we are far from deciphering the "histone code." Fortunately, recent advances in proteomics, mass spectrometry and the availability of new antibodies can help us advance our understanding of the role histone modifications play in plants and other organisms.

The cataloging of histone modifications is just the initial step in trying to understand the complexities of the functional interactions between histone modifications. Such catalog may provide initial candidates to begin functional characterization and information on crosstalk of modifications that may point to a biological role. Unfortunately, the profiling of histone modifications in plants is lagging behind from other organisms such as mammals. There have been few studies focused in identifying PTMs in plant histones and most aimed at newly discovered modifications have been done using whole proteomic approaches, which have bias against histones (see Sect. 8.3.2). In addition, previous analysis of histone PTMs did not include the search for many novel types of modifications that were not yet discovered at that time. Thus, future work should aim to provide a more extensive catalog of histone PTMs in plants, especially of the newly identified modifications. Improvements in sample preparation, development of more sensitive mass spectrometry approaches, novel algorithms for database search and the availability of pan antibodies for specific modifications should all allow for a comprehensive identification of histone modifications even those present at very low levels.

Histone PTMs do not occur alone but often in complex, non-random combinations with numerous others (Taverna et al. 2007b). Thus, one of the major goals in the field is to determine the functional meaning of the combination of modifications that do exist in vivo. So far in plants, all the analysis of histone modifications have made use of endoproteases that cleave the protein into small peptides analyzed by mass spectrometry, an approach called bottom-up (Moradian et al. 2014). Despite the fact that some information of co-occurring histone

PTMs have been obtained with this approach (see Sect. 8.2.2), this have been limited to neighboring residues and additional information regarding long-range combination of modifications is lost when the protein is cleaved. Thus, in order to obtain more data on co-occurring histone modifications in plant histones, other strategies such as top-down and middle-down mass spectrometry need to be applied. Middle-down mass spectrometry methods could be used to detect combinations of modifications in the histone tails whereas top-down mass spectrometry could be applied to PTMs in whole-proteins (Moradian et al. 2014; Molden and Garcia 2014). These methods will provide important information on combinations of histone modifications naturally occurring in plants and help us to start unraveling their functions.

As discussed in Sect. 8.2.2, quantitative proteomics methods are amenable to the analysis of changes in histone modifications between two different genetic backgrounds. Novel quantitative approaches, like LC/MRM-MS, may be used to measure changes in hundreds of histone modifications between wild-type plants and knock-out mutants for "writers" or "erasers" and help us to determine their target PTMs (Liebler and Zimmerman 2013; Chen et al. 2015). Similarly, quantitative differences in histone modifications could be analyzed in different tissues, organs and during plant development using this approach. Another interesting application would be to measure the crosstalk of modifications that occur when the deletion of a "writer" or an "eraser" cause a PTM to be depleted or enriched. This type of experiments could provide additional information of already known histone PTMs and their interplay with other modifications.

While determining the complete repertoire of histone modifications, their combinations and quantitative changes during development may provide clues about their functional significance, these approaches are usually not enough to obtain a complete picture of their biological meaning (Janssen et al. 2017). To this end, other complementary approaches need to be applied in order to gain additional insights. Among these techniques, chromatin immunoprecipitation (ChIP) followed by microarray hybridization (ChIP-chip) or high-throughput sequencing (ChIP-Seq) are one of the most informative techniques used because it allow us to determine the genome-wide distribution of histone modifications (Schmitz and Zhang 2011). In addition, by combining ChIP-chip or ChIP-Seq with transcriptomic profiling of RNAs, using high-throughput RNA sequencing (RNA-Seq), a correlation between enrichment or depletion of histone modifications and gene activity could be obtained along the whole genome (Schmitz and Zhang 2011).

The integrated genome-wide profiling of several histone PTMs, using ChIP-chip, suggest that they may be combined to produce only few chromatin states in Arabidopsis and other metazoans (Roudier et al. 2011). This low combinatorial complexity may also be reflected in the non-random co-occurrence of histone modifications mention above (Taverna et al. 2007b). This apparent simplicity of chromatin organization may work in our favor when determining the functions of histone modifications. However, the complete understanding of how histone PTMs relate to biological function will imply to combine genomic, transcriptomic and proteomic data into a coherent unit. This is a challenging task given the different layers of information that need to be integrated in order to understand how these individual parts work as a single biological system. However, in spite of the challenges, the integration of "omics" data is becoming one of the new frontiers in biological research.

References

Allfrey VG, Faulkner R, Mirsky AE (1964) Acetylation and methylation of histones and their possible role in the regulation of RNA synthesis. Proc Natl Acad Sci U S A 51:786–794

Andrews FH, Shinsky SA, Shanle EK, Bridgers JB, Gest A, Tsun IK, Krajewski K, Shi X, Strahl BD, Kutateladze TG (2016) The Taf14 YEATS domain is a reader of histone crotonylation. Nat Chem Biol 12(6):396–398. https://doi.org/10.1038/nchembio.2065

Bao X, Wang Y, Li X, Li XM, Liu Z, Yang T, Wong CF, Zhang J, Hao Q, Li XD (2014) Identification of 'erasers' for lysine crotonylated histone marks using a chemical proteomics approach. Elife 3:e02999. https://doi.org/10.7554/eLife.02999

Berger SL (2007) The complex language of chromatin regulation during transcription. Nature 447(7143):407–412. https://doi.org/10.1038/nature05915

Bergmuller E, Gehrig PM, Gruissem W (2007) Characterization of post-translational modifications of histone H2B-variants isolated from Arabidopsis thaliana. J Proteome Res 6(9):3655–3668. https://doi.org/10.1021/pr0702159

Bernatavichute YV, Zhang X, Cokus S, Pellegrini M, Jacobsen SE (2008) Genome-wide association of histone H3 lysine nine methylation with CHG DNA methylation in Arabidopsis thaliana. PLoS One 3(9):e3156. https://doi.org/10.1371/journal.pone.0003156

Berr A, Shafiq S, Shen WH (2011) Histone modifications in transcriptional activation during plant development. Biochim Biophys Acta 1809(10):567–576. https://doi.org/10.1016/j.bbagrm.2011.07.001

Bourbousse C, Ahmed I, Roudier F, Zabulon G, Blondet E, Balzergue S, Colot V, Bowler C, Barneche F (2012) Histone H2B monoubiquitination facilitates the rapid modulation of gene expression during Arabidopsis photomorphogenesis. PLoS Genet 8(7):e1002825. https://doi.org/10.1371/journal.pgen.1002825

Britton LM, Newhart A, Bhanu NV, Sridharan R, Gonzales-Cope M, Plath K, Janicki SM, Garcia BA (2013) Initial characterization of histone H3 serine 10 O-acetylation. Epigenetics 8(10):1101–1113. https://doi.org/10.4161/epi.26025

Cai Q, Fu L, Wang Z, Gan N, Dai X, Wang Y (2014) alpha-N-methylation of damaged DNA-binding protein 2 (DDB2) and its function in nucleotide excision repair. J Biol Chem 289(23):16046–16056. https://doi.org/10.1074/jbc.M114.558510

Cerutti H, Casas-Mollano JA (2009) Histone H3 phosphorylation: universal code or lineage specific dialects? Epigenetics 4(2):71–75

Chen T, Muratore TL, Schaner-Tooley CE, Shabanowitz J, Hunt DF, Macara IG (2007) N-terminal alpha-methylation of RCC1 is necessary for stable chromatin association and normal mitosis. Nat Cell Biol 9(5):596–603. https://doi.org/10.1038/ncb1572

Chen LT, Luo M, Wang YY, Wu K (2010) Involvement of Arabidopsis histone deacetylase HDA6 in ABA and salt stress response. J Exp Bot 61(12):3345–3353. https://doi.org/10.1093/jxb/erq154

Chen J, Gao J, Peng M, Wang Y, Yu Y, Yang P, Jin H (2015) In-gel NHS-propionate derivatization for histone post-translational modifications analysis in Arabidopsis thaliana. Anal Chim Acta 886:107–113. https://doi.org/10.1016/j.aca.2015.06.019

Colak G, Pougovkina O, Dai L, Tan M, Te Brinke H, Huang H, Cheng Z, Park J, Wan X, Liu X, Yue WW, Wanders RJ, Locasale JW, Lombard DB, de Boer VC, Zhao Y (2015) Proteomic and biochemical studies of lysine malonylation suggest its malonic aciduria-associated regulatory role in mitochondrial function and fatty acid oxidation. Mol Cell Proteomics 14(11):3056–3071. https://doi.org/10.1074/mcp.M115.048850

Cottrell JS (2011) Protein identification using MS/MS data. J Proteome 74(10):1842–1851. https://doi.org/10.1016/j.jprot.2011.05.014

Dai X, Otake K, You C, Cai Q, Wang Z, Masumoto H, Wang Y (2013) Identification of novel alpha-n-methylation of CENP-B that regulates its binding to the centromeric DNA. J Proteome Res 12(9):4167–4175. https://doi.org/10.1021/pr400498y

Dai L, Peng C, Montellier E, Lu Z, Chen Y, Ishii H, Debernardi A, Buchou T, Rousseaux S, Jin F, Sabari BR, Deng Z, Allis CD, Ren B, Khochbin S, Zhao Y (2014) Lysine 2-hydroxyisobutyrylation is a widely distributed active histone mark. Nat Chem Biol 10(5):365–370. https://doi.org/10.1038/nchembio.1497

Desrosiers R, Tanguay RM (1988) Methylation of Drosophila histones at proline, lysine, and arginine residues during heat shock. J Biol Chem 263(10):4686–4692

Dumortier H, Muller S (2007) Histone autoantibodies. In: Gershwin ME, Meroni PL (eds) Autoantibodies, 2nd edn. Elsevier, Burlington, VA, pp 169–176. https://doi.org/10.1016/B978-044452763-9/50026-3

Dutta A, Abmayr SM, Workman JL (2016) Diverse activities of histone acylations connect metabolism to chromatin function. Mol Cell 63(4):547–552. https://doi.org/10.1016/j.molcel.2016.06.038

Earley K, Lawrence RJ, Pontes O, Reuther R, Enciso AJ, Silva M, Neves N, Gross M, Viegas W, Pikaard CS (2006) Erasure of histone acetylation by Arabidopsis HDA6 mediates large-scale gene silencing in nucleolar dominance. Genes Dev 20(10):1283–1293. https://doi.org/10.1101/gad.1417706

Feng S, Jacobsen SE (2011) Epigenetic modifications in plants: an evolutionary perspective. Curr Opin Plant Biol 14(2):179–186. https://doi.org/10.1016/j.pbi.2010.12.002

Feng Q, Wang H, Ng HH, Erdjument-Bromage H, Tempst P, Struhl K, Zhang Y (2002) Methylation of H3-lysine 79 is mediated by a new family of HMTases without a SET domain. Curr Biol 12(12):1052–1058

Feng S, Jacobsen SE, Reik W (2010) Epigenetic reprogramming in plant and animal development. Science 330(6004):622–627. https://doi.org/10.1126/science.1190614

Fischle W, Wang Y, Allis CD (2003) Binary switches and modification cassettes in histone biology and beyond. Nature 425(6957):475–479. https://doi.org/10.1038/nature02017

Fuchs J, Demidov D, Houben A, Schubert I (2006) Chromosomal histone modification patterns--from conservation to diversity. Trends Plant Sci 11(4):199–208. https://doi.org/10.1016/j.tplants.2006.02.008

Garcia BA, Hake SB, Diaz RL, Kauer M, Morris SA, Recht J, Shabanowitz J, Mishra N, Strahl BD, Allis CD, Hunt DF (2007a) Organismal differences in

post-translational modifications in histones H3 and H4. J Biol Chem 282(10):7641–7655. https://doi.org/10.1074/jbc.M607900200
Garcia BA, Mollah S, Ueberheide BM, Busby SA, Muratore TL, Shabanowitz J, Hunt DF (2007b) Chemical derivatization of histones for facilitated analysis by mass spectrometry. Nat Protoc 2(4):933–938. https://doi.org/10.1038/nprot.2007.106
Goudarzi A, Zhang D, Huang H, Barral S, Kwon OK, Qi S, Tang Z, Buchou T, Vitte AL, He T, Cheng Z, Montellier E, Gaucher J, Curtet S, Debernardi A, Charbonnier G, Puthier D, Petosa C, Panne D, Rousseaux S, Roeder RG, Zhao Y, Khochbin S (2016) Dynamic competing histone H4 K5K8 acetylation and butyrylation are hallmarks of highly active gene promoters. Mol Cell 62(2):169–180. https://doi.org/10.1016/j.molcel.2016.03.014
Hartl M, Fussl M, Boersema PJ, Jost JO, Kramer K, Bakirbas A, Sindlinger J, Plochinger M, Leister D, Uhrig G, Moorhead GB, Cox J, Salvucci ME, Schwarzer D, Mann M, Finkemeier I (2017) Lysine acetylome profiling uncovers novel histone deacetylase substrate proteins in Arabidopsis. Mol Syst Biol 13(10):949. https://doi.org/10.15252/msb.20177819
Hazzalin CA, Mahadevan LC (2005) Dynamic acetylation of all lysine 4-methylated histone H3 in the mouse nucleus: analysis at c-fos and c-jun. PLoS Biol 3(12):e393. https://doi.org/10.1371/journal.pbio.0030393
He D, Wang Q, Li M, Damaris RN, Yi X, Cheng Z, Yang P (2016) Global proteome analyses of lysine acetylation and succinylation reveal the widespread involvement of both modification in metabolism in the embryo of germinating rice seed. J Proteome Res 15(3):879–890. https://doi.org/10.1021/acs.jproteome.5b00805
Huang H, Sabari BR, Garcia BA, Allis CD, Zhao Y (2014) SnapShot: histone modifications. Cell 159(2):458–458.e451. https://doi.org/10.1016/j.cell.2014.09.037
Huang H, Lin S, Garcia BA, Zhao Y (2015) Quantitative proteomic analysis of histone modifications. Chem Rev 115(6):2376–2418. https://doi.org/10.1021/cr500491u
Huang H, Zhang D, Wang Y, Perez-Neut M, Han Z, Zheng YG, Hao Q, Zhao Y (2018) Lysine benzoylation is a histone mark regulated by SIRT2. Nat Commun 9(1):3374. https://doi.org/10.1038/s41467-018-05567-w
Janssen KA, Sidoli S, Garcia BA (2017) Recent achievements in characterizing the histone code and approaches to integrating epigenomics and systems biology. Methods Enzymol 586:359–378. https://doi.org/10.1016/bs.mie.2016.10.021
Janzen CJ, Fernandez JP, Deng H, Diaz R, Hake SB, Cross GA (2006) Unusual histone modifications in Trypanosoma brucei. FEBS Lett 580(9):2306–2310. https://doi.org/10.1016/j.febslet.2006.03.044
Jenuwein T, Allis CD (2001) Translating the histone code. Science 293(5532):1074–1080. https://doi.org/10.1126/science.1063127
Jiang D, Wang Y, Wang Y, He Y (2008) Repression of FLOWERING LOCUS C and FLOWERING LOCUS T by the Arabidopsis Polycomb repressive complex 2 components. PLoS One 3(10):e3404. https://doi.org/10.1371/journal.pone.0003404
Jiang D, Borg M, Lorkovic ZJ, Montgomery SA, Osakabe A, Yelagandula R, Axelsson E, Berger F (2020) The evolution and functional divergence of the histone H2B family in plants. PLoS Genet 16(7):e1008964. https://doi.org/10.1371/journal.pgen.1008964
Jin W, Wu F (2016) Proteome-wide identification of lysine succinylation in the proteins of tomato (Solanum lycopersicum). PLoS One 11(2):e0147586. https://doi.org/10.1371/journal.pone.0147586
Johnson L, Mollah S, Garcia BA, Muratore TL, Shabanowitz J, Hunt DF, Jacobsen SE (2004) Mass spectrometry analysis of Arabidopsis histone H3 reveals distinct combinations of post-translational modifications. Nucleic Acids Res 32(22):6511–6518. https://doi.org/10.1093/nar/gkh992
Karmodiya K, Krebs AR, Oulad-Abdelghani M, Kimura H, Tora L (2012) H3K9 and H3K14 acetylation co-occur at many gene regulatory elements, while H3K14ac marks a subset of inactive inducible promoters in mouse embryonic stem cells. BMC Genomics 13:424. https://doi.org/10.1186/1471-2164-13-424
Kebede AF, Nieborak A, Shahidian LZ, Le Gras S, Richter F, Gomez DA, Baltissen MP, Meszaros G, Magliarelli HF, Taudt A, Margueron R, Colome-Tatche M, Ricci R, Daujat S, Vermeulen M, Mittler G, Schneider R (2017) Histone propionylation is a mark of active chromatin. Nat Struct Mol Biol 24(12):1048–1056. https://doi.org/10.1038/nsmb.3490
Kouzarides T (2007) Chromatin modifications and their function. Cell 128(4):693–705. https://doi.org/10.1016/j.cell.2007.02.005
Li Y, Sabari BR, Panchenko T, Wen H, Zhao D, Guan H, Wan L, Huang H, Tang Z, Zhao Y, Roeder RG, Shi X, Allis CD, Li H (2016) Molecular coupling of histone crotonylation and active transcription by AF9 YEATS domain. Mol Cell 62(2):181–193. https://doi.org/10.1016/j.molcel.2016.03.028
Liebler DC, Zimmerman LJ (2013) Targeted quantitation of proteins by mass spectrometry. Biochemistry 52(22):3797–3806. https://doi.org/10.1021/bi400110b
Liu J, Wang G, Lin Q, Liang W, Gao Z, Mu P, Li G, Song L (2018a) Systematic analysis of the lysine malonylome in common wheat. BMC Genomics 19(1):209. https://doi.org/10.1186/s12864-018-4535-y
Liu K, Yuan C, Li H, Chen K, Lu L, Shen C, Zheng X (2018b) A qualitative proteome-wide lysine crotonylation profiling of papaya (Carica papaya L.). Sci Rep 8(1):8230. https://doi.org/10.1038/s41598-018-26676-y
Liu S, Xue C, Fang Y, Chen G, Peng X, Zhou Y, Chen C, Liu G, Gu M, Wang K, Zhang W, Wu Y, Gong Z (2018c) Global involvement of lysine crotonylation in protein modification and transcription regulation in rice. Mol Cell Proteomics 17(10):1922–1936. https://doi.org/10.1074/mcp.RA118.000640

Liu S, Liu G, Cheng P, Xue C, Zhou Y, Chen X, Ye L, Qiao Z, Zhang T, Gong Z (2019) Genome-wide profiling of histone lysine butyrylation reveals its role in the positive regulation of gene transcription in rice. Rice (N Y) 12(1):86. https://doi.org/10.1186/s12284-019-0342-6

Lorkovic ZJ, Park C, Goiser M, Jiang D, Kurzbauer MT, Schlogelhofer P, Berger F (2017) Compartmentalization of DNA damage response between heterochromatin and euchromatin is mediated by distinct H2A histone variants. Curr Biol 27(8):1192–1199. https://doi.org/10.1016/j.cub.2017.03.002

Loureiro J, Rodriguez E, Dolezel J, Santos C (2006) Comparison of four nuclear isolation buffers for plant DNA flow cytometry. Ann Bot 98(3):679–689. https://doi.org/10.1093/aob/mcl141

Lu Y, Xu Q, Liu Y, Yu Y, Cheng ZY, Zhao Y, Zhou DX (2018) Dynamics and functional interplay of histone lysine butyrylation, crotonylation, and acetylation in rice under starvation and submergence. Genome Biol 19(1):144. https://doi.org/10.1186/s13059-018-1533-y

Luger K, Mader AW, Richmond RK, Sargent DF, Richmond TJ (1997) Crystal structure of the nucleosome core particle at 2.8 A resolution. Nature 389(6648):251–260. https://doi.org/10.1038/38444

Mahrez W, Arellano MS, Moreno-Romero J, Nakamura M, Shu H, Nanni P, Kohler C, Gruissem W, Hennig L (2016) H3K36ac is an evolutionary conserved plant histone modification that marks active genes. Plant Physiol 170(3):1566–1577. https://doi.org/10.1104/pp.15.01744

Malik HS, Henikoff S (2003) Phylogenomics of the nucleosome. Nat Struct Biol 10(11):882–891. https://doi.org/10.1038/nsb996

Mandava V, Fernandez JP, Deng H, Janzen CJ, Hake SB, Cross GA (2007) Histone modifications in Trypanosoma brucei. Mol Biochem Parasitol 156(1):41–50. https://doi.org/10.1016/j.molbiopara.2007.07.005

Meng X, Xing S, Perez LM, Peng X, Zhao Q, Redona ED, Wang C, Peng Z (2017) Proteome-wide analysis of lysine 2-hydroxyisobutyrylation in developing rice (Oryza sativa) seeds. Sci Rep 7(1):17486. https://doi.org/10.1038/s41598-017-17756-6

Meng X, Lv Y, Mujahid H, Edelmann MJ, Zhao H, Peng X, Peng Z (2018) Proteome-wide lysine acetylation identification in developing rice (Oryza sativa) seeds and protein co-modification by acetylation, succinylation, ubiquitination, and phosphorylation. Biochim Biophys Acta 1866(3):451–463. https://doi.org/10.1016/j.bbapap.2017.12.001

Mersfelder EL, Parthun MR (2006) The tale beyond the tail: histone core domain modifications and the regulation of chromatin structure. Nucleic Acids Res 34(9):2653–2662. https://doi.org/10.1093/nar/gkl338

Molden RC, Garcia BA (2014) Middle-down and top-down mass spectrometric analysis of co-occurring histone modifications. Curr Protoc Protein Sci 77:23.27.21–23.27.28. https://doi.org/10.1002/0471140864.ps2307s77

Moradian A, Kalli A, Sweredoski MJ, Hess S (2014) The top-down, middle-down, and bottom-up mass spectrometry approaches for characterization of histone variants and their post-translational modifications. Proteomics 14(4–5):489–497. https://doi.org/10.1002/pmic.201300256

Moraes I, Yuan ZF, Liu S, Souza GM, Garcia BA, Casas-Mollano JA (2015) Analysis of histones H3 and H4 reveals novel and conserved post-translational modifications in sugarcane. PLoS One 10(7):e0134586. https://doi.org/10.1371/journal.pone.0134586

Mujahid H, Meng X, Xing S, Peng X, Wang C, Peng Z (2018) Malonylome analysis in developing rice (Oryza sativa) seeds suggesting that protein lysine malonylation is well-conserved and overlaps with acetylation and succinylation substantially. J Proteome 170:88–98. https://doi.org/10.1016/j.jprot.2017.08.021

Nomoto M, Kyogoku Y, Iwai K (1982) N-trimethylalanine, a novel blocked N-terminal residue of Tetrahymena histone H2B. J Biochem 92(5):1675–1678

Pandey R, Muller A, Napoli CA, Selinger DA, Pikaard CS, Richards EJ, Bender J, Mount DW, Jorgensen RA (2002) Analysis of histone acetyltransferase and histone deacetylase families of Arabidopsis thaliana suggests functional diversification of chromatin modification among multicellular eukaryotes. Nucleic Acids Res 30(23):5036–5055

Picchi GF, Zulkievicz V, Krieger MA, Zanchin NT, Goldenberg S, de Godoy LM (2017) Post-translational modifications of trypanosoma cruzi canonical and variant histones. J Proteome Res 16(3):1167–1179. https://doi.org/10.1021/acs.jproteome.6b00655

Plazas-Mayorca MD, Zee BM, Young NL, Fingerman IM, LeRoy G, Briggs SD, Garcia BA (2009) One-pot shotgun quantitative mass spectrometry characterization of histones. J Proteome Res 8(11):5367–5374. https://doi.org/10.1021/pr900777e

Roudier F, Ahmed I, Berard C, Sarazin A, Mary-Huard T, Cortijo S, Bouyer D, Caillieux E, Duvernois-Berthet E, Al-Shikhley L, Giraut L, Despres B, Drevensek S, Barneche F, Derozier S, Brunaud V, Aubourg S, Schnittger A, Bowler C, Martin-Magniette ML, Robin S, Caboche M, Colot V (2011) Integrative epigenomic mapping defines four main chromatin states in Arabidopsis. EMBO J 30(10):1928–1938. https://doi.org/10.1038/emboj.2011.103

Sabari BR, Zhang D, Allis CD, Zhao Y (2017) Metabolic regulation of gene expression through histone acylations. Nat Rev Mol Cell Biol 18(2):90–101. https://doi.org/10.1038/nrm.2016.140

Sabari BR, Tang Z, Huang H, Yong-Gonzalez V, Molina H, Kong HE, Dai L, Shimada M, Cross JR, Zhao Y, Roeder RG, Allis CD (2018) Intracellular crotonyl-CoA stimulates transcription through p300-catalyzed histone crotonylation. Mol Cell 69(3):533. https://doi.org/10.1016/j.molcel.2018.01.013

Schmitz RJ, Zhang X (2011) High-throughput approaches for plant epigenomic studies. Curr Opin Plant Biol 14(2):130–136. https://doi.org/10.1016/j.pbi.2011.03.010

Schubert D, Primavesi L, Bishopp A, Roberts G, Doonan J, Jenuwein T, Goodrich J (2006) Silencing by plant Polycomb-group genes requires dispersed trimethylation of histone H3 at lysine 27. EMBO J 25(19):4638–4649. https://doi.org/10.1038/sj.emboj.7601311
Shechter D, Dormann HL, Allis CD, Hake SB (2007) Extraction, purification and analysis of histones. Nat Protoc 2(6):1445–1457. https://doi.org/10.1038/nprot.2007.202
Shu H, Nakamura M, Siretskiy A, Borghi L, Moraes I, Wildhaber T, Gruissem W, Hennig L (2014) Arabidopsis replacement histone variant H3.3 occupies promoters of regulated genes. Genome Biol 15(4):R62. https://doi.org/10.1186/gb-2014-15-4-r62
Sidoli S, Bhanu NV, Karch KR, Wang X, Garcia BA (2016) Complete workflow for analysis of histone post-translational modifications using bottom-up mass spectrometry: from histone extraction to data analysis. J Vis Exp 111. https://doi.org/10.3791/54112
Sikorskaite S, Rajamaki ML, Baniulis D, Stanys V, Valkonen JP (2013) Protocol: optimised methodology for isolation of nuclei from leaves of species in the solanaceae and rosaceae families. Plant Methods 9:31. https://doi.org/10.1186/1746-4811-9-31
Sims RJ III, Reinberg D (2008) Is there a code embedded in proteins that is based on post-translational modifications? Nat Rev Mol Cell Biol 9(10):815–820. https://doi.org/10.1038/nrm2502
Smith-Hammond CL, Swatek KN, Johnston ML, Thelen JJ, Miernyk JA (2014) Initial description of the developing soybean seed protein Lys-N(epsilon)-acetylome. J Proteome 96:56–66. https://doi.org/10.1016/j.jprot.2013.10.038
Stock A, Clarke S, Clarke C, Stock J (1987) N-terminal methylation of proteins: structure, function and specificity. FEBS Lett 220(1):8–14
Strahl BD, Allis CD (2000) The language of covalent histone modifications. Nature 403(6765):41–45. https://doi.org/10.1038/47412
Stroud H, Otero S, Desvoyes B, Ramirez-Parra E, Jacobsen SE, Gutierrez C (2012) Genome-wide analysis of histone H3.1 and H3.3 variants in Arabidopsis thaliana. Proc Natl Acad Sci U S A 109(14):5370–5375. https://doi.org/10.1073/pnas.1203145109
Struhl K (1993) Gene expression: chromatin and transcription factors: who's on first? Curr Biol 3(4):220–222
Sun H, Liu X, Li F, Li W, Zhang J, Xiao Z, Shen L, Li Y, Wang F, Yang J (2017) First comprehensive proteome analysis of lysine crotonylation in seedling leaves of Nicotiana tabacum. Sci Rep 7(1):3013. https://doi.org/10.1038/s41598-017-03369-6
Swygert SG, Peterson CL (2014) Chromatin dynamics: interplay between remodeling enzymes and histone modifications. Biochim Biophys Acta 1839(8):728–736. https://doi.org/10.1016/j.bbagrm.2014.02.013
Tan F, Zhang K, Mujahid H, Verma DP, Peng Z (2011a) Differential histone modification and protein expression associated with cell wall removal and regeneration in rice (Oryza sativa). J Proteome Res 10(2):551–563. https://doi.org/10.1021/pr100748e
Tan M, Luo H, Lee S, Jin F, Yang JS, Montellier E, Buchou T, Cheng Z, Rousseaux S, Rajagopal N, Lu Z, Ye Z, Zhu Q, Wysocka J, Ye Y, Khochbin S, Ren B, Zhao Y (2011b) Identification of 67 histone marks and histone lysine crotonylation as a new type of histone modification. Cell 146(6):1016–1028. https://doi.org/10.1016/j.cell.2011.08.008
Taverna SD, Li H, Ruthenburg AJ, Allis CD, Patel DJ (2007a) How chromatin-binding modules interpret histone modifications: lessons from professional pocket pickers. Nat Struct Mol Biol 14(11):1025–1040. https://doi.org/10.1038/nsmb1338
Taverna SD, Ueberheide BM, Liu Y, Tackett AJ, Diaz RL, Shabanowitz J, Chait BT, Hunt DF, Allis CD (2007b) Long-distance combinatorial linkage between methylation and acetylation on histone H3 N termini. Proc Natl Acad Sci U S A 104(7):2086–2091. https://doi.org/10.1073/pnas.0610993104
Tessarz P, Kouzarides T (2014) Histone core modifications regulating nucleosome structure and dynamics. Nat Rev Mol Cell Biol 15(11):703–708. https://doi.org/10.1038/nrm3890
Voigt P, LeRoy G, Drury WJ III, Zee BM, Son J, Beck DB, Young NL, Garcia BA, Reinberg D (2012) Asymmetrically modified nucleosomes. Cell 151(1):181–193. https://doi.org/10.1016/j.cell.2012.09.002
Waterborg JH (2000) Steady-state levels of histone acetylation in Saccharomyces cerevisiae. J Biol Chem 275(17):13007–13011
Waterborg JH, Robertson AJ, Tatar DL, Borza CM, Davie JR (1995) Histones of Chlamydomonas reinhardtii. Synthesis, acetylation, and methylation. Plant Physiol 109(2):393–407
Webb KJ, Lipson RS, Al-Hadid Q, Whitelegge JP, Clarke SG (2010) Identification of protein N-terminal methyltransferases in yeast and humans. Biochemistry 49(25):5225–5235. https://doi.org/10.1021/bi100428x
Wu T, Yuan T, Tsai SN, Wang C, Sun SM, Lam HM, Ngai SM (2009) Mass spectrometry analysis of the variants of histone H3 and H4 of soybean and their post-translational modifications. BMC Plant Biol 9:98. https://doi.org/10.1186/1471-2229-9-98
Xie Z, Dai J, Dai L, Tan M, Cheng Z, Wu Y, Boeke JD, Zhao Y (2012) Lysine succinylation and lysine malonylation in histones. Mol Cell Proteomics 11(5):100–107. https://doi.org/10.1074/mcp.M111.015875
Xie X, Kang H, Liu W, Wang GL (2015) Comprehensive profiling of the rice ubiquitome reveals the significance of lysine ubiquitination in young leaves. J Proteome Res 14(5):2017–2025. https://doi.org/10.1021/pr5009724
Xiong Y, Peng X, Cheng Z, Liu W, Wang GL (2016) A comprehensive catalog of the lysine-acetylation targets in rice (Oryza sativa) based on proteomic analyses. J Proteome 138:20–29. https://doi.org/10.1016/j.jprot.2016.01.019

Xue C, Liu S, Chen C, Zhu J, Yang X, Zhou Y, Guo R, Liu X, Gong Z (2018) Global proteome analysis links lysine acetylation to diverse functions in oryza sativa. Proteomics 18(1). https://doi.org/10.1002/pmic.201700036

Xue C, Qiao Z, Chen X, Cao P, Liu K, Liu S, Ye L, Gong Z (2020) Proteome-wide analyses reveal the diverse functions of lysine 2-hydroxyisobutyrylation in oryza sativa. Rice (N Y) 13(1):34. https://doi.org/10.1186/s12284-020-00389-1

Yelagandula R, Stroud H, Holec S, Zhou K, Feng S, Zhong X, Muthurajan UM, Nie X, Kawashima T, Groth M, Luger K, Jacobsen SE, Berger F (2014) The histone variant H2A.W defines heterochromatin and promotes chromatin condensation in Arabidopsis. Cell 158(1):98–109. https://doi.org/10.1016/j.cell.2014.06.006

Yu Z, Ni J, Sheng W, Wang Z, Wu Y (2017) Proteome-wide identification of lysine 2-hydroxyisobutyrylation reveals conserved and novel histone modifications in Physcomitrella patens. Sci Rep 7(1):15553. https://doi.org/10.1038/s41598-017-15854-z

Yuan ZF, Lin S, Molden RC, Garcia BA (2014) Evaluation of proteomic search engines for the analysis of histone modifications. J Proteome Res 13(10):4470–4478. https://doi.org/10.1021/pr5008015

Zacharaki V, Benhamed M, Poulios S, Latrasse D, Papoutsoglou P, Delarue M, Vlachonasios KE (2012) The Arabidopsis ortholog of the YEATS domain containing protein YAF9a regulates flowering by controlling H4 acetylation levels at the FLC locus. Plant Sci 196:44–52. https://doi.org/10.1016/j.plantsci.2012.07.010

Zhang K, Sridhar VV, Zhu J, Kapoor A, Zhu JK (2007) Distinctive core histone post-translational modification patterns in Arabidopsis thaliana. PLoS One 2(11):e1210. https://doi.org/10.1371/journal.pone.0001210

Zhang N, Zhang L, Shi C, Tian Q, Lv G, Wang Y, Cui D, Chen F (2017) Comprehensive profiling of lysine ubiquitome reveals diverse functions of lysine ubiquitination in common wheat. Sci Rep 7(1):13601. https://doi.org/10.1038/s41598-017-13992-y

Zhao D, Guan H, Zhao S, Mi W, Wen H, Li Y, Zhao Y, Allis CD, Shi X, Li H (2016) YEATS2 is a selective histone crotonylation reader. Cell Res 26(5):629–632. https://doi.org/10.1038/cr.2016.49

Zhen S, Deng X, Wang J, Zhu G, Cao H, Yuan L, Yan Y (2016) First comprehensive proteome analyses of lysine acetylation and succinylation in seedling leaves of brachypodium distachyon L. Sci Rep 6:31576. https://doi.org/10.1038/srep31576

Zheng Y, Thomas PM, Kelleher NL (2013) Measurement of acetylation turnover at distinct lysines in human histones identifies long-lived acetylation sites. Nat Commun 4:2203. https://doi.org/10.1038/ncomms3203

Zhou H, Finkemeier I, Guan W, Tossounian MA, Wei B, Young D, Huang J, Messens J, Yang X, Zhu J, Wilson MH, Shen W, Xie Y, Foyer CH (2018) Oxidative stress-triggered interactions between the succinyl- and acetyl-proteomes of rice leaves. Plant Cell Environ 41:1139. https://doi.org/10.1111/pce.13100

Current Challenges in Plant Systems Biology

9

Danilo de Menezes Daloso and Thomas C. R. Williams

Abstract

Plants, as biological systems, are organized and regulated by a complex network of interactions from the genetic to the morphological level and suffer substantial influence from the environment. Reductionist approaches have been widely used in plant biology but have failed to reveal the mechanisms by which plants can growth under adverse conditions. It seems likely, therefore, that to understand the complexity of plant metabolic responses it is necessary to adopt non-reductionist approaches such as those from systems biology. Although such approaches seem methodologically complex to perform and difficult to interpret, they have been successfully applied in both metabolic and gene expression networks in a wide range of microorganisms and more recently in plants. Given the advance of techniques that allow complex analysis of plant cells, high quantities of data are currently generated and are available for in silico analysis and mathematical modeling. It is increasingly recognized, therefore, that the use of different methods such as graph analysis and dynamic network modeling are needed to better understand this abundance of information. However, before these practical advances, one of the main challenges currently in plant biology is to change the paradigm from the classical reductionism to the systemic level, which requires not only scientific but also educational changes.

Keywords

Biological network · Genome-scale metabolic models · Mathematical modeling · Metabolic fluxes · Systems biology

9.1 Introduction

As sessile organisms, land plants are constantly subjected to changing environmental conditions that trigger local and systemic responses throughout the plant and ultimately have a significant impact upon plant survival, growth, and yield. Changes in the climate are likely to lead to increased incidence of several stresses including drought, temperature extremes, flooding and salinity that are in turn expected to reduce plant productivity. Therefore, plant stress responses represent a high priority area in plant science research, with one important aim being the production of stress-tolerant varieties of important crops species. Plant responses to stress are, however, highly complex,

D. de Menezes Daloso (✉)
Departamento de Bioquímica e Biologia Molecular, Universidade Federal do Ceará, Fortaleza, Brasil
e-mail: daloso@ufc.br

T. C. R. Williams
Departamento de Botânica, Universidade de Brasília, Brasília, Brasil

F. V. Winck (ed.), *Advances in Plant Omics and Systems Biology Approaches*, Advances in Experimental Medicine and Biology 1346, https://doi.org/10.1007/978-3-030-80352-0_9

often involving large-scale alterations in metabolism, signaling and gene expression at the cellular level, together with changes in whole plant physiology and morphology. Under stress conditions, the changes that occur simultaneously in these multiple interconnected levels of cellular organization can make it difficult to understand the mechanisms responsible for stress tolerance as well as to identify and generate stress-tolerant genotypes (Bertolli et al. 2014). This problem is to some extent exacerbated by the relatively limited application of systems biology in plant science. The knowledge concerning the structure, function, and modulation of complex plant networks, such as the plant metabolic network, is therefore limited when compared to that for microorganisms (Jeong et al. 2000; Stelling et al. 2002; Almaas et al. 2004; Bruggeman and Westerhoff 2007).

Reductionist approaches have generally focused on understanding individual cellular components, their chemical composition and their biological functions under different environmental conditions (Palsson 2006; Lorenz et al. 2009). However, given that the way in which a complex network operates and responds to changing conditions may not be immediately apparent from the individual properties of its parts and the fact that the properties of individual parts cannot always be understood outside of a network context, reductionist approaches do not allow a complete characterization of plant phenotypes (Souza et al. 2016). In contrast, systemic analyses overcome these limitations, leading to new interpretations of biological process (Barabási and Oltvai 2004; Friboulet and Thomas 2005). Systems biology approaches have been applied to understand the dynamics of the relationship between plant cells and their surrounding environment (Medeiros et al. 2015). Although such approaches may seem methodologically complex to perform and difficult to understand, a relatively simple network model for plant physiological studies has been proposed (Sato et al. 2010; Bertolli et al. 2013) together with more complex networks used in conjunction with transcriptomic, proteomic, metabolomic, and metabolic flux data. In this chapter, we discuss the current challenges faced when applying systems biology to plants, with a focus on the plant metabolic network and the responses of plants to stress conditions.

The maintenance of biological systems occurs through modulation of complex regulatory networks working at different scales. As biological systems, plants have the capacity to modulate their metabolic networks according to ambient conditions. This occurs to optimize growth and development under constraining environmental conditions (Amzallag 2001; Barabási and Oltvai 2004). The severity of such constraints, the genetic background of the organism, its individual history, and its phenotypic plasticity determines survival or death under non-favorable conditions (Pastori and Foyer 2002; Daloso 2014; Souza and Lüttge 2015). Furthermore, plant tolerance to one or more environmental stress factors depends on a complex interaction network able to create responses at different levels of the plant (Buescher et al. 2012; Nakashima et al. 2014) and has been associated with a complex of genes co-adapted to the environmental condition (Graham et al. 1993) as well as to an "internal memory" of the genotype (Trewavas 2005; Tafforeau et al. 2006; Virlouvet and Fromm 2015). Therefore, given that different individuals can respond differently to a stress factor, which hampers the distinction of susceptible from tolerant genotypes by reductionism approaches, the use of systems biology methods enables not only the integration of different levels of the organism but also the possibility to obtain insights into emergent mechanisms used by a genotype to survive and grow under stress conditions.

Large-scale data generated through omic technologies have been used to obtain unprecedented levels of details regarding the structure and organization of biological networks and interactions between their different components (Hyduke and Palsson 2010; Fernie and Stitt 2012; Amaral and Souza 2017). Furthermore, the advance of data production by omic platforms enables the establishment of genome-scale metabolic models that predict metabolic responses of single cells, organs, and plant individuals under different conditions (Poolman et al. 2009; de Oliveira

Dal'Molin et al. 2010; Williams et al. 2010; Blazier and Papin 2012; Arnold and Nikoloski 2014; Robaina-Estévez et al. 2017). The ultimate goal of these models is to increase our ability to predict plant behavior under field conditions which may ultimately help to select genotypes with greater yield and/or stress tolerance. However, one of the current challenges is to fill the gap between plant responses obtained under controlled and under natural environmental conditions, in which the models created for and from plants grown under controlled conditions must be able to predict the responses of plants in the field. Here we provide a brief historical perspective of plant systems biology and highlight the current technical and theoretical challenges in this field. We conclude by providing a perspective on how systems biology approaches may be used to improve our understanding of plant behavior.

9.2 From the Genomic to the Plant Systems Biology Era

Plant biology has changed dramatically in the last two decades. Much of our knowledge of plant molecular biology comes from studies using *Arabidopsis thaliana* (L.) Heynh. as a model. After the sequencing of its genome (The Arabidopsis Genome Initiative 2000) and the establishment of a protocol to easily insert transgenes (Clough and Bent 1998), which enable the launch of stock centers of mutants (e.g., SALK, SAIL, GABI), Arabidopsis became and remains by far the most thoroughly investigated plant on earth. Making use of the large amount of data obtained through genome sequencing, however, is a nontrivial task which has required greater participation of computational biologists and bioinformaticians in plant biology, and allowed questions about the definition of a gene, the organization of genomes and the significance of repeated sequences to be tackled. However, after obtaining an understanding of the basic concepts of genome organization, another significant challenge became apparent, given that following sequencing 31% of Arabidopsis genes were unclassified and only 9% of the genes identified had been experimentally characterized (The Arabidopsis Genome Initiative 2000). The elucidation of the function of these uncharacterized genes is one of the most important objectives of the post-genomic or functional genomic era.

The current era of functional genomics has been characterized by the collection of large amounts of data using different "omic" platforms. Transcriptomic and proteomic platforms were initially developed for plants with sequenced genomes, however techniques such as microarrays are now being replaced by RNA sequencing methods (RNA-Seq) which do not necessarily require a previously sequenced genome and hence increase the number of species that can be investigated (Wang et al. 2009). In parallel, advances in mass spectrometry (MS)-based platforms have not only contributed to the development of more sophisticated proteomic techniques but have also led to the inclusion of metabolomics amongst the platforms available for the production of large-scale plant biology data (Roessner et al. 2000). Crucially, the combination of these omic platforms with reverse genetics has been a powerful tool in functional genomics, in which the expression of a gene or a set of genes is altered, and the plant phenotype is observed through the eyes of multiple large-scale omic platforms. This approach has greatly increased our understanding of the function of genes that were previously only poorly characterized. However, there are two principal reasons to believe that this approach is unlikely to ever reveal the function of all genes within any given plant. Firstly, plant metabolism and plant signaling are highly plastic systems, with a great deal of redundancy, meaning that perturbations in a single gene can often be compensated by other gene products. Secondly, it is likely that the function of many genes simply cannot be understood from a reductionist point of view alone; they have functions that manifest themselves at the level of organization and operation of biological systems.

In order to overcome the limitations imposed by the use of reductionists approaches, the application of network principles from the general systems theory to biological organisms has changed substantially our view of biological organisms, and led to what might be called a systems biology era (Kitano 2002). Remarkable studies in systems biology using microbes as model organisms showed that the modulation and the behavior of metabolic networks are similar to that observed in complex nonbiological networks (Jeong et al. 2000; Wagner and Fell 2001). This indicates that biological organisms such as plants can be interpreted using network principles. Thus, the recent visualization and interpretation of large-scale omic-based data have followed the principles of network biology (Barabási and Oltvai 2004), which is one of the main contributions of the systems biology to plant biology (Toubiana et al. 2013). In the next sections, we discuss the technical and methodological limitations and challenges associated to the investigation of plants using the main omic approaches named transcriptomic, proteomic, and metabolomic.

9.3 Data Production for Plant Systems Biology: Technical Limitations of Omic Approaches

Systems biology aims to understand the structure and function of biological systems as a whole rather than their components in isolation. Omic platforms, therefore, make important contributions to this field, as they may yield the data necessary to build the models needed to investigate complex systems, reveal previously unidentified interactions between network components and generate the experimental data necessary to test system-scale hypotheses. Despite the sophisticated technology currently available for such analyses, technical limitations still prevent us from detecting all biochemical and molecular responses that plants are capable of manifesting. One of the main challenges in plant systems biology is the generation of large-scale datasets, with proteomic and metabolomic datasets deserving a particular consideration, given that current coverage of proteins and metabolites is well below that known to exist in the plant kingdom. In contrast, transcriptomics requires less improvement in terms of coverage of analytes. Analyses such as microarrays and RNA-Seq can be used to detect expression of almost the entire set of Arabidopsis genes. However, the success of current transcriptomic approaches varies according to the complexity of the genome of the plant species, influenced by the size of the genome, the number of genes, chromosomes, and the genome ploidy. It is noteworthy that the relationship between number of genes is not directly proportional to the number of chromosomes or the size of the genome of a plant species, but it seems reasonable to assume that the application of microarray or RNA-Seq platforms is more difficult and requires more extensive in silico analysis for plants with larger genomes (Wang et al. 2009).

In contrast to transcriptomic, substantial improvements are needed in proteomic and metabolomic approaches. The acquisition of large-scale protein and metabolite data has mainly made use of chromatographic techniques coupled to mass spectrometry (MS), meaning that most of the technical limitations of proteomic and metabolomic platforms are related to the capacity of the chromatography to separate the analytes and the ability of the MS to detect and identify them. This become even more complex in proteins that have been subjected to a posttranslational modification (Friso and van Wijk 2015). In this case, besides detecting and identifying the protein, the posttranslational modification must be identified typically requiring special protocols and high-resolution MS platforms. Further improvement of MS-based approaches therefore seems to be a key area in which technical advances could benefit plant systems biology, aiming to increase the number of proteins detected as well as to improve the methods used to identity posttranslational modifications at large scale. This will help plant biologists to understand how metabolic pathways are regulated at the posttranslational level and permit significant improvements to be made to current metabolic models (Fernie 2012), particularly with regards their capacity to predict the effects of stress conditions.

In a similar manner to proteomics, one of the current limitations of metabolomics is related to the capacity of different platforms to identify and quantify the massive diversity of primary and secondary metabolites that plants can produce. This is in part related to the restricted availability of standards necessary for the establishment of MS-based libraries; while it has been estimated that plant kingdom contains approximately 200,000 metabolites, the mass spectral libraries only contain a small fraction of this number (Weckwerth 2003; Saito and Matsuda 2010). Inevitably though, the diversity and complexity of the chemical structures of plant metabolites means that no single analytical platform will ever be able to uncover all of cellular metabolism. For example, a well-established gas chromatography coupled to a time of flight MS (GC-TOF-MS) metabolomic platform is suitable only for polar metabolites, restricting its use to the study of primary metabolism (Lisec et al. 2006). Alternative platforms for molecules separation such as liquid chromatography (LC) and capillary electrophoresis (CE) can also be coupled to different types of MS instruments that have varying mass analyzers (e.g., Quadrupole, Orbitrap, Triple Quadrupole, ToF, qToF) and have seen significant application in plant metabolomics (Tolstikov and Fiehn 2002; Arrivault et al. 2009; Urakami et al. 2010). The simultaneous use of these platforms substantially increases the number of metabolites detected allowing the creation of more detailed picture of metabolism which is of fundamental importance in corroborating the predictions that can be obtained from genome-scale metabolic models (Fernie 2012; Medeiros et al. 2015). It is noteworthy that great advances have been observed in this field thanks to the collaboration between companies, researchers and the different metabolomic and MS societies. Thus, although metabolomics is currently far from being able to detect and identify the entire set of plant metabolites, the complex puzzle of metabolism has been substantially improved in the last decade and trends suggest that the combination of MS-based metabolomic methods and the mathematical modeling approaches that we discuss below represent a powerful strategy for predictive metabolic engineering (Sweetlove et al. 2003, 2017; Nikoloski et al. 2015).

Plant metabolomics is also complicated by the fact that most plant cells contains three compartmented genomes and numerous organelles that interact through a complex signaling network (Sweetlove and Fernie 2013), making the identification of metabolic responses at the subcellular level extremely challenging. While gene expression can be analyzed at the subcellular level through the origin of the transcript or subcellular location of the protein identified, the investigation of metabolic responses at the subcellular level is more complex. In contrast to mRNA and proteins, where the identity of them is given by their sequence, metabolites found in different organelles have identical chemical structures. Thus, in order to determine metabolite levels with subcellular resolution, the different organelles must be physically separated and the accumulation of metabolites in different subcellular fractions determined. In this vein, the non-aqueous fractionation (NAF) methods can be used to perform this type of separation (Krueger et al. 2014; Medeiros et al. 2017) and thus to overcome this limitation of metabolomics.

The rapid turnover of certain metabolites also presents a challenge to plant metabolomics. Despite methodologies for labeling and rapid freezing of whole Arabidopsis plants and maize leaves that have been established (Szecowka et al. 2013; Arrivault et al. 2017), changes in metabolic fluxes and metabolite pools are faster than the methods commonly used to freeze and analyze plant materials. Thus, another challenge in metabolomics is to improve methods available for observation of the dynamics and compartmentalization of metabolism (Sweetlove and Fernie 2013). Recent reports have carried out real time in situ metabolic analysis through the use of Förster Resonance Energy Transfer (FRET) sensors (De Col et al. 2017; Rizza et al. 2017; Wagner et al. 2019; Nietzel et al. 2020). However, this analysis requires previous transformation of the plant material, limiting the number of metabolites that can be investigated at once and hence the volume of data that can be obtained on the responses of the whole metabolome. Taken together, these facts indicate that metabolomics must be improved to achieve higher or total coverage as it is obtained through transcriptomic and proteomic platforms. However, thanks to the advances made in MS and cooperation between research groups with expertise in different metabolomic platforms most of these challenges should be at least partially overcome in the near future.

9.4 Methodological Limitations of Omic Approaches: From Alive vs. Dead Comparisons to More Dose and Dynamic Analyses

Another challenge in the use of omic approaches in plant systems biology originates from the way in which most experiments are carried out. Most plant studies using omic platforms have been carried out on plant material collected at a single time point, in which a control sample (e.g., non-stressed or wild-type plant) is compared to a plant subjected to "a non-normal growth condition" (e.g., stressed plants) or to a genetically modified plant. This sort of strategy often ignores the dynamics of the responses and assumes that the picture taken by a particular omic platform reveals the main response of the organism under the condition analyzed. This is the opposite to one of the main principles of systems biology, a scientific area that aim to understand the dynamics of the relationships between the different elements that compose an organism (Kitano 2002; Friboulet and Thomas 2005). Special care should be taken when an omic platform is used to analyze gene expression through measurement of the mRNA or protein levels as changes at these levels are not always correlated with changes at the metabolic level. This may occur, in certain cases, due to a temporal difference between changes in gene expression and in metabolism. For instance, the metabolic responses to oxidative stress are much faster than changes in the level of mRNA accumulation (Lehmann et al. 2009), as turnover of metabolites occurs at the milli-second scale (Buziol et al. 2002). Thus, the ideal and the challenge is to analyze the metabolic state of the cell in different time points in ultra-short time scale (Nöh et al. 2007) in order to understand the dynamics of cellular metabolic responses.

Given the higher stability of mRNA and protein abundance compared to metabolite content and taking into account the velocity of biochemical reactions and post translational modifications that underpin metabolic fluxes (Szecowka et al. 2013; Ma et al. 2014; Arrivault et al. 2017), it is particularly important, therefore, to consider the dynamics of the responses rather than a single time point comparison between treated and non-treated samples in order to construct a complete picture of plant responses. Additionally, in plant stress experiments different doses of the stressor, such as different concentrations of a heavy metal or different levels of drought, are needed (Greenham et al. 2017). This strategy may provide substantial information regarding the dynamics of the responses under different levels of stress which is fundamental to understand plant acclimation under adverse conditions. The data produced from this sort of experiment will certainly be useful in the building and curation of dynamic models of metabolism. This is of pivotal importance given that steady-state-based modeling is well established while dynamic modeling still needs profound improvement. In the next section, we provide a brief overview on the transition between the general systems theory from physics to systems biology and argue the changes required in plant science for the complete establishment of plant systems biology.

9.5 Toward a Systemic Plant Biology

As discussed above, many of the advances in plant biology in the last decades were driven by the advent of analytical tools that enable large-scale analysis of gene expression and metabolism. However, systems biology did not originate with large-scale data analysis and instead has its origins in the general systems theory originally proposed by Ludwig von Bertalanffy (von Bertalanffy 1968). It is clear that the large-scale omic platforms have substantially contributed to systems biology by providing an unprecedented quantity and quality of data from different levels of organismal organization. Indeed, these platforms have helped to confirm one of the main principles from General Systems Theory which is that organisms cannot easily be described as the sum of their parts; emergent properties may be detected from the study of the whole system and cannot be identified when looking at the elements of the system in isolation (Souza et al. 2016). Thus, omic platforms facilitate a shift from a theoretical to a practical application of this systems biology principle. This does not, however, mean that systems biology depends entirely on use of omic platforms, nor does it mean that any large-scale exercise in data accumulation represents systems biology. Indeed, systems biology approaches have been applied to the analysis of biological sub-networks without the use of omic approaches, especially in other areas of plant biology such as plant ecology (Odum 1983). Therefore, it is necessary to avoid the use of the term systems biology when referring to works that have only used omic platforms.

Omic platforms have allowed plant biologists to produce more data than it can readily be analyzed and interpreted. This has transformed plant biology into a data-rich multidisciplinary field, where expertise in mathematics and computational science is increasingly recognized as fundamental to a plant biologist. In this vein, an important current challenge in plant systems biology is the shift from a reductionism paradigm to a more systemic view of plant science, something that will require profound theoretical and practical changes in plant science. First, the formation of plant biologists and biotechnologists must be dramatically changed in the (under)graduate programs. It is urgently needed to increase the number of mathematicians and computational scientists in these programs. The challenge is to form professionals that have at least basic skills of mathematical modeling and computational science with a complete background of biology. To achieve this, it is first necessary to recognize systems biology as the future of plant science. Efforts from universities and the different plant biology societies are needed to overcome the reductionism view of plant science at global scale. It is important to mention that this strategy has already been adopted in several universities as well as and it will not abolish science made at reductionist level. Several important areas of our society depend on the basic science produced under the eyes of the classical reductionism. However, the well-recognized power of systems biology in identifying and predicting responses already shown in ecology, medicine, social, and sexual networks must be also applied in plant biology.

Beyond these philosophical and educational changes, it is important to mention that the experimental design, the elaboration of hypotheses and currently accepted theories in plant biology may also be influenced by systems biology principles. Firstly, a "design-build-test-learn" cycle with the help of mathematical models and bioinformatics tools may be incorporated in the design of new experiments (Gutierrez et al. 2005; Sweetlove et al. 2017) (Fig. 9.1). Given the amount of omics data available for in silico analysis and the number of models already created to predict signaling and metabolic responses (Li et al. 2006; Hills et al. 2012; Chatterjee et al. 2017; Lima et al. 2017; Robaina-Estévez et al. 2017; Zuniga et al. 2017; Christopher et al. 2019; Shameer et al. 2019; Benes et al. 2020; Vallarino et al. 2020) it seems likely that the classical way of generating hypothesis, based solely on a previous experimental observation, is becoming obsolete.

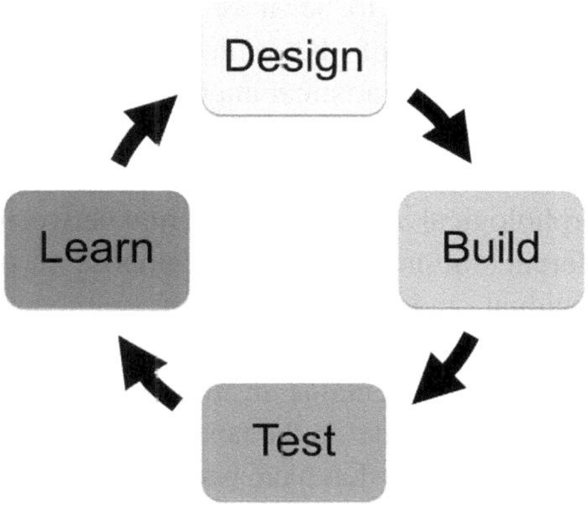

Fig. 9.1 The design-build-test-learn cycle

Previous observations are certainly important, but before a hypothesis is experimentally tested, a previous in silico analysis may provide great insights into that hypothesis as well, aiding in the conceiving novel experimental design. In turn, the results obtained from such experiments may help to improve in silico models and hence their power to predict plant responses, closing the "design-build-test-learn" cycle (Fig. 9.1).

The adoption of the general systems theory in plant biology allows different perspective that may change many paradigms within plant biology. For instance, the idea that plants tend to direct their metabolism toward homeostasis, a metabolic state far from the dynamic of chaos, can be substantially changed after observing the dynamics of plant responses using systems biology tools. Indeed, nonlinear chaotic dynamic has been already observed in the dynamics of sap flow (Souza et al. 2004b), stomatal conductance (Souza et al. 2004a), CO_2 fixation in CAM plants (Lüttge and Beck 1992), and in glycolysis (Nielsen et al. 1997). Furthermore, the analysis and interpretation of data acquired from plants may change using systems biology tools. For instance, the visualization of data in systems biology is often carried out through networks. In biology, this approach has been used since Barabási's works when it was demonstrated that biological networks have similar structures as nonbiological networks (Jeong et al. 2000, 2001). In this context, network parameters such as node, hub, hub-like nodes, density and degree of connectivity of the network will prove to be more important for the comprehension of the biological phenomena than univariate statistical analysis such as average and variance analysis. Therefore, the reconstruction of networks from the genetic to the morphological level, the integration between the different sub-networks that compose a plant individual and the creation of mathematical modeling tools that can predict the behavior of these networks remains an important challenge for the next decade. In the next sections, we discuss the methods that have been used to analyze metabolic networks in plants and highlight the challenges that plant biologists are facing.

9.6 Reconstructing Plant Metabolic Networks: The Challenge to Reach an Integrative Nonsteady State Large-Scale Modeling Platform

Large-scale gene expression data from different tissues under several environmental conditions are available for Arabidopsis plants, which has led to well-established gene expression and co-expression networks (Toubiana et al. 2013; Ruprecht et al. 2017). However, large-scale protein-protein interaction and metabolic networks have received greater attention in yeast, bacteria, *C. elegans*, mammalians and other organisms rather than in plants. The identification, analysis and modeling of these networks represent important goals of plant systems biology. Crucially though, we must also focus on studying the interactions between networks at different levels of organization, as current studies and models tend to be limited to a specific level, such as gene networks (Thum et al. 2008; Urano et al. 2009; Usadel et al. 2009; Ruprecht et al. 2017), protein-protein interaction networks (Jones et al. 2014; Zhang et al. 2017) or metabolic networks (Williams et al. 2010; Arnold and Nikoloski 2014; Robaina Estévez and Nikoloski 2015). Thus, one of the great challenges is to integrate these different networks and models. With this in hand we will be better able to understand the behavior of the entire plant network and predict plant responses in the field and thus optimize growth and yield of crop plants under adverse conditions. In this section we focus particularly on metabolic network modeling given that this seems to be the most challenging task to achieve in plant systems biology.

Since the first work published using a large-scale plant metabolomic platform (Roessner et al. 2000), the number of metabolite profiling studies has increased substantially, thanks to the use and evolution of GC-MS and LC-MS platforms. However, snapshot analyses of plant metabolism may only poorly reflect the flow of material through the metabolic network and hence yield little insight into how metabolic fluxes respond to environmental, genetic or developmental perturbations. Indeed, theoretical and experimental analyses indicate that alterations in metabolic flux may lead to increases, decreases or no effect at all of metabolite abundances—a problem compounded by the fact that often, due to subcellular compartmentation, it is not possible to directly measure the intermediates of any given metabolic pathway. Methods for the quantification and prediction of metabolic fluxes, therefore, represent important systems biology tools. While radiolabeling remains a sensitive method for the investigation of specific metabolic processes (Kruger et al. 2017), fluxes in plant metabolism at the network level are studied using either metabolic flux analysis (MFA) or stoichiometric modeling. In the case of MFA, plant material is initially labeled using stable isotopes such as ^{13}C, ^{18}O, or ^{15}N (Silva et al. 2016), either individually or in combination (António et al. 2016), until metabolic and isotopic steady state is reached and further measurement of isotope incorporation using MS or nuclear magnetic resonance (NMR) spectroscopy is done.

Medium-scale metabolic network models have been used to interpret the isotope labeling data and, in this manner, determine metabolic fluxes. Such studies in plants have revealed the operation of novel metabolic processes (Schwender et al. 2004) and the effects of growth conditions (Williams et al. 2008; Allen and Young 2013) and mutations (Lonien and Schwender 2009) on fluxes. They have also produced insights into reactions contributing to cofactor and carbon balances, information that is difficult to obtain using other methods (Allen et al. 2009; Masakapalli et al. 2013). Furthermore, advancements in both analytical methods and theoretical tools should lead to the expansion of the use of these methods, and have also begun to make the study of photoautotrophic metabolism possible through the use of instationary flux analysis (Szecowka et al. 2013; Ma et al. 2014; Arrivault et al. 2017). Photosynthetic metabolism cannot readily be investigated using steady state-based flux analysis, and it is anticipated that instationary analysis should see wider use in the near future. Spatial resolution of metabolic fluxes, both at the cellular and subcellular level remains a challenge for isotope labeling-based methods, though highly compartmented cellular models have been used in some studies (Masakapalli et al. 2010) and methods have also been proposed for measurement of cell type specific fluxes (Rossi et al. 2017).

While MFA continues to prove a highly valuable tool for the study of plant metabolism it has two significant drawbacks. Firstly, not all experimental systems can readily be labeled using stable isotopes to obtained meaningful data. Secondly, and perhaps more importantly in terms of systems biology, MFA is not a tool for the prediction of fluxes, as the flux distribution within the metabolic network is fitted to the experimental isotope labeling data. This means that MFA cannot be used to directly predict the effects of environmental or genetic alterations on metabolic fluxes, although it can be used to formulate hypotheses regarding the operation of the plant metabolic network that can be tested through further experimentation. Stoichiometric modeling, on the other hand, can be used to predict how fluxes may respond to perturbations in the metabolic network. Here, metabolic models are built from simple stoichiometric descriptions of individual reactions. Such models may therefore be extremely large, and indeed so-called genome-scale models aim to include all reactions for which genes encoding the respective enzyme have been annotated in the genome. Extensive curation is typically needed to fill gaps in pathways and place reactions in the correct subcellular location. In this way, genome-scale models have been produced for Arabidopsis (Poolman et al. 2009; de Oliveira Dal'Molin et al. 2010; Williams et al. 2010; Arnold and Nikoloski 2014) and crops including rice (Mohanty et al. 2016; Chatterjee et al. 2017), corn (Seaver et al. 2015), and tomato (Yuan et al. 2016). The construction of additional models for further crop plants represents an important goal in metabolic modeling. Once built, constraints are introduced and flux balance analysis (FBA) carried out allowing the use of the model to test and generate hypotheses about the operation of the metabolic network. The constraints often take the form of measurements of biomass composition, production or consumption that fix the rates of entry and removal of material from the model. This approach was, for example, used to investigate changes in metabolism during tomato fruit ripening (Colombié et al. 2017). However, a number of methods also exist for the integration of large-scale gene expression data into metabolic models (Robaina Estévez and Nikoloski 2014) and these approaches are beginning to see significant use. For example, gene expression data for guard cells and mesophyll cells was recently used to produce specific models for these two cell types and in this way uncovers differences in their predicted metabolism. Remarkably, part of the predictions generated by these models was successfully confirmed by ^{13}C-labeling kinetic analysis (Robaina-Estévez et al. 2017), highlighting the power of systems biology approaches for the establishment of hypothesis to be experimentally assessed following the idea of the "design-build-test-learn" cycle (Fig. 9.1).

Given the successes obtained using stoichiometric models to explore the metabolism of single cell types under constant conditions, the next logical step is to use such models to explore the larger scale temporal and spatial organization of plants. For example, a duplicated genome-scale model of Arabidopsis, where one sub-model represents daytime metabolism and second nighttime metabolism, correctly predicted certain known features of C3 and CAM leaf metabolism, including diurnal patterns of starch accumulation and degradation (Cheung et al. 2014; Shameer et al. 2018). This approach could in theory be extended to even more complex biological metabolic rhythms. Ultimately, it seems likely that stoichiometric metabolic models will need to be integrated with other types of model, including those representing signaling processes, gene expression and whole plant development and growth, in a similar manner to that use to construct a "digital *Arabidopsis*" based on a functional-structural model, carbon dynamic model, photothermal model and photoperiodism model (Chew et al. 2014, 2016). Genome-scale stoichiometric models could also be integrated with the sophisticated crop models that are used to predict yield under field conditions. Such models often include parameters, such as carbon conversion efficiencies (Setiyono et al. 2010) that can be determined using genome-scale metabolic models, potentially helping to bridge the gap between gene function and whole plant growth and development.

Given the advances achieved in understanding metabolic networks in microorganisms, it seems likely that our comprehension of the function and key points of regulation of plant metabolic networks will be improved through use of both MFA and mathematical modeling coupled with sophisticated laboratory experiments. For example, a recent adaptive laboratory evolution study using whole-genome sequencing and MFA revealed how *Escherichia coli* adapt its metabolism to overcome the negative effects of a mutation in phosphoglucose isomerase (PGI), an important glycolytic enzyme (Long et al. 2017). The application of such an approach in plant science will certainly provide great insight into how plants adapt their metabolism under adverse conditions. The challenge will be to transfer the knowledge obtained from steady state MFA and FBA to a dynamic view of the entire plant metabolic network and from studies with model plants such as Arabidopsis in the laboratory to crop plants in the field. In the next section we will explore this last challenge.

9.7 From Arabidopsis to Crops, from the Phytotron to the Field: The Challenge for Crop Yield Improvement

The higher degree of complexity of plant cells compared to animals is mainly due to their capacity to quickly respond to changing environmental conditions. For instance, specialized cells such as the guard cells found at leaf epidermis integrate a number of different environmental and different endogenous signals to ultimately determine the appropriate degree of stomatal opening (Assmann and Jegla 2016). Furthermore, plants possess a complex hormonal and signaling network that responds to pathogen infection and herbivore attack together with the capacity to trigger more rapid responses on repeated exposure to a given stress condition (Ding et al. 2012; Virlouvet and Fromm 2015). This plasticity is of pivotal importance given the sessile nature of plants, in which a plant must adapt their metabolism under adverse conditions in order to grow and reproduce. In this sense, it is almost impossible to identify markers and understand how plants respond to a specific environmental condition by looking at the parts that make up a plant (Bertolli et al. 2014). This has tremendous importance in plant breeding. For example, a number of quantitative trait loci (QTL) and genes have been identified that confer increased drought tolerance (Tuberosa and Salvi 2006), and countless articles indicating that manipulation of a particular gene confers higher drought tolerance in Arabidopsis in a laboratory setting. However, we are failing to transfer the technology from the laboratory to the field, i.e., there are few commercially growth plants that exploit this knowledge (Nuccio et al. 2018). Why have we had so little success? There are multiple reasons for this lack of success (Flexas 2016; Nunes-Nesi et al. 2016), however, one of these is the use of reductionist approaches to identify plant stress metabolic responses as described above. Furthermore, experiments with combinations of stresses and measurements of the dynamics of the response are scarce, especially under field conditions, a major problem considering that different stresses often occur simultaneously. Therefore, the main challenge in molecular plant breeding is to fulfill the gap between the knowledge obtained in plants growing under controlled conditions to the field. As perspective, a recent study have fulfilled this gap in which the metabolite profiling of maize roots grown under controlled conditions could predict the hybrid performance in the field (Lima et al. 2017), indicating that this challenge could be overcome in the near future through the use of omic technologies and systems biology tools.

9.8 Concluding Remarks and Future Perspectives

The next decade will likely be marked by considerable advances in both analytical omic tools and the development of new techniques for modeling plant networks. Technical advances will require increased links between industries and research institutes. This is already reality in several countries, but more groups with different aims in plant science need to be integrated. Advances in modeling of plant systems will require the increased participation of mathematicians and computational scientists in plant biology research. This will help current plant biologists analyze and interpret omic data using systems biology tools, favor a change from reductionist to a more systemic view of plant biology and contribute to the training of a new generation of plant biologists able to use mathematical and modeling tools as basic research skills.

Development of software and the application of multivariate statistical analysis have contributed substantially to the analysis of large-scale omic data. The current challenge is to improve cross scale statistical analysis and create cross scale models to integrate different levels of observation and thus to have a better picture of the responses at whole plant network level. Furthermore, there is also a need to make the analysis of metabolic flux a routine part of plant biology, which has long been neglected (Fernie et al. 2005). While several pieces of software have been used for MFA and FBA of plants these analyses are relatively little used, especially when compared to transcriptomic, proteomic, and metabolomics methods. This may be due the fact that network flux analysis is the most recently established omic platform and suffers from difficult to perform experiments and complex data analysis. In order to increase the number of fluxomic studies and their contribution to plant systems biology, the main challenges are (1) advances in analytical techniques discussed above, (2) development of methods for short time scale MFA in whole plants under different conditions; and (3) improvements in software for the calculation and analysis of isotope enrichment. Taken together, overcoming the challenges discussed in this chapter will certainly change plant biology and will improve our understanding of the plant growth, development and responses to a changing environment. This, in turn, will increase the power of predictive metabolic engineering and provide improved perspectives for the increasing crop yield in the near future.

This cycle was originally proposed by Lee J. Sweetlove, Jens Nielsen, and Alisdair R. Fernie as a strategy to be adopted for engineering plant central metabolism (Sweetlove et al. 2017). Here we suggest the adoption of this cycle as a strategy to be used in plant systems biology studies in general. This idea is also influenced by the principles of the previously proposed systems biology cycle (Gutierrez et al. 2005), in which the establishment of hypothesis is aided by predictions made using systemic tools such as mathematical models and bioinformatics analyses. By adopting such an experimental approach, the design of the experiment, the organism to be tested, and the execution of the experiment per se will be based not only in previous experimental observations but, fundamentally, also in modeling predictions and simulations. The knowledge obtained from such experiments (the learn part of the cycle) is of fundamental importance not only to increase our understanding regarding the biological phenomena under investigation such as plant response to stress conditions but also to curate the models being used. In turn, well-established models will help the elaboration of critical hypothesis to be tested and ultimately speed up plant metabolic engineering and the breeding of stress-tolerant genotypes.

References

Allen DK, Young JD (2013) Carbon and nitrogen provisions alter the metabolic flux in developing soybean embryos. Plant Physiol 161:1458–1475. https://doi.org/10.1104/pp.112.203299

Allen DK, Ohlrogge JB, Shachar-Hill Y (2009) The role of light in soybean seed filling metabolism. Plant J 58:220–234. https://doi.org/10.1111/j.1365-313X.2008.03771.x

Almaas E, Kovács B, Vicsek T et al (2004) Global organization of metabolic fluxes in the bacterium Escherichia coli. Nature 427:839–843. https://doi.org/10.1038/nature02289

Amaral MN, Souza GM (2017) The challenge to translate OMICS data to whole plant physiology: the context matters. Front Plant Sci 8:8–11. https://doi.org/10.3389/fpls.2017.02146

Amzallag GN (2001) Data analisys in plant physiology: are we missing the reality? Plant Cell Environ 24:881–890

António C, Päpke C, Rocha M et al (2016) Regulation of primary metabolism in response to low oxygen availability as revealed by carbon and nitrogen isotope redistribution. Plant Physiol 170:43–56. https://doi.org/10.1104/pp.15.00266

Arnold A, Nikoloski Z (2014) Bottom-up metabolic reconstruction of arabidopsis and its application to determining the metabolic costs of enzyme production. Plant Physiol 165:1380–1391. https://doi.org/10.1104/pp.114.235358

Arrivault S, Guenther M, Ivakov A et al (2009) Use of reverse-phase liquid chromatography, linked to tandem mass spectrometry, to profile the Calvin cycle and other metabolic intermediates in Arabidopsis rosettes at different carbon dioxide concentrations. Plant J 59:824–839. https://doi.org/10.1111/j.1365-313X.2009.03902.x

Arrivault S, Obata T, Szecówka M et al (2017) Metabolite pools and carbon flow during C4 photosynthesis in maize: 13CO 2 labeling kinetics and cell type fractionation. J Exp Bot 68:283–298. https://doi.org/10.1093/jxb/erw414

Assmann SM, Jegla T (2016) Guard cell sensory systems: recent insights on stomatal responses to light, abscisic acid, and CO2. Curr Opin Plant Biol 33:157–167

Barabási A, Oltvai ZN (2004) Network biology: understanding the cell's functional organization. Nat Rev Genet 5:101–113. https://doi.org/10.1038/nrg1272

Benes B, Guan K, Lang M, Long SP, Lynch JP, Marshall-Colón A, Peng B, Schnable J, Sweetlove L, Turk M (2020) Multiscale computational models can guide experimentation and targeted measurements for crop improvement. Pkant J 103:21–31. https://doi.org/10.1111/tpj.14722

von Bertalanffy L (1968) General system theory. George Braziller, New York, NY

Bertolli S, Vítolo H, Souza G (2013) Network connectance analysis as a tool to understand homeostasis of plants under environmental changes. Plants 2:473–488. https://doi.org/10.3390/plants2030473

Bertolli SC, Mazzafera P, Souza GM (2014) Why is it so difficult to identify a single indicator of water stress in plants? A proposal for a multivariate analysis to assess emergent properties. Plant Biol 16:578–585. https://doi.org/10.1111/plb.12088

Blazier AS, Papin JA (2012) Integration of expression data in genome-scale metabolic network reconstructions. Front Physiol 3:299

Bruggeman FJ, Westerhoff HV (2007) The nature of systems biology. Trends Microbiol 15:45–50. https://doi.org/10.1016/j.tim.2006.11.003

Buescher JM, Liebermeister W, Jules M et al (2012) Global network reorganization during dynamic adaptations of Bacillus subtilis metabolism. Science (80-) 335:1099–1103. https://doi.org/10.1126/science.1206871

Buziol S, Bashir I, Baumeister A et al (2002) New bioreactor-coupled rapid stopped-flow sampling technique for measurements of metabolite dynamics on a subsecond time scale. Biotechnol Bioeng 80:632–636. https://doi.org/10.1002/bit.10427

Chatterjee A, Huma B, Shaw R, Kundu S (2017) Reconstruction of oryza sativa indica genome scale metabolic model and its responses to varying RuBisCO activity, light intensity, and enzymatic cost conditions. Front Plant Sci 8:1–18. https://doi.org/10.3389/fpls.2017.02060

Cheung CYM, Poolman MG, Fell DA et al (2014) A diel flux balance model captures interactions between light and dark metabolism during day-night cycles in C3 and crassulacean acid metabolism leaves. Plant Physiol 165:917–929. https://doi.org/10.1104/pp.113.234468

Chew YH, Wenden B, Flis A et al (2014) Multiscale digital *Arabidopsis* predicts individual organ and whole-organism growth. Proc Natl Acad Sci 111:E4127–E4136. https://doi.org/10.1073/pnas.1410238111

Chew YH, Seaton DD, Millar AJ (2016) Multi-scale modelling to synergise plant systems biology and crop science. Field Crops Res 202:77–83. https://doi.org/10.1016/j.fcr.2016.02.012

Christopher T, Williams R, Moreira TB, Shaw R, Ganguly O, Luo X, Kim S, Gabriel L, Coelho F, Yue C et al (2019) A genome-scale metabolic model of soybean (Glycine max) highlights metabolic fluxes during seedling growth. Plant Physiol 180:1912

Clough SJ, Bent AF (1998) Floral dip: a simplified method for Agrobacterium-mediated transformation of *Arabidopsis thaliana*. Plant J 16:735–743. https://doi.org/10.1046/j.1365-313X.1998.00343.x

Colombié S, Beauvoit B, Nazaret C et al (2017) Respiration climacteric in tomato fruits elucidated by constraint-based modelling. New Phytol 213:1726–1739. https://doi.org/10.1111/nph.14301

Daloso DM (2014) The ecological context of bilateral symmetry of organ and organisms. Nat Sci 6:184–190. https://doi.org/10.4236/ns.2014.64022

De Col V, Fuchs P, Nietzel T et al (2017) ATP sensing in living plant cells reveals tissue gradients and stress dynamics of energy physiology. elife 6:1–29. https://doi.org/10.7554/eLife.26770

Ding Y, Fromm M, Avramova Z (2012) Multiple exposures to drought "train" transcriptional responses in Arabidopsis. Nat Commun 3:740. https://doi.org/10.1038/ncomms1732

Fernie AR (2012) Grand challenges in plant systems biology: closing the circle(s). Front Plant Sci 3:1–4. https://doi.org/10.3389/fpls.2012.00035

Fernie AR, Stitt M (2012) On the discordance of metabolomics with proteomics and transcriptomics: coping with increasing complexity in logic, chemistry, and network interactions scientific correspondence. Plant Physiol 158:1139–1145. https://doi.org/10.1104/pp.112.193235

Fernie AR, Geigenberger P, Stitt M (2005) Flux an important, but neglected, component of functional genomics. Curr Opin Plant Biol 8:174–182. https://doi.org/10.1016/j.pbi.2005.01.008

Flexas J (2016) Genetic improvement of leaf photosynthesis and intrinsic water use efficiency in C3 plants: why so much little success? Plant Sci 251:155. https://doi.org/10.1016/j.plantsci.2016.05.002

Friboulet A, Thomas D (2005) Systems biology - an interdisciplinary approach. Biosens Bioelectron 20:2404–2407. https://doi.org/10.1016/j.bios.2004.11.014

Friso G, van Wijk KJ (2015) Update: post-translational protein modifications in plant metabolism. Plant Physiol 169:01378.2015. https://doi.org/10.1104/pp.15.01378

Graham JH, Emlen JM, Freeman DC (1993) Developmental stability and its applications in ecotoxicology. Ecotoxicology 2:175–184. https://doi.org/10.1007/BF00116422

Greenham K, Guadagno CR, Gehan MA et al (2017) Temporal network analysis identifies early physiological and transcriptomic indicators of mild drought in brassica rapa. elife 6:1–26. https://doi.org/10.7554/eLife.29655

Gutierrez RA, Shasha DE, Coruzzi GM (2005) Systems biology for the virtual plant. Plant Physiol 138:550–554. https://doi.org/10.1104/pp.104.900150

Hills A, Chen Z-H, Amtmann A et al (2012) OnGuard, a computational platform for quantitative kinetic modeling of guard cell physiology. Plant Physiol 159:1026–1042. https://doi.org/10.1104/pp.112.197244

Hyduke DR, Palsson BØ (2010) Towards genome-scale signalling-network reconstructions. Nat Rev Genet 11:297–307. https://doi.org/10.1038/nrg2750

Jeong H, Tombor B, Albert R et al (2000) The large-scale organization of metabolic networks. Nature 407:651–654. https://doi.org/10.1038/35036627

Jeong H, Mason SP, Barabasi A-L, Oltvai ZN (2001) Lethality and centrality in protein networks. Nature 411:41–42. https://doi.org/10.1038/35075138

Jones AM, Xuan Y, Xu M et al (2014) Border control--a membrane-linked interactome of Arabidopsis. Science (80-) 344:711–716. https://doi.org/10.1126/science.1251358

Kitano H (2002) Systems biology: a brief overview. Science (80-) 295:1662–1664. https://doi.org/10.1126/science.1069492

Krueger S, Steinhauser D, Lisec J, Giavalisco P (2014) Analysis of subcellular metabolite distributions within arabidopsis thaliana leaf tissue: a primer for subcellular metabolomics. In: Sanchez-Serrano J, Salinas J (eds) Arabidopsis protocols. Methods in molecular biology, vol 1062. Humana Press, Totowa, NJ, pp 575–596

Kruger NJ, Masakapalli SK, Ratcliffe RG (2017) Assessing metabolic flux in plants with radiorespirometry. In: Jagadis Gupta K (ed) Plant respiration and internal oxygen. Methods in molecular biology, vol 1670. Humana Press, New York, NY, pp 1–16

Lehmann M, Schwarzländer M, Obata T et al (2009) The metabolic response of Arabidopsis roots to oxidative stress is distinct from that of heterotrophic cells in culture and highlights a complex relationship between the

levels of transcripts, metabolites, and flux. Mol Plant 2:390–406. https://doi.org/10.1093/mp/ssn080
Li S, Assmann SM, Albert R (2006) Predicting essential components of signal transduction networks: a dynamic model of guard cell abscisic acid signaling. PLoS Biol 4:1732–1748. https://doi.org/10.1371/journal.pbio.0040312
Lima FA, Westhues M, Cuadros-Inostroza Á et al (2017) Metabolic robustness in young roots underpins a predictive model of maize hybrid performance in the field. Plant J 90:319–329. https://doi.org/10.1111/tpj.13495
Lisec J, Schauer N, Kopka J et al (2006) Gas chromatography mass spectrometry–based metabolite profiling in plants. Nat Protoc 1:387–396. https://doi.org/10.1038/nprot.2006.59
Long CP, Gonzalez JE, Feist AM et al (2017) Dissecting the genetic and metabolic mechanisms of adaptation to the knockout of a major metabolic enzyme in *Escherichia coli*. Proc Natl Acad Sci 115:201716056. https://doi.org/10.1073/pnas.1716056115
Lonien J, Schwender J (2009) Analysis of metabolic flux phenotypes for two arabidopsis mutants with severe impairment in seed storage lipid synthesis. Plant Physiol 151:1617–1634. https://doi.org/10.1104/pp.109.144121
Lorenz DR, Cantor CR, Collins JJ (2009) A network biology approach to aging in yeast. Proc Natl Acad Sci U S A 106:1145–1150. https://doi.org/10.1073/pnas.0812551106
Lüttge U, Beck F (1992) Endogenous rhythms and chaos in crassulacean acid metabolism. Planta 188:28. https://doi.org/10.1007/BF00198936
Ma F, Jazmin LJ, Young JD, Allen DK (2014) Isotopically nonstationary 13C flux analysis of changes in Arabidopsis thaliana leaf metabolism due to high light acclimation. Proc Natl Acad Sci U S A 111:16967–16972. https://doi.org/10.1073/pnas.1319485111
Masakapalli SK, Le Lay P, Huddleston JE et al (2010) Subcellular flux analysis of central metabolism in a heterotrophic arabidopsis cell suspension using steady-state stable isotope labeling. Plant Physiol 152:602–619. https://doi.org/10.1104/pp.109.151316
Masakapalli SK, Kruger NJ, Ratcliffe RG (2013) The metabolic flux phenotype of heterotrophic Arabidopsis cells reveals a complex response to changes in nitrogen supply. Plant J 74:569–582. https://doi.org/10.1111/tpj.12142
Medeiros DB, Daloso DM, Fernie AR et al (2015) Utilizing systems biology to unravel stomatal function and the hierarchies underpinning its control. Plant Cell Environ 38:1457–1470. https://doi.org/10.1111/pce.12517
Medeiros DB, Barros K, Barros JA et al (2017) Impaired malate and fumarate accumulation due the mutation of tonoplast dicarboxylate transporter has little effects on stomatal behaviour. Plant Physiol 175:00971.2017. https://doi.org/10.1104/pp.17.00971
Mohanty B, Kitazumi A, Cheung CYM et al (2016) Identification of candidate network hubs involved in metabolic adjustments of rice under drought stress by integrating transcriptome data and genome-scale metabolic network. Plant Sci 242:224–239. https://doi.org/10.1016/j.plantsci.2015.09.018
Nakashima K, Yamaguchi-Shinozaki K, Shinozaki K (2014) The transcriptional regulatory network in the drought response and its crosstalk in abiotic stress responses including drought, cold, and heat. Front Plant Sci 5:170. https://doi.org/10.3389/fpls.2014.00170
Nielsen K, SØrensen PG, Hynne F (1997) Chaos in glycolysis. J Theor Biol 186:303–306. https://doi.org/10.1006/jtbi.1996.0366
Nietzel T, Mostertz J, Ruberti C et al (2020) Redox-mediated kick-start of mitochondrial energy metabolism drives resource-efficient seed germination. Proc Natl Acad Sci U S A 117:741–751
Nikoloski Z, Perez-Storey R, Sweetlove LJ (2015) Inference and prediction of metabolic network fluxes. Plant Physiol 169:1443–1455. https://doi.org/10.1104/pp.15.01082
Nöh K, Grönke K, Luo B et al (2007) Metabolic flux analysis at ultra short time scale: isotopically non-stationary 13C labeling experiments. J Biotechnol 129:249–267. https://doi.org/10.1016/j.jbiotec.2006.11.015
Nuccio ML, Paul M, Bate NJ et al (2018) Where are the drought tolerant crops?An assessment of more than two decades of plant biotechnology effort in crop improvement. Plant Sci 273:110. https://doi.org/10.1016/j.plantsci.2018.01.020
Nunes-Nesi A, Nascimento VL, Silva FMO et al (2016) Natural genetic variation for morphological and molecular determinants of plant growth and yield. J Exp Bot 67:2989–3001. https://doi.org/10.1093/jxb/erw124
Odum HT (1983) Systems ecology: an introduction. Wiley, New York, NY
de Oliveira Dal'Molin CG, Quek L-E, Palfreyman RW et al (2010) AraGEM, a genome-scale reconstruction of the primary metabolic network in arabidopsis. Plant Physiol 152:579–589. https://doi.org/10.1104/pp.109.148817
Palsson B (2006) Systems biology: properties of reconstructed networks. Cambridge University Press, Cambridge. https://doi.org/10.1017/CBO9780511790515
Pastori GM, Foyer CH (2002) Common components, networks, and pathways of cross-tolerance to stress. The central role of "redox" and abscisic acid-mediated controls. Plant Physiol 129:460–468. https://doi.org/10.1104/pp.011021
Poolman MG, Miguet L, Sweetlove LJ, Fell DA (2009) A genome-scale metabolic model of arabidopsis and some of its properties. Plant Physiol 151:1570–1581. https://doi.org/10.1104/pp.109.141267
Rizza A, Walia A, Lanquar V et al (2017) In vivo gibberellin gradients visualized in rapidly elongating tissues. Nat Plants 3:803–813. https://doi.org/10.1038/s41477-017-0021-9
Robaina Estévez S, Nikoloski Z (2014) Generalized framework for context-specific metabolic model

extraction methods. Front Plant Sci 5:491. https://doi.org/10.3389/fpls.2014.00491

Robaina Estévez S, Nikoloski Z (2015) Context-specific metabolic model extraction based on regularized least squares optimization. PLoS One 10:e0131875. https://doi.org/10.1371/journal.pone.0131875

Robaina-Estévez S, Daloso DM, Zhang Y et al (2017) Resolving the central metabolism of Arabidopsis guard cells. Sci Rep 7:8307. https://doi.org/10.1038/s41598-017-07132-9

Roessner U, Wagner C, Kopka J et al (2000) Simultaneous analysis of metabolites in potato tuber by gas chromatography - mass spectrometry. Plant J 23:131–142

Rossi MT, Kalde M, Srisakvarakul C et al (2017) Cell-type specific metabolic flux analysis: a challenge for metabolic phenotyping and a potential solution in plants. Metabolites 7:59. https://doi.org/10.3390/metabo7040059

Ruprecht C, Vaid N, Proost S et al (2017) Beyond genomics: studying evolution with gene coexpression networks. Trends Plant Sci 22:298–307. https://doi.org/10.1016/j.tplants.2016.12.011

Saito K, Matsuda F (2010) Metabolomics for functional genomics, systems biology, and biotechnology. Annu Rev Plant Biol 61:463–489. https://doi.org/10.1146/annurev.arplant.043008.092035

Sato AM, Catuchi TA, Ribeiro RV, Souza GM (2010) The use of network analysis to uncover homeostatic responses of a drought-tolerant sugarcane cultivar under severe water deficit and phosphorus supply. Acta Physiol Plant 32:1145–1151. https://doi.org/10.1007/s11738-010-0506-x

Schwender J, Goffman F, Ohlrogge JB, Shachar-Hill Y (2004) Rubisco without the Calvin cycle improves the carbon efficiency of developing green seeds. Nature 432:779–782. https://doi.org/10.1038/nature03145

Seaver SMD, Bradbury LMT, Frelin O et al (2015) Improved evidence-based genome-scale metabolic models for maize leaf, embryo, and endosperm. Front Plant Sci 6:142. https://doi.org/10.3389/fpls.2015.00142

Setiyono TD, Cassman KG, Specht JE et al (2010) Simulation of soybean growth and yield in near-optimal growth conditions. Field Crop Res 119:161–174. https://doi.org/10.1016/j.fcr.2010.07.007

Shameer S, Baghalian K, Cheung CYM et al (2018) Computational analysis of the productivity potential of CAM. Nat Plants 4:165–171. https://doi.org/10.1038/s41477-018-0112-2

Shameer S, Ratcliffe RG, Sweetlove LJ (2019) Leaf energy balance requires mitochondrial respiration and export of chloroplast NADPH in the light. Plant Physiol 180:1947–1961

Silva WB, Daloso DM, Fernie AR et al (2016) Can stable isotope mass spectrometry replace radiolabelled approaches in metabolic studies? Plant Sci 249:59–69. https://doi.org/10.1016/j.plantsci.2016.05.011

Souza GM, Lüttge U (2015) Stability as a phenomenon emergent from plasticity–complexity–diversity in eco-physiology. Prog Bot 76:211–239

Souza GM, De Oliveira RF, Cardoso VJM (2004a) Temporal dynamics of stomatal conductance of plants under water deficit: can homeostasis be improved by more complex dynamics? Braz Arch Biol Technol 47:423–431. https://doi.org/10.1590/S1516-89132004000300013

Souza GM, Ribeiro RV, Santos MG et al (2004b) Approximate Entropy as a measure of complexity in sap ow temporal dynamics of two tropical tree species under water de cit. An Acad Bras Cienc 76:625–630

Souza GM, Prado CHBA, Ribeiro RV et al (2016) Toward a systemic plant physiology. Theor Exp Plant Physiol 28:341–346. https://doi.org/10.1007/s40626-016-0071-9

Stelling J, Klamt S, Bettenbrock K et al (2002) Metabolic network structure determines key aspects of functionality and regulation. Nature 420:190–193. https://doi.org/10.1038/nature01166

Sweetlove LJ, Fernie AR (2013) The spatial organization of metabolism within the plant cell. Annu Rev Plant Biol 64:723–746. https://doi.org/10.1146/annurev-arplant-050312-120233

Sweetlove LJ, Last RL, Fernie AR (2003) Perspectives in systems biology predictive metabolic engineering: a goal for systems biology 1. Plant Physiol 132:420–425. https://doi.org/10.1104/pp.103.022004.At

Sweetlove LJ, Nielsen J, Fernie AR (2017) Engineering central metabolism – a grand challenge for plant biologists. Plant J 90:749–763. https://doi.org/10.1111/tpj.13464

Szecowka M, Heise R, Tohge T et al (2013) Metabolic fluxes in an illuminated Arabidopsis rosette. Plant Cell 25:694–714. https://doi.org/10.1105/tpc.112.106989

Tafforeau M, Verdus MC, Norris V et al (2006) Memory processes in the response of plants to environmental signals. Plant Signal Behav 1:9–14. https://doi.org/10.4161/psb.1 1.2164

The Arabidopsis Genome Initiative A (2000) Analysis of the genome sequence of the flowering plant Arabidopsis thaliana. Nature 408:796–815. https://doi.org/10.1038/35048692

Thum KE, Shin MJ, Gutiérrez RA et al (2008) An integrated genetic, genomic and systems approach defines gene networks regulated by the interaction of light and carbon signaling pathways in Arabidopsis. BMC Syst Biol 2:31. https://doi.org/10.1186/1752-0509-2-31

Tolstikov VV, Fiehn O (2002) Analysis of highly polar compounds of plant origin: combination of hydrophilic interaction chromatography and electrospray ion trap mass spectrometry. Anal Biochem 301:298–307. https://doi.org/10.1006/abio.2001.5513

Toubiana D, Fernie AR, Nikoloski Z, Fait A (2013) Network analysis: tackling complex data to study plant metabolism. Trends Biotechnol 31:29–36. https://doi.org/10.1016/j.tibtech.2012.10.011

Trewavas A (2005) Green plants as intelligent organisms. Trends Plant Sci 10:413–419. https://doi.org/10.1016/j.tplants.2005.07.005

Tuberosa R, Salvi S (2006) Genomics-based approaches to improve drought tolerance of crops. Trends

Plant Sci 11:405–412. https://doi.org/10.1016/j.tplants.2006.06.003
Urakami K, Zangiacomi V, Yamaguchi K, Kusuhara M (2010) Quantitative metabolome profiling of Illicium anisatum by capillary electrophoresis time-of-flight mass spectrometry. Biomed Res 31:161–163. https://doi.org/10.2220/biomedres.31.161
Urano K, Maruyama K, Ogata Y et al (2009) Characterization of the ABA-regulated global responses to dehydration in Arabidopsis by metabolomics. Plant J 57:1065–1078. https://doi.org/10.1111/j.1365-313X.2008.03748.x
Usadel B, Obayashi T, Mutwil M et al (2009) Co-expression tools for plant biology: opportunities for hypothesis generation and caveats. Plant Cell Environ 32:1633–1651. https://doi.org/10.1111/j.1365-3040.2009.02040.x
Vallarino G, Fernie AR, Ratcliffe RG, Shameer S, Sweetlove LJ (2020) Flux balance analysis of metabolism during growth by osmotic cell expansion and its application to tomato fruits. Plant J 103:68–82. https://doi.org/10.1111/tpj.14707
Virlouvet L, Fromm M (2015) Physiological and transcriptional memory in guard cells during repetitive dehydration stress. New Phytol 205:596–607. https://doi.org/10.1111/nph.13080
Wagner A, Fell DA (2001) The small world inside large metabolic networks. Proc R Soc B Biol Sci 268:1803–1810. https://doi.org/10.1098/rspb.2001.1711
Wagner S, Steinbeck J, Fuchs P, Lichtenauer S, Elsässer M, Schippers JHM, Nietzel T, Ruberti C, Van Aken O, Meyer AJ et al (2019) Multiparametric real-time sensing of cytosolic physiology links hypoxia responses to mitochondrial electron transport. New Phytol 224:1668
Wang Z, Gerstein M, Snyder M (2009) RNA-seq: a revolutionary tool for transcriptomics. Nat Rev Genet 10:57–63. https://doi.org/10.1038/nrg2484. RNA-Seq
Weckwerth W (2003) Metabolomics in systems biology. Annu Rev Plant Biol 54:669–689. https://doi.org/10.1146/annurev.arplant.54.031902.135014
Williams TCR, Miguet L, Masakapalli SK et al (2008) Metabolic network fluxes in heterotrophic Arabidopsis cells: stability of the flux distribution under different oxygenation conditions. Plant Physiol 148:704–718. https://doi.org/10.1104/pp.108.125195
Williams TCR, Poolman MG, Howden AJM et al (2010) A genome-scale metabolic model accurately predicts fluxes in central carbon metabolism under stress conditions. Plant Physiol 154:311–323. https://doi.org/10.1104/pp.110.158535
Yuan H, Cheung CYM, Poolman MG et al (2016) A genome-scale metabolic network reconstruction of tomato (Solanum lycopersicum L.) and its application to photorespiratory metabolism. Plant J 85:289–304. https://doi.org/10.1111/tpj.13075
Zhang Y, Beard KFM, Swart C et al (2017) Protein-protein interactions and metabolite channelling in the plant tricarboxylic acid cycle. Nat Commun 8:15212. https://doi.org/10.1038/ncomms15212
Zuniga C, Levering J, Antoniewicz MR, Guarnieri MT, Betenbaugh MJ, Zengler K (2017) Predicting dynamic metabolic demands in the photosynthetic eukaryote Chlorella vulgaris. Plant Physiol 176:450

Contribution of Omics and Systems Biology to Plant Biotechnology

10

Ronaldo J. D. Dalio, Celso Gaspar Litholdo Jr, Gabriela Arena, Diogo Magalhães, and Marcos A. Machado

Abstract

The development of modern genetic engineering approaches and high throughput technologies in biological research, besides the holistic view of systems biology, have triggered the progress of biotechnology to address plant productivity and stress adaptation. Indeed, plant biotechnology has the potential to overcome many problems we currently face that impair our agriculture, such as diseases and pests, environmental pressures, or climate change. The system biology field encompasses the identification of the general principles and patterns found in living systems, by studying the molecular diversity and integrate this knowledge in complex models of regulatory networks. The "omics," which comprises but not limited to genomic, transcriptomic, proteomic, epigenomic, and metabolomic studies in entire plants, allow a better understanding of plant system biology and further contribute to biotechnology development. In this chapter, we provided an overview on omic studies for the searching and identification of metabolites and proteins employed by microorganisms to develop biotechnological products. Moreover, we present an overview of the central aspects of small RNA as regulators of gene expression connecting system networks and the potential application into plant biotechnology.

Keywords

Plant biotechnology · Plant-microbe interaction · Effectors · Small RNAs · Omics

10.1 Introduction

10.1.1 Plants Have Shaped Human Life History on Earth

The energy of sunlight converted by algae and plants to carbohydrates and other organic molecules is fundamental for life in our planet. Our society has developed along with the improvement of our capacity to cultivate and store plants as a main source of food through agriculture. Climate change, diseases and pests have reduced the sources of energy and, besides suitable agricultural-land area, represent the main obstacles for the optimal production and yield in agriculture nowadays. Furthermore, the population

R. J. D. Dalio (✉) · D. Magalhães
Centro de Citricultura Sylvio Moreira, Laboratório de Biotecnologia, Instituto Agronômico, Cordeirópolis, SP, Brasil

IdeeLab Biotecnologia, Piracicaba, SP, Brasil

C. G. Litholdo Jr · G. Arena · M. A. Machado
Centro de Citricultura Sylvio Moreira, Laboratório de Biotecnologia, Instituto Agronômico, Cordeirópolis, SP, Brasil
e-mail: marcos@ccsm.br

F. V. Winck (ed.), *Advances in Plant Omics and Systems Biology Approaches*, Advances in Experimental Medicine and Biology 1346, https://doi.org/10.1007/978-3-030-80352-0_10

growth rate is rapidly increasing, which raises concerns about food security in a near future. To balance the equation, many people are trusting on the development of plant biotechnology.

Indeed, plant biotechnology have the potential to overcome many problems we currently face that impair our agriculture, such as diseases and pests, environmental pressures or climate change, to cite a few examples. However, it is not yet known if the rate of plant biotechnology development will cope with the always-growing needs for food. Besides increasing plant productivity and resistance against biotic and abiotic stresses, plant biotechnology is also crucial to the development of the much needed second and third generation biofuels.

In this scenario, "omics" and plant system biology emerges as fundamental knowledge to understand, not only the physiology of a single plant, but also to extrapolate this information to more complex natural and anthropogenic ecosystems, which in turn have the potential to accelerate the development of plant biotechnology. Besides the biotechnological products that have arisen from genetic manipulation of organisms, such as genetic modified organisms (GMO), antibiotics and vaccines, the modern biotechnology provides advances in the study of omics, and consequently to the system biology field. This emerging field, which is closely related to synthetic biology, encompasses the identification of the general principles and patterns found in living and engineered systems, along with the study of the molecular diversity of living organisms, to finally, integrate this knowledge in complex models of the regulatory networks (Breitling 2010).

Several new methods for DNA sequencing known as "next-generation" or "second-generation" sequencing were developed around the year 2000, and expand enormously the genomic information available nowadays, which comprises hundreds of organisms. Together with transcriptomics, proteomics, epigenomics, and metabolomics studies that are now facilitated by high-throughput methodologies and bioinformatics analyses, the enormous growth in omics studies now makes systems biology expand in biological research.

In the next sections, it will be discussed the characterization of metabolites and proteins employed by plant-associated beneficial microorganisms, and also plant susceptibility genes that are targeted by pathogen effectors, in order to develop biotechnological products. Additionally, the posttranscriptional regulatory role of small RNAs, representing another layer of gene expression regulation, is presented. Finally, our perspectives for the contribution of omics and systems biology to advance plant biotechnology are further discussed.

10.2 Development

10.2.1 Plant–Microbe Interaction: Effectors, Omics and Strategies for Plant Breeding

Plants are in constant interaction with microbes in the environment. The nature of those relationships might range from no obvious interactions (not compatible), beneficial (mutualistic) to harmful (pathogenic), which also can be influenced by changes in environmental conditions. In almost all cases, microbes utilize effectors to modulate host physiology aiming to establish successful colonization. In this topic we will discuss the effector-based strategies employed by both beneficial and pathogenic microbes, the omic tools to identify effectors and plant targets, and the biotechnological approaches to engineer plants with higher productivity and resistance.

10.2.2 Effectors from Beneficial Microorganisms

Mutualistic microbes, which provide essential biochemical products/processes to host plants, are mainly associated with the root system and usually referred to as plant growth-promoting bacteria (PGPB) and fungi (PGPF) (Pieterse et al. 2014). These beneficial organisms can improve plant growth and development by using both direct and indirect mechanisms. Probably the most well-known beneficial relationship between plants and microorganisms is the interactions of Rhizobium and other nitrogen-fixing bacteria with plants colonized by these bacteria.

Direct mechanisms, such as nitrogen fixation, phosphorous solubilization and production of growth-promoting compounds as plant regulators (auxin, cytokinin, gibberellin), refer to a directly induction of plant growth and development by microbe-associated molecules (Olanrewaju et al. 2017). The production of auxins by beneficial microbes has been greatly explored due to the numerous positive effects that this versatile hormone can cause, for instance by regulating cell division and cell enlargement to provide growth of roots, stem and leaves (Vanneste and Friml 2009). The indole-3-acetic acid (IAA) produced by the plant-associated microorganisms can stimulate root development if the plant IAA concentration is insufficient, or causes contrary effect to inhibit root growth in cases where the concentration of the hormone is optimal (Spaepen et al. 2007). In *Triticum aestivum*, the IAA content produced by strains belonging mainly to *Bacillus* and *Pseudomonas* species increased the number of tillers, the spike length and seed weight, demonstrating the potential of this hormone to increase plant growth and yield (Ali et al. 2009). Cytokinin is also produced by soil microorganisms capable to work as a plant growth regulator (PGR) (Arkhipova et al. 2007). Cytokinins content can cause beneficial effects on plant growth and yield, by acting in a lot of biological processes, including cell division, cell enlargement, tissue expansion, stomatal opening and shoot growth (Weyens et al. 2009). For example, treatment of *Platycladus orientalis* (oriental thuja) seedling with cytokinin produced by *Bacillus subtilis* increased drought stress tolerance thus improving plant health (Liu et al. 2013).

Indirect mechanisms, such as production of antibiotics, quorum quenching and induced systemic resistance (ISR), refer to an indirectly induction of plant growth and development by the inhibition of pathogens attack (Olanrewaju et al. 2017). Bacteria from the genera *Pseudomonas* and *Bacillus* have been shown to produce a large variety of effectors with antimicrobial properties, such as ecomycins, 2,4 Diacetyl Phloroglucinol (DAPG), Phenazine-1-carboxylic acid (PCA), subtilin, TasA, and sublancin (Goswami et al. 2016). Beneficial microbes can also inhibit infection of phytopathogenic bacteria by disrupting their communication (Olanrewaju et al. 2017). In response to fluctuations in cell population density, quorum-sensing bacteria synthesize extracellular signaling molecules, called autoinducers, which triggers gene expression regulations in proximal bacterial cells. By using quorum sensing, bacteria can regulate a diverse array of physiological activities, such as biofilm formation and virulence, in a coordinated action within bacterial population (Miller and Bassler 2001). Some beneficial PGPBs produce lactonase enzymes that degrade pathogen-produced autoinducer, thus disrupting quorum sensing and preventing bacterial pathogens from inhibiting plant growth (Olanrewaju et al. 2017).

Indirect promotion of plant growth by beneficial microbes can also be achieved by triggering the ISR, a plant priming for defense against subsequent attacks from a broad spectrum of pathogens and herbivores. Induced resistance is triggered not locally at the site of contact with the mutualistic microbe but also systemically in plant parts that were not exposed to the inducer. Both PGPB and PGPF in the rhizosphere have been described to stimulate plant health by triggering the plant immune system (Pieterse et al. 2014). For instance, pioneer studies reported that plants with root system colonized by a PGPB strain of *Pseudomonas fluorescens* had a higher production of antimicrobial phytoalexins and enhanced resistance to the pathogen *Fusarium oxysporum* (Van Peer et al. 1991).

Similarly, the colonization of cucumber roots by *Pseudomonas* and *Serratia* PGPB strains resulted in reduced anthracnose disease symptoms caused by *Colletotrichum orbiculare* (Wei et al. 1991). Since then, numerous studies had reported the ability of plant growth-promoting microbes to induce ISR and enhance plant health (Pieterse et al. 2014). Many microbial effectors responsible for the onset of ISR have been described. Examples from PGPB include antibiotics, homoserine lactones, iron-regulated siderophores, lipopolysaccharides-containing cell wall and flagella. Volatiles such as 2R,3R-butanediol and C13 synthesized by *B. subtilis* and *Paenibacillus polymyxa*, respectively, also elicit ISR. ISR-inducing effectors from PGPF include enzymatic proteins, such as xylanases and celulases (Pieterse et al. 2014).

Besides effectors that induce systemic plant defenses, beneficial microbes might also deliver effectors that suppress local plant defenses to help the establishment of mutualistic interactions with the host. Some effectors, employed by PGPB and PGPF to overcome plant immune responses, have been described, e.g., the SP7 from *Rhizophagus intraradices* (Kloppholz et al. 2011). Suppression of plant defenses is a mechanism also typically exerted by effectors from pathogenic microbes to achieve successful infection, and it will be addressed in the following subtopic.

10.2.3 Effectors of Plant-Pathogens

During the co-evolution of plants and pathogens, plants have developed a multilayered immune system to self-protect while adapted pathogens acquired mechanisms to overcome its defenses. At the cell surface, plants carry pattern recognition receptors (PRRs) to recognize conserved molecules associated to pathogens/microbes (pathogen/microbe-associated molecular patterns—PAMPs/MAMPs) and elicit the so-called pattern-triggered immunity (PTI). To counteract PTI, specialized pathogens deliver effector proteins that suppress the plant defense signaling and induce an effector-triggered susceptibility (ETS). As a counter-counter-defense strategy, plants have evolved proteins coded by resistance genes (R genes) to sense the effectors or their effects in plant cells, triggering the effector-triggered immunity (ETI) (Fig. 10.1) (Jones and Dangl 2006).

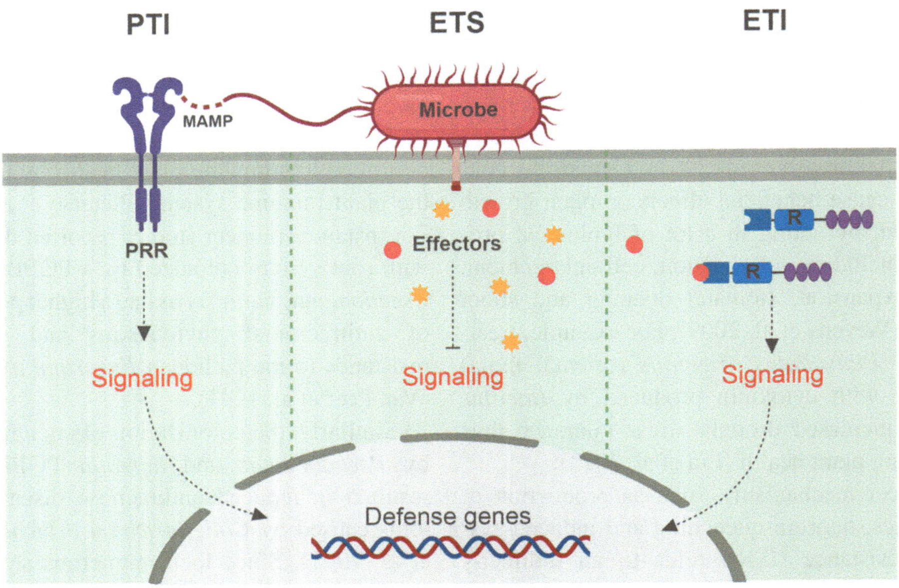

Fig. 10.1 Multilayered plant immune system. Plants carry pattern recognition receptors (PRRs) that recognize pathogen/microbe—associated molecular patterns (PAMPs/MAMPs) and elicit the pattern-triggered immunity (PTI). Adapted pathogens have acquired effector proteins that are delivered in the host cell to suppress PTI, inducing an effector-triggered susceptibility (ETS). As a counter-counter-defense strategy, plants have acquired resistance (R) proteins that recognize the effectors or their effects in plant cells, triggering the effector-triggered immunity (ETI)

Most R genes encode members of a family of nucleotide-binding leucine-rich repeat (NLR) receptors that recognize specific pathogen effectors. The first described R gene, Pto, was identified more than 20 years ago in tomato conferring resistance to strains of *Pseudomonas syringae* carrying specific effectors (former called as avirulence genes) (Martin et al. 1993; Scofield et al. 1996). Since then, several R genes have been identified in distinct plant species, as promoters of resistance to all kinds of pathogens. Classic examples include the tobacco gene N that confers resistance to tobacco mosaic virus (TMV) (Whitham et al. 1994), the Arabidopsis RPS2 and RPM1 that recognize effectors from *P. syringae* (Bent et al. 1994; Grant et al. 1995), and the tomato Cf-2 and Cf-9 that promotes resistance to *Cladosporium fulvum* (Jones et al. 1994; Dixon et al. 1996). When pathogens attempt to overcome plant defenses by delivering effector molecules, R genes-encoded proteins might recognize either the effectors itself or plant affected proteins, triggering a signaling cascade that culminate in the plant resistance.

Along with suppression of plant defenses, pathogen effectors might exploit the so-called plant susceptibility genes (S genes) that facilitates the infection process or supports compatibility with a pathogen (Zaidi et al. 2018). Proteins coded by S genes might assist pathogen in several steps of the establishment of a compatible interaction such as host recognition, penetration, proliferation and spread. The best-known example of an S gene is the *Mildew resistance locus O* (*Mlo*) that encodes a membrane-associated protein required for powdery mildew fungal penetration of host epidermal cells. Besides *Mlo*, the rice *SWEET* genes were identified as susceptibility genes to bacterial blight (Zhou et al. 2015). The associated pathogen, *Xanthomonas oryzae*, encodes transcription activator-like (TAL) effectors that recognize specific regions (effector binding elements, EBE) in the promoter of the *SWEET* genes and induce their expression (Zhou et al. 2015). Because *SWEET* genes encode sugar transporters, they likely promote susceptibility to bacterial blight by triggering sugar release to the apoplast and thus providing nutrient to the pathogen (Blanvillain-Baufumé et al. 2016). In fact, several S genes targeted by *Xanthomonas* spp. TAL effectors have been identified (Hutin et al. 2015). Another S gene recently characterized is the citrus LOB1, which support host susceptibility to citrus canker disease, caused by *Xanthomonas citri* subsp *citri* (Hu et al. 2014). Like *X. oryzae*, *X. citri* also uses its TAL effectors to bind EBEs in the promoter of LOB1 and induce its expression (Hu et al. 2014). Even though its biological role remains to be determined, induction of LOB1 using custom-designed TAL effectors leads to similar citrus canker symptoms (Zhang et al. 2017), highlighting its central role in the development of the disease.

10.2.4 Omics as Tools to Identify Microbe Effectors and Plant Targets

The increasing advances in omics technologies are boosting the discovery of microbial effectors in a rapid and efficient manner. Next-generation sequencing technologies are used to sequence microbe genomes, allowing in silico prediction of effectors. Putative effectors can be predicted from sequence datasets by detecting features associated to secreted proteins, such as the presence of a signal peptide, the absence of transmembrane and membrane anchorage domains, and small sequence size/length (Dalio et al. 2017). Genome sets from different strains of the same microbe can be compared by searching for core effectors, known to be important for the microbe colonization and, hence, less subjected to mutations that could help them to escape from introduced sources of plant resistance (Dangl et al. 2013). Following such strategies, sets of effectors of several microorganisms have been disclosed with high efficiency (Vleeshouwers and Oliver 2014). Further proteomic and transcriptomic data, from microbe upon contact with plant signals, help to select secreted proteins potentially involved in the host–microbe interaction.

Once the most promising effector candidates are selected, their biological activity can be validated by transient or stable gene expression in plants (Dalio et al. 2017). The set of transcriptomic, proteomic, metabolomic, and phenomic data from transformed plants compared to wild type shall settle the status of the predicted protein as a true effector and can contribute to the elucidation of plant modifications imposed by effector activity. Further approaches to validate effector function is knocking out or knocking down the effector gene-by-gene editing or silencing (Dalio et al. 2017). In such cases, the obtainment of omics data from both microbe with the disrupted gene and colonized test-plant are also useful to demonstrate that the function of the effector is compromised.

Subsequently, the identification of effectors has facilitated the discovery of corresponding plant target genes. "Effectoromics" studies have been successful in identifying a growing list of effectors and their corresponding R and S genes (Dangl et al. 2013; Vleeshouwers and Oliver 2014). By using the functionally validated effector, plants can be screened for proteins that directly interact with the effector. For instance, candidate targets for effector manipulation can be elucidated using yeast two-hybrid screening, which has been applied at genomic scale, or pull-down assays followed by proteomics identification of interacting proteins (Dalio et al. 2017). Irrespective if those or other approaches are employed, the searching for targeted R or S genes directed by effector-based screens provide higher throughput and more straightforward phenotypes than pathogen-based screens (Dalio et al. 2017; Vleeshouwers and Oliver 2014). Similar strategy could be employed to identify plant targets from beneficial microbe effectors.

This effector-rationalized approach (Dangl et al. 2013) was used to search for the source of *Phytophthora infestans* resistance in the potato "Sarpo Mira," one of the few cultivars reported to retain field resistance to late blight for several years (Rietman et al. 2012). A collection of core effectors was predicted from *P. infestans* genome and expressed in potato leaves. The induced resistance response to specific effectors, whose corresponding R genes were mostly known, enabled the dissection of R genes that confers late blight resistance in "Sarpo Mira" genotype (Rietman et al. 2012). Similar strategies can be used to provide breeding programs with R genes for deployment in susceptible genotypes. A different approach relied on the TAL effectors from *Xanthomonas* species, which binds EBE regions in the promoter of S genes, inducing their expression and facilitating pathogen infection. Using the knowledge on EBE regions, an engineered R gene was produced by adding EBE regions to the promoter of *Xa27* gene and deployed in rice (Hummel et al. 2012). The synthetic R gene was successfully activated by TAL effectors, conferring rice resistance to both bacterial blight and bacterial leaf streak (Hummel et al. 2012). Besides deploying engineered R genes, S genes targeted by TAL effectors have been edited to generate resistant genotypes. Using an effector-rationalized approach, the discovery of *Xanthomonas* TAL binding sites combined with transcriptomic data has led to the discovery of several S genes for different *Xanthomonas*/host interaction (Hutin et al. 2015). The identified S genes are greatly increasing the knowledge on *Xanthomonas*-causing diseases and have now been used as targets for gene editing to confer resistance to such diseases (Hutin et al. 2015; Li et al. 2012).

10.2.5 Biotechnology Approaches for Genetic Engineering Plants to Improve Productivity and Disease Resistance

Crops have been selected for higher yield and disease resistance throughout the history of agriculture (Table 10.1). Traditional breeding methods allowed the introgression of interesting traits well before the comprehension of the molecular mechanisms involved in plant–microbe interactions. Currently, the elucidation of effector-targeted plant genes is used by breeding programs to develop crop varieties with higher levels of resistance and productivity. Combined with conventional time-consuming breeding techniques, technologies based on genetic engineering (Fig. 10.2) have been used to improve and speed up the process of developing high-yield and durable disease-resistant crop varieties (Table 10.1).

Table 10.1 Historical scientific events that have developed the modern biotechnology

Years	Scientist/pioneer/discoverer	Innovative events
8500 BC	Southwest Asians	Emergence of plant and animal domestication
1675	Anton Van Leeuwenhoek	Discovery of microorganisms by "The Father of Microbiology"
1862–1885	Louis Pasteur	Discoveries of the principles of vaccination, microbial fermentation, and pasteurization
1865	Gregor Mendel	Establishment of the principles of genetics and theories of heredity by "The Father of Genetics"
1919	Károly Ereky	Creation of the term biotechnology
1928	Ludwig von Bertalanffy	Proposition of the general systems theory, one of the precursors of systems biology
1929	Alexander Fleming	Purification of penicillin from the fungus *Penicillium notatum*
1930	George Beadle and Edward Tatum	Confirmation that genes direct the production of proteins
1944	Oswald Avery	Identification of DNA as the material of which genes and chromosomes are made and transmit the genetic information
1953	Francis Crick, Maurice Wilkins, and James Watson	Revelation of the structure of DNA molecule
1961	François Jacob and Jacques Monod	Elucidation of the control of enzyme expression levels as the result of regulation of DNA transcription
1967	Har Gobing Khorana and Marshall Niremberg	Elucidation of the genetic code
1972	Paul Berg	Development of recombinant DNA techniques—"the emergence of genetic engineering"
1976	Walter Fiers	Sequencing of the first complete genome of bacteriophage
1976	Herbert Boyer and Robert Swanson	Establishment of the first biotechnology company, the Genentech
1977	Frederick Sanger	Determination of the first DNA sequence
1978	Werner Arber, Daniel Nathans, and Hamilton Smith	Isolation of restriction enzymes from bacteria
1982	Richard Palmiter	Generation of the first genetic modified organism (GMO)
1985	Kary Banks Mullis	Development of the polymerase chain reaction (PCR) technique
1986	Thomas H. Roderick	Creation of the term genomics
1986	USA and France	Establishment of the first field trials of transgenic tobacco resistant to herbicide
1994	USA	Approval of the first GMO to be commercially available, a transgenic tomato
1998	Washington Uni and Sanger Institute	Sequencing of the first complete animal genome, of the *Caenorhabditis elegans*
1998	Craig Mello and Andrew Fire	Elucidation of the mechanism of RNA interference (RNAi) in animals
1998	Peter Waterhouse and Ming-Bo Wang	Discovery that the double-stranded RNA (dsRNA) induces the RNAi in plants
2000s	Roche, ABI, and Solexa/Illumina technologies	Development of high throughput-sequencing technologies
2000s	Several	Emergence of modern systems biology approaches, "the age of systems"
2000	Arabidopsis Genome Initiative	Sequencing of the first complete plant genome, of the *Arabidopsis thaliana*
2002	Koichi Tanaka, John Fenn, and Kurt Wüthrich	Recognition for the development of identification and structure analyses for proteomics
2008	Solexa/Illumina technologies	Development of RNA-seq for modern transcriptomic studies
2012	Jennifer Doudna and Emmanuelle Charpentier	Development of a precise gene editing technology using CRISPR-Cas9
2017	David Liu and Feng Zhang	Development of a flexible RNA base editing technology using CRISPR-Cas13

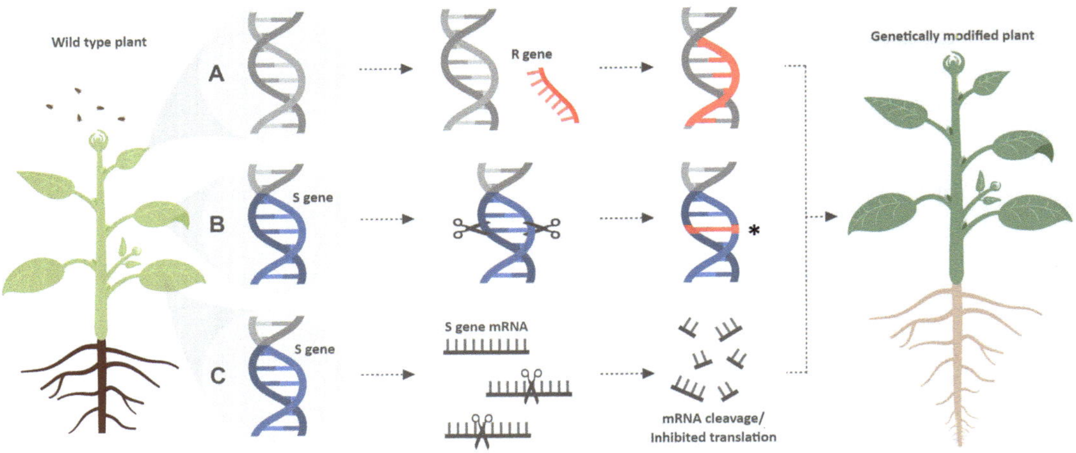

Fig. 10.2 Genetic engineering approaches to generate plants with improved disease resistance. (**a**) Transgeny to introduce genes of interest in wild type plants, including resistance (R) genes. (**b**) Gene editing to disrupt native genes such as susceptibility (S) genes. Sire-directed gene editing can be achieved using tools such as CRISPR/Cas9 system, that introduce double-stranded DNA breaks and triggers an error-prone DNA repair pathway, resulting in indels mutations (*). (**c**) Gene silencing to disrupt the function of undesirable genes (e.g., S genes). RNA interference can promote the cleavage or translation inhibition of mRNAs, knocking down the expression of the target gene

Transgenic approaches have been used to introduce genes of interest in plant species, including dominant R genes for disease resistance (Dangl et al. 2013). For instance, transgenic tomatoes with field-level resistance to bacterial spot disease were produced by transferring the R gene *Bs2* from pepper (Horvath et al. 2012). Similarly, the gene *RB* from potato wild relatives was introduced in the cultivated potato by transgeny and generated increased resistance to late blight (Halterman et al. 2008). Showing that R genes from non-hosts can effectively promote plant resistance, the maize R gene *Rxo1* was used to generate a transgenic rice, conferring resistance to bacterial streak (Zhao et al. 2005). The downside of using dominant R genes to generate plant resistance is that they usually present a short life in the field due to the adaptative potential of the corresponding pathogen effectors (Dangl et al. 2013). On the one hand, stacking multiple R genes simultaneously should provide more durable resistance since multiple effector genes would have to suffer mutation to evade resistance (Dangl et al. 2013). On the other hand, enhancing plant resistance by disrupting S genes rather than expressing R genes is an attractive approach.

Along with the introduction of foreign genes by conventional transgeny, the disruption of native gene functions might be achieved by gene silencing or editing. Gene silencing can be activated by the presence of double-stranded RNAs (dsRNA) and results in the cleavage or translation inhibition of RNAs. Briefly, dsRNA triggers their own cleavage by Dicer nucleases, producing small interfering RNAs (siRNA), which in turn are recruited by RNA-induced silencing complexes (RISC) that target RNAs with sequence homology to the incorporated siRNA (Kamthan et al. 2015). Transgenic plants with constructs designed to produce siRNA contain in the dsRNA a sequence to target gene, a technology known as RNA interference (RNAi). With the employment of engineered siRNA, RNAi can be used to manipulate gene expression and suppress undesirable traits such as susceptibility to pathogens.

The RNAi system have been applied as a strategy to control plant insects (Galdeano et al. 2017; Mao et al. 2007), viruses (Fuentes et al. 2016; Niu et al. 2006) and other attackers by directly targeting the pathogen/herbivore genes. Another approach is the use of RNAi to target plant S genes. Such strategy was used to silence the potato *SYR1* gene, resulting in reduced formation of papillae components in response to infection with *P. infestans* and increased resistance to late blight (Eschen-Lippold et al. 2012). In another example, RNAi was employed to silence *SSI2* (suppressor of salicylate insensitivity of npr1-5) gene in rice, a negative regulator of plant defenses, conferring resistance to fungal blast and bacterial leaf blight diseases (Jiang et al. 2009). We will discuss more about sRNAs in the next chapter section.

Besides gene silencing, an increasing approach within molecular plant breeding is the use of site-directed genome editing. One of the most revolutionary tools within gene editing techniques is the CRISPR-Cas9 system. With CRISPR/Cas9 tool, double-stranded DNA breaks can be introduced at specific genome regions by a site-specific nuclease, leading to the activation of DNA repair pathways. In the absence of a repair template, the non-homologous end-joining (NHEJ) pathway repairs the DNA in an error-prone process that often causes insertions or deletions around the DNA breaks, generating mutated alleles (Zaidi et al. 2018). Though recently developed, CRISPR/Cas9 system has already been applied in several economically important crops such as rice (Jiang et al. 2013), maize (Char et al. 2017), tomato (Brooks et al. 2014), and sweet orange (Jia and Wang 2014). CRISPR quickly became successful due to its high simplicity, efficiency, specificity and versatility (Bortesi and Fischer 2015; Zaidi et al. 2018). The major advantage of the gene editing, however, is the possibility to generate genetically modified cultivars that lack transgenes in the final line and thus can be exempted from GMO legislation and are more likely to be accepted by the public. Without transgenes or other foreign DNA sequences, some genome-edited plants using CRISPR already evaded regulation by USDA and are reaching market in record time (Waltz 2018).

In the context of developing disease resistance, S genes are promising targets for gene editing, since their mutation can limit the ability of a pathogen to cause disease. By using gene editing, the S gene *LOB1* was successfully modified in grapefruit, generating plants without symptoms of *Xanthomonas citri* bacterial infection (Jia et al. 2017). Likewise, CRISPR/Cas9 was used to edit the rice S gene *SWEET13*, resulting in resistance to bacterial blight (Zhou et al. 2015). Regarding fungal pathogens, gene editing disabled multiple homeoalleles of *MLO* gene in wheat, conferring heritable broad-spectrum resistance to powdery mildew (Wang et al. 2014). Resistance to potyviruses was obtained in *Arabidopsis* (Pyott et al. 2016) and cucumber (Chandrasekaran et al. 2016) by disrupting the S gene *eIF4E* (*eukaryotic translation initiation factor E*), which codes for a protein essential to the viral infection cycle. The CRISPR/Cas9 system was also employed in tomato to inactivate *DMR6* (*downy mildew resistance 6*), an S gene involved in the homeostasis of the defense hormone salicylic acid, generating plants with high levels of resistance to a wide variety of pathogens (Thomazella et al. 2016). The results obtained so far using CRISPR technology have proven that mutation on S genes can generate plant resistance to several diseases. Ongoing studies are focusing in obtaining final lines that do not contain foreign DNA to facilitate consumer acceptance. The use of genome editing to mutate S genes is emerging as a revolutionary approach to provide a transgene-free, long term, and efficient control measure of plant diseases.

Apart from breeding strategies to obtain genetically engineered plant, another biotechnology strategy used to improve plant health is the use of heterologous expression systems to produce molecules of interest. This approach can be used to large-scale production of effectors from mutualistic microbes that stimulate plant growth or disease resistance. Heterologous expression of microbe-associated quorum quenching molecules have been explored aiming disruption of biofilm-forming phytopathogenic bacteria (Kalia 2015). For instance, the *aiiA* gene coding for lactonase effectors from distinct *Bacillus* species was engineered into *Lysobacter enzymogenes* and *E. coli*, resulting in reduced virulence of *Pectobacterium carotovorum* on Chinese cabbage (Qian et al. 2010) and attenuated soft rot symptoms of *Erwinia carotovora* in potato, respectively (Pan et al. 2008). Similar strategy of recombinant protein systems can be used to synthesize other effectors from beneficial microbes in a commercial scale to increase plant health such as growth-promoting hormones, hydrolytic enzymes, siderophores, or antibiotics.

The improved identification of microbe effectors using omics technologies is providing valuable resources for plant breeding programs. Knowledge of microbe effectors and their target plant genes can be applied in combination with biotechnology techniques to speed up the development of plant varieties with higher productivity and durable disease resistance.

10.2.6 Gene Expression Regulation by Small Noncoding RNAs

The comprehension of system biology depends on a wide data collection, integration and analysis of biological molecules, focusing on interactions and emerging properties. In this context, the small noncoding-RNAs (sRNAs) has appeared, in the last couple of decades, as active and essential regulatory molecules for protein-coding gene expression, influencing several interconnected biochemical pathways. Therefore, the identification of sRNAs and characterization of their interactive network, including the discovery of sRNA target genes and associated biochemical pathways is crucial for the application of systems biology to plant biotechnology.

In this section, we provide an overview of the central aspects of endogenous sRNAs, mostly microRNAs (miRNAs), function during plant development and the evolutionary history of *MIRNA* genes. MiRNAs have been shown to act as posttranscriptional regulators, directing several essential processes in the plant, and miRNA-based technology is also a target for plant engineering to achieve high yields, quality and stress resistance. The applications of sRNAs e miRNAs research on plant biotechnology and the importance to incorporate these regulatory molecules into systems biology are further discussed.

10.2.6.1 Biological Roles of Plant miRNAs

Expansion of the miRNA regulatory system is associated with requirements for additional endogenous control of genomic information (Mattick 2004). The remarkable and constant expansion of miRNAome coincides with the major morphological innovations present in the animal bilaterians, vertebrates, and placental mammals, where many tissue- and organ-specific miRNA/target regulatory associations could have been fundamental to the emergence of complex bodies. This is reflected in the strong correlation between the number of *MIR* families contained in an organism and its position in the hierarchy of the animal kingdom (Hertel et al. 2006; Sempere et al. 2006). Moreover, there is a correlation between the number of target genes regulated by a miRNA and the age of a *MIR* gene. In animals, the number of targets of an individual miRNA also appears to increase over evolutionary time, with the more phylogenetically ancient miRNAs having more target genes than young miRNAs (Brennecke et al. 2005).

In plants, analyses of the miRNAome of common ancestors have suggested that only a few *MIR* genes are highly conserved across the entire kingdom (Cuperus et al. 2011; Nozawa et al. 2010; Ma et al. 2010). The sRNAs derived from conserved *MIR* families represent the most abundant miRNAs in a particular miRNAome, as a result of a moderate to high levels of *MIR* gene expression (Axtell 2008; Cuperus et al. 2011). These conserved miRNAs are usually derived from multi-gene *MIR* families, containing identical or highly similar mature miRNA sequences (Jones-Rhoades 2012). A high level of functional redundancy is noticed among members of the same *MIR* family (Sieber et al. 2007; Allen et al. 2007). However, the expansion of *MIR* gene families, combined with the occurrence of mutations outside the mature miRNA sequence, would provide diversification of the spatiotemporal expression in different *MIR* family members (Li and Mao 2007).

The development of multicellular organisms depends on complex regulatory networks that integrate endogenous and environmental signals. The signaling effectors in this process include phytohormones, peptides, transcription factors, and sRNAs, which are globally interconnected over long and short distances within the plant, acting in a spatiotemporal manner (Sparks et al. 2013).

Phytohormones are important mediators throughout plant development, perceiving and transmitting the internal and external cues, and whose signaling pathways are under constant cross-talk mechanisms to adjust the responses (Vanstraelen and Benková 2012). A close relationship between miRNAs and phytohormones has been seen in several studies, showing intersections in their pathways and feedback mechanisms where *MIR* genes respond to hormones which in turn regulate several genes involved in hormonal signaling pathways (Liu and Chen 2009; Liu et al. 2009; Curaba et al. 2014). Tissue- or stage-specific miRNA accumulation often plays a central role affecting, directly or indirectly, the expression of genes to adjust the transcriptome in accordance with the development requirements, in a highly dynamic regulatory network (Válóczi et al. 2006; Meng et al. 2011). The fine-tuning regulation of plant development by miRNA has been revealed from the characterization of several plant mutants, either impaired in steps of the miRNA biogenesis, displaying pleiotropic developmental defects, or impaired in particular *MIR* genes and targets, leading to more specific developmental defects (Mallory and Vaucheret 2006). However, pleiotropic defects have been also observed, mostly in cases where a miRNA has several targets.

During seed development, miR160 and miR167 regulation of the auxin-related transcription factors, *ARF17* and *ARF6/8*, affect embryo development, seed production and germination rates (Mallory et al. 2005; Todesco et al. 2010). The gibberellin (GA)- and abscisic acid (ABA)-regulated transcription factors *MYELOBLASTOSIS* (*MYB*) GAMYB-like genes *MYB33/65* are regulated by miR159, affecting seed size and fertility (Allen et al. 2007). In the early stages of embryogenesis, miR165/166 and miR394 seem to be essential for stem cell differentiation and shoot apical meristem (SAM) maintenance. MiR165/166 regulates the *HD-ZIPIII* transcription factors to define the vascular cell types in the roots and maintain cell pluripotency in the SAM, via association with AGO10 (Carlsbecker et al. 2010; Zhu et al. 2011) whereas miR394 is required for stem cell differentiation and targets an F-Box encoding gene *LCR* (Knauer et al. 2013; Litholdo et al. 2016). Although these studies did not show any phytohormone relationship with these miRNAs regulation, hormones, such as auxin and cytokinin could be contributing to this cell differentiation processes (Knauer et al. 2013; Leibfried et al. 2005).

During leaf development, miR165/166 and miR394 also play an important role. MiR165/166, in conjunction with miR390/ta-siRNAs, determine the abaxial-adaxial leaf polarity (Nogueira et al. 2007), and miR394 influences leaf shape and curvature, which is suggested to involve auxin signaling (Song et al. 2012). The miR393 regulation of the auxin receptors *TIR1* and *AUXIN SIGNALING F-BOX* (*AFB*s) genes also mediates some auxin-related aspects of leaf development (Si-Ammour et al. 2011). Moreover, miR164 and miR319 are important regulators during leaf initiation, growth and differentiation, by targeting *CUC1/CUC2* and *TCP* transcription factors genes, respectively (Pulido and Laufs 2010).

During the plant life cycle, miR156 and miR172 are the main players regulating the transition from juvenile to adult vegetative phase, and from the vegetative to reproductive phase (Spanudakis and Jackson 2014). Both miRNAs regulate transcription factors, including 11 genes encoding SPL protein regulated by miR156, and six *AP2*-like genes regulated by miR172. Interestingly, both miRNAs show opposite expression pattern during phase changes, mediated by integrated and coordinated transcriptional activation of their pathways (Wu et al. 2009). Additionally, miR159, miR319, and miR390 also regulate flowering time, implying that GA and auxin might coordinate the regulation of these miRNAs.

During root development, the regulation of *HD-ZIPIII* genes by miR165/166 modulates lateral root initiation, vascular tissue differentiation and nitrogen-fixing nodule development (Boualem et al. 2008; Carlsbecker et al. 2010; Miyashima et al. 2011). Another auxin-dependent process in the roots involves miR160 regulation of the transcription factors *ARF10*/*ARF16*, and the miR390-triggered production of ta-siRNAs, targeting *ARF4* (Wang et al. 2005; Yoon et al. 2010). Moreover, miR828-triggered ta-siRNAs target members of *MYB* transcription factors, playing a role in root hair patterning and anthocyanin production (Luo et al. 2012; Xia et al. 2012).

Interestingly, the complex network of interactions between miRNAs and hormonal signaling pathways integrates plant development and stress response signals. For instance, the miR393 regulation of the F-Box genes *TIR1* and *AFB2* mediates the auxin-dependent root development in response to ABA-related drought stress (Chen et al. 2012). Moreover, the metabolism of some inorganic nutrients depends on the gene regulation mediated by mobile miRNAs, such as miR395, miR398, and miR399, which are responsive to starvation of sulfur, copper/zinc, and phosphate, respectively (Kawashima et al. 2009; Yamasaki et al. 2007; Bari et al. 2006). All the *MIR* genes exemplified in this section are conserved among several evolutionary distant plant species, demonstrating the crucial roles of these conserved miRNAs in fundamental and ubiquitous aspects of plant development. However, non-conserved *MIR*s has also been uncovered, such as miR824 that is only found in Brassicaceae yet plays a role in development. It regulates the conserved transcription factor *AGAMOUS-LIKE16* (*AGL16*), which is important for normal stomata development (Kutter et al. 2007). This suggests that non-conserved miRNAs can emerge and acquire developmental functions in a restricted number of species.

The majority of *MIR* loci identified in a specific miRNAome have been found to be young, non-conserved microRNAs (Jones-Rhoades 2012; Axtell 2013). It has been assumed that most of the recently evolved *MIR* genes are short-lived, imprecisely processed, and functionally irrelevant. This is mostly due to the lack of identified and/or validated target genes and therefore some of the non-conserved miRNAs are likely to be under neutral selective pressure (Axtell 2008; Jones-Rhoades 2012). However, these assumptions could be the result of restricted spatiotemporal expression pattern of young *MIRs* or their expression being activated only under a particular stress condition. Moreover, it has been suggested that recently evolved miRNAs could have a distinct mode of interaction with their target genes or even in their mode of targeting, which might prevent the identification of the targets using the usual rules and approaches (Axtell 2008, Cuperus et al. 2011).

10.2.7 Recent Applications of Omics and Small RNA Research in Plant Biotechnology

The development of modern genetic engineering approaches and high-throughput technologies in biological research, besides the holistic view of systems biology, have triggered the progress of biotechnology to address plant productivity and stress adaptation (Table 10.1). The introduction of transgenes into plants has been widely and efficiently used for crop breeding, generating genetically modified organism with desired traits. Currently, the available omics information for selection of specific characteristics for breeding has offered a range of opportunities. Due to omics-scale molecular analysis and elucidation of genetic information and interactive networks, the modern biotechnology has the potential to target any traits for breeding, by interfering in one or multiple genes and/or networks.

Small RNA research has many potential applications in the plant biotechnology, aiming to increase food production, disease and pest controls, and to overcome the consequences of climate change (Zhou and Luo 2013; Kamthan et al. 2015; Zhang and Wang 2015, 2016; Liu et al. 2017). Several miRNAs may target multiple genes at a same time and it has been shown that manipulating a single *MIR* gene can significantly interfere in intricated gene networks, to provide an appropriate strategy for crop improvement. For instance, *MIR156*—the sRNA miR156 targets transcription factors-encoding genes, namely SQUAMOSA-promoter binding like proteins (SPL) (Schwab et al. 2005; Wang et al. 2008; Yamaguchi et al. 2009; Lal et al. 2011; Kim et al. 2012), and an increase by more than 100% in plant biomass is observed by overexpressing *MIR156* in different plant species, including Arabidopsis, rice, tomato, and switchgrass (Schwab et al. 2005; Fu et al. 2012; Xie et al. 2012).

MiRNAs also play an important role in plant responses to biotic and abiotic stresses (Ku et al. 2015; Litholdo et al. 2017), and accordingly, the manipulation of miRNAs to increase plant defenses has been applied to several plants, including agricultural crop species (Djami-Tchatchou et al. 2017). The first *MIR* gene revealed to play a role in plant stress responses was the *MIR393*—miR393 regulates the auxin signaling transcription factors and the overexpression of this miRNA leads to inhibition of bacterial growth (Navarro et al. 2006). Transgenic plants overexpressing miR7696 and miR396 also confers enhanced resistance to rice blast infection and cyst nematode infection in Arabidopsis, respectively (Campo et al. 2013; Hewezi et al. 2008). For abiotic stresses, the increased abundance of miR169 in transgenic tomato plants enhanced drought tolerance, by regulating target genes involved in stomatal opening, transpiration rate, and therefore, leaf water loss (Zhang et al. 2011). MiR319 has been shown to confer resistance to different environmental conditions, such as cold, salt and drought stress—transgenic rice and creeping bentgrass plants, overexpressing miR319, showed respectively increased tolerance to these conditions (Yang et al. 2013; Zhou et al. 2013).

Besides the manipulation of single miRNA/target genes module, to generate transgenic plants, the miRNA-mediated gene silencing serves also as a biotechnological tool and is currently applied in plant science, to generate mutants of theoretically any gene of interest. Individual genes can be silenced by introducing into plants engineered RNA silencing expression constructs, such as artificial miRNAs to target and inactivate endogenous gene expression (Molesini et al. 2012). This approach can disrupt the production of a specific unwanted compound, for example the caffeine to deliver a decaffeinated coffee plant. Conversely, the expression of endogenous small RNAs can be altered by suppression or overexpression of the mature sRNA sequence to alter plant development and protection (Djami-Tchatchou et al. 2017). The deregulation of specific plant miRNAs, and consequently the target gene(s), can aim numerous purposes, such as an increase in plant biomass, tolerance to biotic and abiotic stresses, fruit maturation control, and production of compounds of interest (Molesini et al. 2012; Sunkar et al. 2012; Zhang 2015).

10.3 Concluding remarks

With this chapter, we provided an overview on omic studies for the searching and identification of metabolites and proteins employed by microorganisms to develop biotechnological products. Additionally, we present an overview of the central aspects of small RNA as regulators of gene expression connecting system networks and the potential application into plant biotechnology. First used to generate virus resistance, several other RNAi strategies have been used for trans-kingdom gene regulation. Double-stranded RNA (dsRNA) produced by plants to target pathogen endogenous gene and reduced virulence has been one of the most successful approach to control insects, nematodes, and more recently, fungi. Host-induced gene silencing (HIGS) by the generation of transgenic plants carrying pathogen-targeting constructs, and spraying dsRNA solution in target organisms are experimentally validated in biotechnology approaches.

The omics, which comprises but not limited to genomic, transcriptomic, proteomic, epigenomic, and metabolomic studies in entire plants, allow a better understanding of plant biology and contribute further to biotechnology development. Recent methodological advances are enabling biological analyses of single-cells to provide opportunities to enhance our understanding of plant biology as a system (Libault et al. 2017).

During the last decade, the discovery of regulatory small RNAs altered the perception that only protein-coding genes are the players in gene regulatory network, since sRNAs emerged as central players in the transcriptional and posttranscriptional gene expression. The change in paradigm altered the way system biology is comprehended and how we can use this regulatory mechanism to improve biotechnology toolbox.

References

Ali B, Sabri AN, Ljung K, Hasnain S (2009) Auxin production by plant associated bacteria: impact on endogenous IAA content and growth of Triticum aestivum L. Lett Appl Microbiol 48:542–547

Allen RS, Li J, Stahle MI, Dubroué A, Gubler F, Millar AA (2007) Genetic analysis reveals functional redundancy and the major target genes of the Arabidopsis miR159 family. Proc Natl Acad Sci 104:16371–16376

Arkhipova TN, Prinsen E, Veselov SU, Martinenko EV, Melentiev AI, Kudoyarova GR (2007) Cytokinin producing bacteria enhance plant growth in drying soil. Plant Soil 292:305–315

Axtell MJ (2008) Evolution of microRNAs and their targets: are all microRNAs biologically relevant? Biochim Biophys Acta Gene Regul Mech 1779:725–734

Axtell MJ (2013) Classification and comparison of small RNAs from plants. Annu Rev Plant Biol 64:137–159

Bari R, Datt Pant B, Stitt M, Scheible W-R (2006) PHO2, MicroRNA399, and PHR1 Define a Phosphate-signaling pathway in plants. Plant Physiol 141:988–999

Bent AF, Kunkel BN, Dahlbeck D, Brown KL, Schmidt R, Giraudat J et al (1994) RPS2 of Arabidopsis thaliana: a leucine-rich repeat class of plant disease resistance genes. Science 265:1856–1860

Blanvillain-Baufumé S, Reschke M, Solé M, Auguy F, Doucoure H, Szurek B, Meynard D, Portefaix M, Cunnac S, Guiderdoni E, Boch J, Koebnik R (2016) Targeted promoter editing for rice resistance to Xanthomonas oryzae pv. oryzae reveals differential activities for SWEET14 -inducing TAL effectors. Plant Biotechnol J 15:306.

Bortesi L, Fischer R (2015) The CRISPR/Cas9 system for plant genome editing and beyond. Biotechnol Adv 33:41–52

Boualem A, Laporte P, Jovanovic M, Laffont C, Plet J, Combier J-P, Niebel A, Crespi M, Frugier F (2008) MicroRNA166 controls root and nodule development in Medicago truncatula. Plant J 54:876–887

Breitling R (2010) What is systems biology? Front Physiol 1:9

Brennecke J, Stark A, Russell RB, Cohen SM (2005) Principles of microRNA–target recognition. PLoS Biol 3:e85

Brooks C, Nekrasov V, Lippman ZB, Van Eck J (2014) Efficient gene editing in tomato in the first generation using the clustered regularly interspaced short palindromic repeats/CRISPR-Associated9 system. Plant Physiol 166:1292–1297

Campo S, Peris-Peris C, Siré C, Moreno AB, Donaire L, Zytnicki M, Notredame C, Llave C, San SB (2013) Identification of a novel microRNA (miRNA) from rice that targets an alternatively spliced transcript of the Nramp6 (Natural resistance-associated macrophage protein 6) gene involved in pathogen resistance. New Phytol 199:212.

Carlsbecker A, Lee J-Y, Roberts CJ, Dettmer J, Lehesranta S, Zhou J, Lindgren O, Moreno-Risueno MA, Vatén A, Thitamadee S, Campilho A, Sebastian J, Bowman JL, Helariutta Y, Benfey PN (2010) Cell signalling by microRNA165/6 directs gene dose-dependent root cell fate. Nature 465:316–321

Chandrasekaran J, Brumin M, Wolf D, Leibman D, Klap C, Pearlsman M et al (2016) Development of broad virus resistance in non-transgenic cucumber using CRISPR/Cas9 technology. Mol Plant Pathol 17:1140–1153

Char SN, Neelakandan AK, Nahampun H, Frame B, Main M, Spalding MH et al (2017) An Agrobacterium-delivered CRISPR/Cas9 system for high-frequency targeted mutagenesis in maize. Plant Biotechnol J 15:257–268

Chen H, Li Z, Xiong L (2012) A plant microRNA regulates the adaptation of roots to drought stress. FEBS Lett 586:1742–1747

Cuperus JT, Fahlgren N, Carrington JC (2011) Evolution and functional diversification of MIRNA genes. Plant Cell 23:431–442

Curaba J, Singh MB, Bhalla PL (2014) miRNAs in the crosstalk between phytohormone signalling pathways. J Exp Bot 65(6):1425–1438

Dalio RJD, Herlihy J, Oliveira TS, McDowell JM, Machado M (2017) Effector biology in focus: a primer for computational prediction and functional characterization. Mol Plant-Microbe Interact 31:22–33

Dangl JL, Horvath DM, Staskawicz BJ (2013) Pivoting the plant immune system from dissection to deployment. Science 341:746–751

Dixon MS, Jones DA, Keddie JS, Thomas CM, Harrison K, Jones JD (1996) The tomato Cf-2 disease resistance locus comprises two functional genes encoding leucine-rich repeat proteins. Cell 84:451–459

Djami-Tchatchou AT, Sanan-Mishra N, Ntushelo K, Dubery IA (2017) Functional roles of microRNAs in agronomically important plants—potential as targets for crop improvement and protection. Front Plant Sci 8:378

Eschen-Lippold L, Landgraf R, Smolka U, Schulze S, Heilmann M, Heilmann I et al (2012) Activation of defense against Phytophthora infestans in potato by down-regulation of syntaxin gene expression. New Phytol 193:985–996

Fu C, Sunkar R, Zhou C, Shen H, Zhang JY, Matts J, Wolf J, Mann DG, Stewart CN Jr, Tang Y et al (2012)

Overexpression of miR156 in switchgrass (Panicum virgatum L.) results in various morphological alterations and leads to improved biomass production. Plant Biotechnol J 10:443–452

Fuentes A, Carlos N, Ruiz Y, Callard D, Sánchez Y, Ochagavía ME et al (2016) Field trial and molecular characterization of rnai-transgenic tomato plants that exhibit resistance to tomato yellow leaf curl geminivirus. Mol Plant-Microbe Interact 29:197–209

Galdeano DM, Breton MC, Lopes JRS, Falk BW, Machado MA (2017) Oral delivery of double-stranded RNAs induces mortality in nymphs and adults of the Asian citrus psyllid, Diaphorina citri. PLoS One 12:e0171847

Goswami D, Thakker JN, Dhandhukia PC (2016) Portraying mechanics of plant growth promoting rhizobacteria (PGPR): a review-promoting rhizobacteria (PGPR); indole acetic acid (IAA); phosphate solubilization; siderophore production; antibiotic production; induced systematic resistance (ISR); ACC deaminase. Cogent Food Agric 2:1127500

Grant MR, Godiard L, Straube E, Ashfield T, Lewald J, Sattler A et al (1995) Structure of the Arabidopsis RPM1 gene enabling dual specificity disease resistance. Science 269:843–846

Halterman DA, Kramer LC, Wielgus S, Jiang J (2008) Performance of transgenic potato containing the late blight resistance gene RB. Plant Dis 92:339–343

Hertel J, Lindemeyer M, Missal K, Fried C, Tanzer A, Flamm C, Hofacker I, Stadler P, The Students of Computer Labs (2006) The expansion of the metazoan microRNA repertoire. BMC Genomics 7:25

Hewezi T, Howe P, Maier TR, Baum TJ (2008) Arabidopsis small RNAs and their targets during cyst nematode parasitism. Mol Plant-Microbe Interact 21:1622–1634

Horvath DM, Stall RE, Jones JB, Pauly MH, Vallad GE, Dahlbeck D et al (2012) Transgenic resistance confers effective field level control of bacterial spot disease in tomato. PLoS One 7:e42036

Hu Y, Zhang J, Jia H, Sosso D, Li T, Frommer WB et al (2014) Lateral organ boundaries 1 is a disease susceptibility gene for citrus bacterial canker disease. Proc Natl Acad Sci U S A 111:E521–E529

Hummel AW, Doyle EL, Bogdanove AJ (2012) Addition of transcription activator-like effector binding sites to a pathogen strain-specific rice bacterial blight resistance gene makes it effective against additional strains and against bacterial leaf streak. New Phytol 195:883–893

Hutin M, Pérez-Quintero AL, Lopez C, Szurek B (2015) MorTAL Kombat: the story of defense against TAL effectors through loss-of-susceptibility. Front Plant Sci 6:535

Jia H, Wang N (2014) Targeted genome editing of sweet orange using Cas9/sgRNA. PLoS One 9:e93806

Jia H, Zhang Y, Orbović V, Xu J, White FF, Jones JB et al (2017) Genome editing of the disease susceptibility gene CsLOB1 in citrus confers resistance to citrus canker. Plant Biotechnol J 15:817–823

Jiang C-J, Shimono M, Maeda S, Inoue H, Mori M, Hasegawa M et al (2009) Suppression of the rice fatty-acid desaturase gene OsSSI2 enhances resistance to blast and leaf blight diseases in rice. Mol Plant-Microbe Interact 22:820–829

Jiang W, Zhou H, Bi H, Fromm M, Yang B, Weeks DP (2013) Demonstration of CRISPR/Cas9/sgRNA-mediated targeted gene modification in Arabidopsis, tobacco, sorghum and rice. Nucleic Acids Res 41:e188–e188

Jones JDG, Dangl JL (2006) The plant immune system. Nature 444:323–329

Jones DA, Thomas CM, Hammond-Kosack KE, Balint-Kurti PJ, Jones JD (1994) Isolation of the tomato Cf-9 gene for resistance to Cladosporium fulvum by transposon tagging. Science 266:789–793

Jones-Rhoades M (2012) Conservation and divergence in plant microRNAs. Plant Mol Biol 80:3–16

Kalia VC (2015) Quorum sensing vs quorum quenching: a battle with no end in sight. Springer, New York, NY

Kamthan A, Chaudhuri A, Kamthan M, Datta A (2015) Small RNAs in plants: recent development and application for crop improvement. Front Plant Sci 6:208

Kawashima CG, Yoshimoto N, Maruyama-Nakashita A, Tsuchiya YN, Saito K, Takahashi H, Dalmay T (2009) Sulphur starvation induces the expression of microRNA-395 and one of its target genes but in different cell types. Plant J 57:313–321

Kim JJ, Lee JH, Kim W, Jung HS, Huijser P, Ahn JH (2012) The microRNA156-SQUAMOSA PROMOTER BINDING PROTEIN-LIKE3 module regulates ambient temperature responsive flowering via FLOWERING LOCUS T in Arabidopsis. Plant Physiol 159:461–478

Kloppholz S, Kuhn H, Requena N (2011) A secreted fungal effector of Glomus intraradices promotes symbiotic biotrophy. Curr Biol 21:1204–1209

Knauer S, Holt AL, Rubio-Somoza I, Tucker EJ, Hinze A, Pisch M, Javelle M, Timmermans MC, Tucker MR, Laux T (2013) A protodermal miR394 signal defines a region of stem cell competence in the Arabidopsis shoot meristem. Dev Cell 24:125–132

Ku YS, Wong JW, Mui Z, Liu X, Hui JH, Chan TF, Lam HM (2015) Small RNAs in plant responses to abiotic stresses: regulatory roles and study methods. Int J Mol Sci 16(10):24532–24554

Kutter C, Schöb H, Stadler M, Meins F, Si-Ammour A (2007) MicroRNA-mediated regulation of stomatal development in Arabidopsis. Plant Cell 19:2417–2429

Lal S, Pacis LB, Smith HMS (2011) Regulation of the SQUAMOSA PROMOTER-BINDING PROTEIN-LIKE genes/microRNA156 module by the homeodomain proteins PENNYWISE and POUND-FOOLISH in Arabidopsis. Mol Plant 4:1123–1132

Leibfried A, To JPC, Busch W, Stehling S, Kehle A, Demar M, Kieber JJ, Lohmann JU (2005) WUSCHEL controls meristem function by direct regulation of cytokinin-inducible response regulators. Nature 438:1172–1175

Li A, Mao L (2007) Evolution of plant microRNA gene families. Cell Res 17:212–218

Li T, Liu B, Spalding MH, Weeks DP, Yang B (2012) High-efficiency TALEN-based gene editing produces disease-resistant rice. Nat Biotechnol 30:390–392.

Libault M, Pingault L, Zogli P, Schiefelbein J (2017) Plant systems biology at the single-cell level. Trends Plant Sci 11:949–960

Litholdo CG Jr, Schwedersky RP, Hemerly A, Ferreira PCG (2017) The role of microRNAs in plant-pathogen interactions. Revisao Anual de Patologia de Plantas (RAAP) 25:41–58

Litholdo CG Jr, Parker BL, Eamens AL, Larsen MR, Cordwell SJ, Waterhouse PM (2016) Proteomic identification of putative microRNA394 target genes in arabidopsis thaliana identifies major latex protein family members critical for normal development. Mol Cell Proteomics 15(6):2033–2047

Liu Q, Chen Y-Q (2009) Insights into the mechanism of plant development: interactions of miRNAs pathway with phytohormone response. Biochem Biophys Res Commun 384:1–5

Liu Q, Zhang Y-C, Wang C-Y, Luo Y-C, Huang Q-J, Chen S-Y, Zhou H, Qu L-H, Chen Y-Q (2009) Expression analysis of phytohormone-regulated microRNAs in rice, implying their regulation roles in plant hormone signaling. FEBS Lett 583:723–728

Liu F, Xing S, Ma H, Du Z, Ma B (2013) Cytokinin-producing, plant growth-promoting rhizobacteria that confer resistance to drought stress in Platycladus orientalis container seedlings. Appl Microbiol Biotechnol 97:9155–9164

Liu SR, Zhou JJ, Hu CG, Wei CL, Zhang JZ (2017) MicroRNA-mediated gene silencing in plant defense and viral counter-defense. Front Microbiol 8:1801

Luo Q-J, Mittal A, Jia F, Rock C (2012) An autoregulatory feedback loop involving PAP1 and TAS4 in response to sugars in Arabidopsis. Plant Mol Biol 80:117–129

Ma Z, Coruh C, Axtell MJ (2010) Arabidopsis lyrata small RNAs: transient MIRNA and small interfering RNA loci within the Arabidopsis genus. Plant Cell 22:1090–1103

Mallory AC, Vaucheret H (2006) Functions of microRNAs and related small RNAs in plants. Nat Genet 38:S31–S36

Mallory AC, Bartel DP, Bartel B (2005) MicroRNA-directed regulation of Arabidopsis AUXIN RESPONSE FACTOR17 is essential for proper development and modulates expression of early auxin response genes. Plant Cell 17:1360–1375

Mao Y-B, Cai W-J, Wang J-W, Hong G-J, Tao X-Y, Wang L-J et al (2007) Silencing a cotton bollworm P450 monooxygenase gene by plant-mediated RNAi impairs larval tolerance of gossypol. Nat Biotechnol 25:1307–1313

Martin GB, Brommonschenkel SH, Chunwongse J, Frary A, Ganal MW, Spivey R et al (1993) Map-based cloning of a protein kinase gene conferring disease resistance in tomato. Science 262:1432–1436

Mattick JS (2004) RNA regulation: a new genetics? Nat Rev Genet 5:316–323

Meng Y, Shao C, Wang H, Chen M (2011) The regulatory activities of plant microRNAs: a more dynamic perspective. Plant Physiol 157:1583–1595

Millar A, Waterhouse P (2005) Plant and animal microRNAs: similarities and differences. Funct Integr Genom 5:129–135

Miller MB, Bassler BL (2001) Quorum sensing in bacteria. Annu Rev Microbiol 55:165–199

Miyashima S, Koi S, Hashimoto T, Nakajima K (2011) Non-cell-autonomous microRNA165 acts in a dose-dependent manner to regulate multiple differentiation status in the Arabidopsis root. Development 138:2303–2313

Molesini B, Pii Y, Pandolfini T (2012) Fruit improvement using intragenesis and artificial microRNA. Trends Biotechnol 30:80–88

Navarro L, Dunoyer P, Jay F, Arnold B, Dharmasiri N, Estelle M, Voinnet O, Jones JDG (2006) A plant miRNA contributes to antibacterial resistance by repressing auxin signaling. Science 312:436

Niu Q-W, Lin S-S, Reyes JL, Chen K-C, Wu H-W, Yeh S-D et al (2006) Expression of artificial microRNAs in transgenic Arabidopsis thaliana confers virus resistance. Nat Biotechnol 24:1420–1428

Nogueira FTS, Madi S, Chitwood DH, Juarez MT, Timmermans MCP (2007) Two small regulatory RNAs establish opposing fates of a developmental axis. Genes Dev 21:750–755

Nozawa M, Miura S, Nei M (2010) Origins and evolution of microRNA genes in Drosophila species. Genome Biol Evol 2:180–189

Olanrewaju OS, Glick BR, Babalola OO (2017) Mechanisms of action of plant growth promoting bacteria. World J Microbiol Biotechnol 33:197.

Pan J, Huang T, Yao F, Huang Z, Powell CA, Qiu S et al (2008) Expression and characterization of aiiA gene from Bacillus subtilis BS-1. Microbiol Res 163:711–716

Pieterse CMJ, Zamioudis C, Berendsen RL, Weller DM, Van Wees SCM, Bakker PAHM et al (2014) Induced systemic resistance by beneficial Microbes. Annu Rev Phytopathol 52:347–375

Pulido A, Laufs P (2010) Co-ordination of developmental processes by small RNAs during leaf development. J Exp Bot 61:1277–1291

Pyott DE, Sheehan E, Molnar A (2016) Engineering of CRISPR/Cas9-mediated potyvirus resistance in transgene-free Arabidopsis plants. Mol Plant Pathol 17:1276–1288.

Qian G-L, Fan J-Q, Chen D-F, Kang Y-J, Han B, Hu B-S et al (2010) Reducing Pectobacterium virulence by expression of an N-acyl homoserine lactonase gene Plpp-aiiA in Lysobacter enzymogenes strain OH11. Biol Control 52:17–23

Rietman H, Bijsterbosch G, Cano LM, Lee H-R, Vossen JH, Jacobsen E et al (2012) Qualitative and quantitative late blight resistance in the potato cultivar sarpo mira is determined by the perception of five distinct RXLR effectors. Mol Plant-Microbe Interact 25:910–919

Schwab R, Palatnik JF, Riester M, Schommer C, Schmid M, Weigel D (2005) Specific effects of microRNAs on the plant transcriptome. Dev Cell 8:517–527

Scofield S, Tobias C, Rathjen J, Chang J, Lavelle D, Michelmore R et al (1996) Molecular basis of gene-for-gene specificity in bacterial speck disease of tomato. Science 274:2063–2065

Sempere LF, Cole CN, Mcpeek MA, Peterson KJ (2006) The phylogenetic distribution of metazoan microRNAs: insights into evolutionary complexity and constraint. J Exp Zool B Mol Dev Evol 306B:575–588

Si-Ammour A, Windels D, Arn-Bouldoires E, Kutter C, Ailhas J, Meins F, Vazquez F (2011) miR393 and secondary siRNAs regulate expression of the TIR1/AFB2 auxin receptor clade and auxin-related development of Arabidopsis leaves. Plant Physiol 157:683–691

Sieber P, Wellmer F, Gheyselinck J, Riechmann JL, Meyerowitz EM (2007) Redundancy and specialization among plant microRNAs: role of the MIR164 family in developmental robustness. Development 134:1051–1060

Song JB, Huang SQ, Dalmay T, Yang ZM (2012) Regulation of leaf morphology by microRNA394 and its target leaf curling responsiveness. Plant Cell Physiol 53:1669

Spaepen S, Vanderleyden J, Remans R (2007) Indole-3-acetic acid in microbial and microorganism-plant signaling. FEMS Microbiol Rev 31:425–448.

Spanudakis E, Jackson S (2014) The role of microRNAs in the control of flowering time. J Exp Bot 65:365–380

Sparks E, Wachsman G, Benfey PN (2013) Spatiotemporal signalling in plant development. Nat Rev Genet 14:631–644

Sunkar R, Li Y-F, Jagadeeswaran G (2012) Functions of microRNAs in plant stress responses. Trends Plant Sci 17:196–203

Thomazella DPT, Brail Q, Dahlbeck D, Staskawicz BJ (2016) CRISPR-Cas9 mediated mutagenesis of a DMR6 ortholog in tomato confers broad-spectrum disease resistance. bioRxiv:064824

Todesco M, Rubio-Somoza I, Paz-Ares J, Weigel D (2010) A Collection of target mimics for comprehensive analysis of microRNA function in Arabidopsis thaliana. PLoS Genet 6:e1001031

Válóczi A, Várallyay É, Kauppinen S, Burgyán J, Havelda Z (2006) Spatio-temporal accumulation of microRNAs is highly coordinated in developing plant tissues. Plant J 47:140–151

Van Peer R, Niemann GJ, Schippers B (1991) Induced resistance and phytoalexin accumulation in biological control of Fusarium wilt of carnation by Pseudomonas sp. strain WCS417r. Phytopathology 81:728–734

Vanneste S, Friml J (2009) Auxin: a trigger for change in plant development. Cell 136:1005–1016

Vanstraelen M, Benková E (2012) Hormonal interactions in the regulation of plant development. Annu Rev Cell Dev Biol 28:463–487

Vleeshouwers VGAA, Oliver RP (2014) Effectors as tools in disease resistance breeding against biotrophic, hemibiotrophic, and necrotrophic plant pathogens. Mol Plant-Microbe Interact 27:196–206

Waltz E (2018) With a free pass, CRISPR-edited plants reach market in record time. Nat Biotechnol 36:6

Wang J-W, Wang L-J, Mao Y-B, Cai W-J, Xue H-W, Chen X-Y (2005) Control of root cap formation by microRNA-targeted auxin response factors in Arabidopsis. Plant Cell 17:2204–2216

Wang JW, Schwab R, Czech B, Mica E, Weigel D (2008) Dual effects of miR156-targeted SPL genes and CYP78A5/KLUH on plastochron length and organ size in Arabidopsis thaliana. Plant Cell 20:1231–1243

Wang Y, Cheng X, Shan Q, Zhang Y, Liu J, Gao C et al (2014) Simultaneous editing of three homoeoalleles in hexaploid bread wheat confers heritable resistance to powdery mildew. Nat Biotechnol 32:947–951

Wei G, Kloepper JW, Tuzun S (1991) Induction of systemic resistance of cucumber to Colletotrichum orbiculare by select strains of plant-growth promoting rhizobacteria. Phytopathology 81:1508–1512

Weyens N, van der Lelie D, Taghavi S, Newman L, Vangronsveld J (2009) Exploiting plant–microbe partnerships to improve biomass production and remediation. Trends Biotechnol 27:591–598

Whitham S, Dinesh-Kumar SP, Choi D, Hehl R, Corr C, Baker B (1994) The product of the tobacco mosaic virus resistance gene N: similarity to toll and the interleukin-1 receptor. Cell 78:1101–1115

Wu G, Park MY, Conway SR, Wang J-W, Weigel D, Poethig RS (2009) The sequential action of miR156 and miR172 regulates developmental timing in Arabidopsis. Cell 138:750–759

Xia R, Zhu H, An Y-Q, Beers E, Liu Z (2012) Apple miRNAs and tasiRNAs with novel regulatory networks. Genome Biol 13:R47

Xie KB, Shen JQ, Hou X, Yao JL, Li XH, Xiao JH, Xiong LZ (2012) Gradual increase of miR156 regulates temporal expression changes of numerous genes during leaf development in rice. Plant Physiol 158:1382–1394

Yamaguchi A, Wu MF, Yang L, Wu G, Poethig RS, Wagner D (2009) The microRNAregulated SBP-box transcription factor SPL3 Is a direct upstream activator of LEAFY, FRUITFULL, and APETALA1. Dev Cell 17:268–278

Yamasaki H, Abdel-Ghany SE, Cohu CM, Kobayashi Y, Shikanai T, Pilon M (2007) Regulation of copper homeostasis by micro-RNA in Arabidopsis. J Biol Chem 282:16369–16378

Yang C, Li D, Mao D, Liu X, Ji C, Li X, Zhao X, Cheng Z, Chen C, Zhu L (2013) Overexpression of microRNA319 impacts leaf morphogenesis and leads to enhanced cold tolerance in rice (Oryza sativa L.). Plant Cell Environ 36:2207–2218

Yoon EK, Yang JH, Lim J, Kim SH, Kim S-K, Lee WS (2010) Auxin regulation of the microRNA390-dependent transacting small interfering RNA pathway in Arabidopsis lateral root development. Nucleic Acids Res 38:1382–1391

Zaidi SSA, Mukhtar MS, Mansoor S (2018) Genome editing: targeting susceptibility genes for plant disease resistance. Trends Biotechnol 36:898–906

Zhang B (2015) MicroRNA: a new target for improving plant tolerance to abiotic stress. J Exp Bot 66:1749–1761

Zhang B, Wang Q (2015) MicroRNA-based biotechnology for plant improvement. J Cell Physiol 230(1):1–15

Zhang B, Wang Q (2016) MicroRNA, a new target for engineering new crop cultivars. Bioengineered 7(1):7–10

Zhang XH, Zou Z, Gong PJ, Zhang JH, Ziaf K, Li HX, Xiao FM, Ye ZB (2011) Overexpression of microRNA169 confers enhanced drought tolerance to tomato. Biotechnol Lett 33:403–409

Zhang J, Huguet-Tapia JC, Hu Y, Jones J, Wang N, Liu S, White FF (2017) Homologues of CsLOB1 in citrus function as disease susceptibility genes in citrus canker. Mol Plant Pathol 18(6):798–810

Zhao B, Lin X, Poland J, Trick H, Leach J, Hulbert S (2005) A maize resistance gene functions against bacterial streak disease in rice. Proc Natl Acad Sci U S A 102:15383–15388

Zhou M, Luo H (2013) MicroRNA-mediated gene regulation: potential applications for plant genetic engineering. Plant Mol Biol 83(1–2):59–75

Zhou M, Li DY, Li ZG, Hu Q, Yang CH, Zhu LH, Luo H (2013) Constitutive expression of a miR319 gene alters plant development and enhances salt and drought tolerance in transgenic creeping bentgrass. Plant Physiol 161:1375–1391

Zhou J, Peng Z, Long J, Sosso D, Liu B, Eom JS, Huang S, Liu S, Cruz CV, Frommer WB, White FF, Yang B (2015) Gene targeting by the TAL effector PthXo2 reveals cryptic resistance gene for bacterial blight of rice. Plant J 82:632–643

Zhu H, Hu F, Wang R, Zhou X, Sze S-H, Liou LW, Barefoot A, Dickman M, Zhang X (2011) Arabidopsis Argonaute10 specifically sequesters miR166/165 to regulate shoot apical meristem development. Cell 145:242–256

Index

F. V. Winck (ed.), *Advances in Plant Omics and Systems Biology Approaches*, Advances in Experimental Medicine and Biology 1346, https://doi.org/10.1007/978-3-030-80352-0

O

P

MIX
Papier aus verantwortungsvollen Quellen
Paper from responsible sources
FSC® C105338

If you have any concerns about our products,
you can contact us on
ProductSafety@springernature.com

In case Publisher is established outside the EU,
the EU authorized representative is:
Springer Nature Customer Service Center GmbH
Europaplatz 3, 69115 Heidelberg, Germany

Printed by Libri Plureos GmbH
in Hamburg, Germany